Advances in Geographic Information Science

Series editors

Shivanand Balram, Burnaby, Canada
Suzana Dragicevic, Burnaby, Canada

More information about this series at http://www.springer.com/series/7712

Daniel A. Griffith · Yongwan Chun
Denis J. Dean
Editors

Advances in Geocomputation

Geocomputation 2015—The 13th International Conference

Springer

Editors
Daniel A. Griffith
School of Economic, Political
and Policy Sciences
University of Texas at Dallas
Richardson, TX
USA

Denis J. Dean
School of Economic, Political
and Policy Sciences
University of Texas at Dallas
Richardson, TX
USA

Yongwan Chun
School of Economic, Political
and Policy Sciences
University of Texas at Dallas
Richardson, TX
USA

ISSN 1867-2434 ISSN 1867-2442 (electronic)
Advances in Geographic Information Science
ISBN 978-3-319-22785-6 ISBN 978-3-319-22786-3 (eBook)
DOI 10.1007/978-3-319-22786-3

Library of Congress Control Number: 2016955684

Printed on acid-free paper

This Springer imprint is published by Springer Nature
The registered company is Springer International Publishing AG
The registered company address is: Gewerbestrasse 11, 6330 Cham, Switzerland

Preface

In May of 2015, geospatial scholars from around the world came to the University of Texas at Dallas (UTD), in Richardson, TX, for Geocomputation 2015, which was convened by the following Organizing Committee (a mixture of spatial and computer scientists): Denis J. Dean and Daniel A. Griffith (co-chairs), Brian J.L. Berry, Yongwan Chun, Yan Huang, Fang Qiu, Weili Wu, May Yuan, and Kang Zhang. This was an international, peer-reviewed, full-paper conference concerned with (1) enriching geography and the spatial sciences with a toolbox of methods to model and analyze a range of highly complex, often nondeterministic problems; (2) exploring a middle ground from the doubly informed perspective of geography and computer science; and (3) developing truly enabling technology for the quantitative spatial scientist, technology that offers a rich source of computational and representational challenges for the computer scientist. How scholars deal with these developments is a significant and potentially transformative issue, one requiring a conscious attempt to move the research agenda from geographical information systems (GIS) back to geographical analysis and modeling, deemphasizing the technical aspects of GIS (e.g., databases, large monolithic systems that standardized on interfaces, file structures, and query languages) while emphasizing GIScience. Accordingly, the primary aims of this conference were to

- facilitate global networking between institutions and individuals;
- promote active collaboration among researchers from diverse parent disciplines;
- generate a framework allowing newcomers to see their work in an international context;
- act as a constructive forum for interdisciplinary discussion of research ideas;
- create an international focus for current state-of-the-art research;
- provide a mechanism for disseminating the latest innovations and discoveries;
- highlight the benefits and limitations of new computational techniques; and
- indicate fruitful directions for further research.

Scholars from geography, as well as many other disciplines, have gathered for the past two decades at a series of Geocomputation conferences to contribute to this discussion. The following list of some of the notable past keynote speakers gives a notion of the range of these voices: Michael Batty (2011), Keith Clarke (2011), A. Stewart Fotheringham (2001, 2011), Mark Gahegan (2009), Menno-Jan Kraak (2007), Paul Longley (2013), Harvey Miller (2013), Peter Nijkamp (2011), Kirsi Virrantaus (2007), Shaowen Wang (2013), Jo Wood (2011), Xuejun Yang (2013), and Feng Zhao (2013). Those for the 2015 conference were Scott Morehouse, Shashi Shekhar, Dana Tomlin, Paul Torrens, and Michael Batty.

Supplemental funding for the conference was furnished by the US National Science Foundation (BCS-1461259) and the UTD. Proceedings copyediting was furnished by Ms. Kay Steinmetz of the Syracuse University Press.

The Symposium Series, the Venue, and the Conference Program

Geocomputation is conceptualized as the art and science of solving complex spatial problems with computers. For two decades, a loosely organized group of spatial scientists (e.g., http://www.geocomputation.org/steer.html) has been concerned with this broad theme. Since 1996, this interest group has facilitated the organization of an annual-that-has-become-a-biannual conference series: Geocomputation———. These conferences have been hosted solely by academic units at universities around the world (on four continents) and have attracted a diverse set of scholars from geography, computer science, geosciences, and the environmental sciences, as well as from government and other nonacademic organizations. Each conference has attracted between 80 and 300 participants, with a tendency for larger attendances at the more recent conferences. The quality of scholarly content is high: Short papers are submitted well in advance and subjected to a rigorous peer-reviewed process; the resulting series impact, being multidisciplinary and broadly international, has been substantial. The Geocomputation organization's Web site, http://www.geocomputation.org/, hosts all the conference papers, making conference findings openly accessible.

The 13th in this series, Geocomputation 2015, was held on the campus of the UTD, an innovative institution in the heart of North Texas that is part of the prestigious University of Texas System. It originally was the Texas Instruments Research Center and has grown since its founding in 1969 to include 133 degree programs, with cutting-edge curricula serving a variety of undergraduate and graduate student interests. The university continues its original commitment to providing some of the state's most lauded science and engineering programs and also has gained prominence for a breadth of educational paths, from criminology to biomedical engineering to arts and technology. Its Geospatial Information Sciences program is ranked first in Texas, first in both GIScience/Computation and Spatial Analysis/Statistics (according to *Geographical Perspectives* in 2015), and 16th in the nation (based on a 2010 study by Academic Analytics of Stony Brook, NY). This program spans the three Schools of Economic, Political and Policy Sciences, Natural Sciences and Mathematics, and Engineering, providing an ideal environment for hosting Geocomputation 2015.

Geocomputation 2015 was held May 20–23, 2015 (workshops on May 20; conference sessions on May 21–23). Cutting-edge research shared by participants from all over the world included

- new geocomputational algorithms and architectures;
- high-performance and cloud-enabled geocomputation;
- spatiotemporal modeling;
- geostatistics, spatial statistics, and spatial econometrics;
- geosimulation, agent-based modeling, and cellular automata;
- geovisualization and geovisual analytics;
- geospatial knowledge discovery and geospatial data mining;
- geospatial sensor web;
- various geocomputational applications;
- accuracy and uncertainty of geocomputational models;
- teaching geocomputation;
- applications in environmental, ecological, and biological modeling and analysis;
- applications in health and medical informatics; and
- applications in geodemographics, urban studies, criminology, and transportation modeling.

In addition to three peer-reviewed paper tracks resulting in both presentation and poster sessions, Geocomputation 2015 included keynote presentations by leading scholars, technical workshops aimed at training scholars in cutting-edge methodological approaches, and many opportunities for discussion and networking.

Most of the earlier conferences were held in Europe, with the 1997 conference being held in New Zealand, the 2001 and 2009 in Australia, and the 2013 in China, with only the 1999 and 2005 conferences being held in the USA. UTD hosted the conference in the USA for the first time in a decade. For this conference, the Organizing Committee established three concurrent sessions for peer-reviewed, accepted papers presented by general participants, with no competing session for the keynote speakers. The program for Geocomputation 2015 follows:

Workshops (either half-day or full-day)

Wednesday, May 20, 2015

8:00–9:00 A.M.	Breakfast
9:30 A.M.–12:30 P.M.	W1. Support vector machines for spatial and temporal analysis
	W3. Small area population estimation using areal interpolation
12:30–1:30 P.M.	Lunch
2:00–5:30 P.M.	W1. Support vector machines for spatial and temporal analysis

Conference

Wednesday, May 20, 2015

5:00–6:15 P.M.	Conference opens (registration begins); and
6:00–7:30 P.M.	Welcome reception.

Thursday, May 21, 2015

7:30–8:30 A.M.	Breakfast
8:45–9:00 A.M.	Welcoming remarks: Brian J.L. Berry
9:00–9:45 A.M.	Keynote: Shashi Shekhar
9:45–10:30 A.M.	Keynote: Scott Morehouse
10:45–11:15 A.M.	Mid-morning break
11:15 A.M.–12:30 P.M.	Sessions
	A-1. Geocomputation and the Urban Environment (Chair: Brian J.L. Berry) A-2. Algorithms (Chair: Yan Huang) A-3. Accuracy and Uncertainty in Geocomputation (Chair: Daniel A. Griffith)
12:30–1:30 P.M.	Lunch
2:00–3:40 P.M.	Sessions
	B-1. Agent-Based Models (Chair: Weili Wu) B-2. Spatio-Temporal Geocomputation (Chair: May Yuan) B-3. Geocomputation and the Natural Environment (Chair: Fang Qiu)
3:40–4:10 P.M.	Mid-afternoon break
4:10–5:50 P.M.	Sessions
	C-1. Spatial and Geostatistics I (Chair: Yongwan Chun) C-2. Social Media (Chair: May Yuan) C-3. High Performance (HP) and Cloud Computing (Chair: Yan Huang)
6:00–7:30 P.M.	Texas BBQ (Under the Trellis)

Friday, May 22, 2015

7:30–8:30 A.M.	Breakfast
8:45–9:30 A.M.	Keynote: Dana Tomlin
9:30–10:15 A.M.	Keynote: Paul Torrens

10:30–11:00 A.M.	Mid-morning break
11:00 A.M.–12:15 P.M.	Sessions
	D-1. Geometry and Space (Chair: May Yuan) D-2. Geocomputation Simulation I (Chair: Brian J.L. Berry) D-3. Geocomputation and Species Modeling (Chair: Yan Huang)
12:30–1:30 P.M.	Lunch
2:00–3:40 P.M.	Sessions
	E-1. Spatial and Geostatistics II (Chair: Daniel A. Griffith) E-2. Cellular Automata (Chair: Fang Qiu) E-3. Geocomputation Time Modeling (Chair: Yongwan Chun)
3:40–4:10 P.M.	Mid-afternoon break
4:10–5:50 P.M.	Sessions
	F-1. Geocomputation Visualization (Chair: Fang Qiu) F-2. Geocomputation and Movement (Chair: Daniel A. Griffith) F-3. Artificial Intelligence and Data Mining (Chair: Weili Wu)
7:00–9:00 P.M.	Banquet dinner, Hyatt Regency, North Dallas

Saturday, May 23, 2015

7:30–8:30 A.M.	Breakfast
8:45–9:30 A.M.	Keynote: Michael Batty
9:45–10:15 A.M.	Mid-morning break, Clarke Center, common area
10:15–11:30 A.M.	Sessions
	G-1. Geocomputation Simulation II (Chair: Brian J. L. Berry) G-2. Geocomputation, Data Mining, and Big Data (Chair: Yongwan Chun) G-3. Medical Geocomputation (Chair: Weili Wu)
11:50 A.M.–12:30 P.M.	Closing remarks
12:30 P.M.	Lunch

Posters were displayed throughout the conference in the lobby of the meeting facility. Participants were scheduled to present 71 refereed papers and display 13 refereed posters; best paper and poster prizes were awarded. This edited proceedings is one outcome of Geocomputation 2015.

Richardson, TX, USA
May 2016

Daniel A. Griffith
Yongwan Chun
Denis J. Dean

Contents

Introduction

Daniel A. Griffith, Yongwan Chun and Denis J. Dean

At Geocomputation 2015, a number of researchers presented their work orally or as a poster. Oral presentations included five keynote lectures by the following internationally renowned people in the field that the conference featured: Michael Batty, Scott Morehouse, Shashi Shekhar, Dana Tomlin, and Paul Torrens. The research expertise of these distinguished participants demonstrates interest in geocomputation from a wide variety of fields, including geography, computer science, landscape architecture, and software engineering. In their presentations, these keynote speakers delivered discussion and insights about fundamental concepts, technical advances, and cutting-edge applications in geocomputation. While four of them delivered research-oriented speeches from an academic perspective, Scott Morehouse of Esri, Inc. addressed issues in geocomputation from an industry viewpoint.

Two of these five keynote speeches are included in this volume. Shashi Shekhar's presentation—which resulted in his co-authored paper here with Eftelioglu, Jiang and Tang—focuses on visions and challenges in spatial computing. He emphasizes that a number of research questions need to be investigated to materialize the transformative potential of spatial computing, and addresses how subareas of spatial computing play a role in spatial computing, which comprises spatial database management, spatial data mining, geographic information systems (GIS), global positioning systems (GPS), and spatial statistics. His chapter in this volume emphasizes the relevance of spatial computing to issues pertaining to food, energy, and water resources, referred to as the FEW nexus. Dana Tomlin highlights two

D.A. Griffith (✉) · Y. Chun · D.J. Dean
School of Economic, Political and Policy Sciences, University of Texas at Dallas, Richardson, TX 75080, USA
e-mail: dagriffith@utdallas.edu

Y. Chun
e-mail: ywchun@utdallas.edu

D.J. Dean
e-mail: denis.dean@utdallas.edu

D.A. Griffith et al. (eds.), *Advances in Geocomputation*,
Advances in Geographic Information Science,
DOI 10.1007/978-3-319-22786-3_1

fundamental data models in GIS—vector and raster—in his keynote speech, stating "Though more recent advances…have since mitigated many of these practical differences [between vector and raster], perceptual differences nonetheless remain." As a developer of Map Algebra, a raster-based cartographic modeling language, Dana discusses how a raster-based view would be improved by employing a number of examples and graphics.

The other three keynote speakers also delivered skillful presentations about aspects of geocomputation, but, unfortunately, versions of these presentations do not appear in this volume. Michael Batty presented a new geocomputational approach to define cities using percolation theory. Specifically, he examined clusters of street networks that are created by disconnecting street segments whose length is longer than a threshold value. He showed its potential with an example from the United Kingdom. Paul Torrens presented a cohesive scheme to enhance and further develop a geosimulation model of human movement and interactions at a fine scale. This scheme has been developed with multiple components, including captures of human movements using a sensor network, immersive 3D representations, and interaction movements of agents. Scott Morehouse delivered a presentation about geographic information models. With a lengthy history in GIS software development at Esri, he discussed various components of GIS, including spatial databases, software engineering, and system architecture.

1 Concluding Comments

This proceedings volume marks the final product of the Geocomputation 2015 conference. The papers for this conference are classified into the following four broad thematic groups, for both presentation at the conference and organization of this volume: (1) spatial data: construction, representation, and visualization; (2) spatial analysis: methods and applications; (3) spatial statistical and geostatistical modeling; and (4) computational challenges and advances in geocomputation: high-performance computation and dynamic simulation. The Conference Organizing Committee (D. Griffith and D. Dean, co-chairs; B. Berry, Y. Chun, Y. Huang, F. Qiu, W. Wu, M. Yuan, and K. Zhang) produced this classification and the allocation of papers to it. Meanwhile, papers were refereed by both Conference Organizing Committee members and the following Conference Scientific Committee members:

Li An	San Diego State University	USA
Itzhak Benenson	Tel Aviv University	Israel
Mark Birkin	University of Leeds	UK
Dan Brown	University of Michigan	USA
Tao Cheng	University College London	UK

(continued)

(continued)

Keith Clarke	University of California, Santa Barbara	USA
Suzana Dragicevic	Simon Fraser University	Canada
Andrew Evans	University of Leeds	UK
Stewart Fotheringham	Arizona State University	USA
Alison Heppenstall	University of Leeds	UK
Mark Horner	Florida State University	USA
Bin Jiang	University of Gävle	Sweden
Brian Lees	University of New South Wales	Australia
Bin Li	Central Michigan University	USA
Songnian Li	Ryerson University	Canada
Feng Lu	Chinese Academy of Sciences	China
Alan Murray	University of California, Santa Barbara	USA
Tomoki Nakaya	Ritsumei University	Japan
David O'Sullivan	University of California, Berkeley	USA
Shih-Lung Shaw	University of Tennessee	USA
Xuan Shi	University of Arkansas	USA
Narushige Shiode	The University of Warwick	UK
Jean-Claude Thill	University of North Carolina, Charlotte	USA
Paul Torrens	University of Maryland	USA
John Wilson	University of Southern California	USA
David Wong	George Mason University	USA
Dawn Wright	Esri Inc.	USA
Ningchuan Xiao	Ohio State University	USA
Chaowei Yang	George Mason University	USA
Qiming Zhou	Hong Kong Baptist University	China

An additional paper reviewer included

Yifei Lou	University of Texas at Dallas	USA

Finally, the authors who contributed to this volume furnished multiple revisions to both written and graphical parts of their papers.

The Nexus of Food, Energy, and Water Resources: Visions and Challenges in Spatial Computing

Emre Eftelioglu, Zhe Jiang, Xun Tang and Shashi Shekhar

Abstract In the coming decades, the increasing world population is expected to increase the demand for food, energy, and water (FEW) resources. In addition, these resources will be under stress due to climate change and urbanization. Previously, more problems were caused by piecemeal approaches analyzing and planning those resources independent of each other. The goal of the FEW nexus approach is to prevent such problems by understanding, appreciating, and visualizing the interconnections and interdependencies of FEW resources at local, regional, and global levels. The nexus approach seeks to use the FEW resources as an interrelated system of systems, but data and modeling constraints make this a challenging task. Also, the lack of complete knowledge and observability of FEW interactions exacerbates the problem. Related work focuses on physical science solutions (e.g., desalination, biopesticides). No doubt these are necessary and worthwhile for FEW resource security. Spatial computing may help domain scientists achieve their goals for the FEW nexus. In this chapter, we describe our vision of spatial computing's role in understanding the FEW nexus from a spatial data life cycle perspective. We provide details of each of the spatial computing components. For each component, we list new technical challenges that are likely to drive future spatial computing research.

Keywords Food · Energy and water nexus · Spatial computing

E. Eftelioglu · Z. Jiang · X. Tang (✉) · S. Shekhar
University of Minnesota, Minneapolis, MN, USA
e-mail: xuntang@cs.umn.edu

E. Eftelioglu
e-mail: emre@cs.umn.edu

Z. Jiang
e-mail: zhe@cs.umn.edu

S. Shekhar
e-mail: shekhar@cs.umn.edu

D.A. Griffith et al. (eds.), *Advances in Geocomputation*,
Advances in Geographic Information Science,
DOI 10.1007/978-3-319-22786-3_2

1 Introduction

Critical resources, such as food, energy, and water (FEW) are under increasing stress due to population growth, urbanization, and climate change (Andrews-Speed et al. 2014; Hoff 2011; NSF 2014). Although a doomsday scenario is unlikely, policy makers and industry need to take required precautions to avoid future FEW scarcities. Therefore, technological breakthroughs are needed to ensure the availability of FEW resources and to meet the needs of the world's future population (NIC 2012).

The FEW resources (NSF 2015; UNU-FLORES 2015; Holger 2011) are inextricably interconnected. This is often referred to as the FEW nexus, implying that constraints (or choke points) on one of the resources may limit the availability of another of the resources. For example, production of food or energy at a location (e.g., the California Central Valley) is limited by the availability of water. Figure 1 shows an illustration of the FEW nexus. The interdependencies can be seen in terms of the following aspects. First, necessary processes in energy production, such as cooling nuclear power plants, energy generation in hydroelectric power plants, and the extraction of coal bed natural gas for thermoelectric power plants, need water. Moreover, biofuel for energy generation needs food (i.e., agricultural products). In addition, energy is needed in water pumping, drainage, desalination, distribution, and fertilizer production for food production. However, sometimes unexpected interactions occur within FEW systems that cause harm and increase vulnerability. Examples are groundwater depletion due to high water use for irrigation purposes, groundwater pollution caused by extreme fertilizer usage in fields, surface water depletion due to energy production, and electricity blackouts due to the high energy requirements of extreme irrigation pumping.

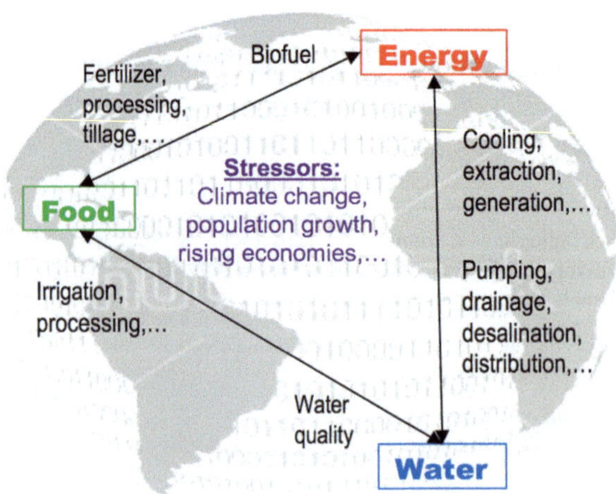

Fig. 1 Interaction of food, energy, and water systems

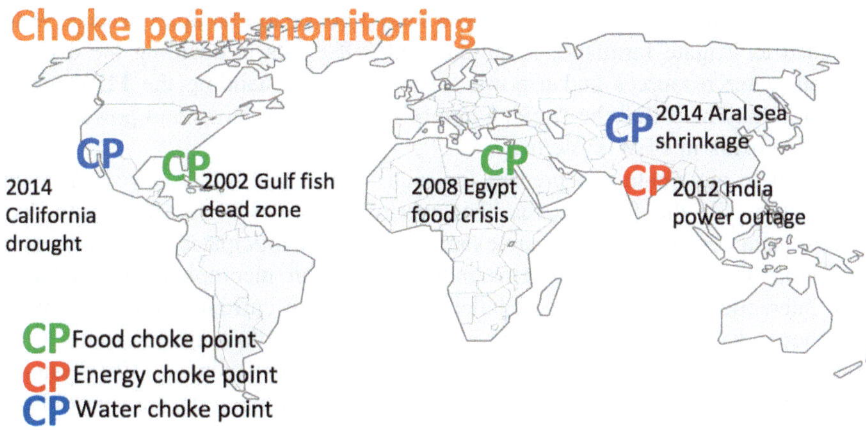

Fig. 2 FEW choke point locations that occurred in the last decade

Significant issues arise from the interdependent and interconnected nature of FEW systems, which traditionally were analyzed and planned independently. Solutions implemented in one sector can have unintended and dangerous consequences in other sectors (NSF 2015). Figure 2 shows examples of such consequences that are shown with a choke point (CP) symbol. One example is the excessive use of fertilizers for crop production in the US Midwest, which caused nutrients to reach the Gulf of Mexico through rivers, triggering the large-scale growth of algae and a subsequent loss of dissolved oxygen in the water; this process killed fish and made the area a dead zone (Rabalais et al. 2002).

Similarly, excessive water usage for agricultural purposes caused the Aral Sea to shrink to less than half of its size in a couple of decades. In July 2012, the electricity subsidy for agricultural water pumps in South Asia during a drought led to power grid failure, triggering the largest blackout on earth (Webber 2015). An even larger cascade of events occurred across the globe in 2008, when major land use changes to boost biofuel production caused a worldwide food crisis, which in turn led to political and economic instability in Bangladesh and Egypt (Bobenrieth and Wright 2009).

The goal of understanding the nexus of FEW (Hoff 2011; Mohtar and Daher 2012; Scott et al. 2015) is to reduce the risks of unintended consequences associated with FEW resources, to improve the quality of life, to create economic opportunities, to build regulations to ensure the resilience and accessibility of FEW systems, to provide equity to people (Scott et al. 2015), and to appreciate interconnections and interdependencies across FEW systems. Better understanding the nexus of FEW also may benefit geodesign (Miller 2012), a technology that helps decide geographic-related resource arrangement. An illustrative example is that farm and water sources should not be too close because fertilizer usage may pollute the water. In contrast, these two resources should not be too far apart because water is needed to irrigate farmland. A similar trade-off exists in deciding the distance between water resources and a power plant. By understanding the FEW nexus, geodesign may help find the "golden points" to locate the farms and power plants, critical elements in FEW systems.

We fail to anticipate and prepare for such interactions because we lack a full understanding of the FEW nexus. The world food price crisis, the South Asia blackout, and ocean dead zones are examples of the unanticipated consequences of attempting to solve FEW problems individually with incomplete knowledge of FEW interactions. We see three challenging areas that currently inhibit scientists and other stakeholders from learning what they need to about these systems. First, limited observability hinders data collection within and across FEW systems. Distributions of underground water and wind energy, for example, are very difficult to observe (Flammini et al. 2014). Second, data management and querying facilities for global FEW observations are inadequate. Without computational support, FEW datasets collected from different sources (e.g., remote sensing imagery, ground sensor observations, river networks, global supply chains) and from different geographical areas cannot be fully exploited together. Third, current research is hampered by the lack of effective data-sharing protocols across sectors and countries. Datasets from individual FEW systems are often in different formats of representation, hindering data sharing across owners (e.g., precision agricultural data from farm owners), governments, and decision makers.

National and international agencies have begun to consider research challenges to understanding the nexus of FEW resources. A report by the Food and Agriculture Organization of the United Nations calls for stakeholder dialogue based on empirical evidence, scenario development, and response options (Flammini et al. 2014). A report by the National Science Foundation (NSF) Mathematics and Physical Sciences Advisory Committee identifies key areas where physical science could address FEW challenges (NSF 2014). Examples include developing desalination technologies to increase sustainable water supplies for agriculture and improving crop protection via biopesticides and genetic techniques. The US Department of Energy also published a report summarizing the challenges and opportunities for understanding the water-energy nexus. This report highlights promising technologies such as advanced materials, water recovery, and cooling technologies. However, all recent efforts address the problem from a social or physical science perspective without addressing geocomputational challenges in collecting, integrating, managing, analyzing, and visualizing spatial data related to the FEW nexus.

In this chapter, we identify key areas in which spatial computing can help achieve an understanding of the FEW nexus. We also list new technical challenges in each area that are likely to drive future spatial computing research. Next, we describe our vision of spatial computing's role in understanding the FEW nexus.

2 A Spatial Computing Vision

We believe that spatial computing has the potential to provide transformative insights into the FEW nexus and to predict future FEW resource CPs similar to weather forecasting. In the early 1900s, weather predictions were limited to same-day forecasts. The use of radar and interpolation techniques after the World War II, however, allowed weather forecasting for three to five days, as well as early warnings before hurricanes and other extreme weather events. Figure 3 is an example of a weather forecast map that displays predictions of near-future weather event (e.g., rainy, sunny, cloudy). Similarly, spatial computing may help monitor or even project future events of either food, energy, or water resource risks (e.g., scarcity, environmental catastrophe) on a map (e.g., event type, location, time), just like weather forecasts.

Our vision is that spatial computing can offer insights into the interactions and interdependencies of the FEW nexus, as well as provide future projections and early warnings. The tools for achieving this vision will be spatial and spatiotemporal nexus data management, analytics, visualization, and decision support. For example, if we could identify and analyze the teleconnection patterns between the food crisis and biofuel production across countries that occurred in 2008, we might be able to avoid such a widespread crisis from happening in the future. Spatial analytics may improve water management at a global scale, rather than at regional or country-wide scales. For example, virtual water trade (Hoekstra and Hung 2002) could relocate water-intensive crops (e.g., cotton) from countries less endowed with water to those with a hydrological advantage (e.g., high rainfall) to leverage spatial variability of FEW resources. Spatial tools (e.g., CPs represented as critical nodes, paths, cut computations) could help support the resilience of individual FEW resources and prioritize system elements for increased redundancy. Another important theory for studying the FEW nexus is the life-cycle assessment (LCA, also known as life-cycle analysis or cradle-to-grave analysis). LCA (UNEP 2016) is a technique that determines and evaluates environmental impacts associated with all

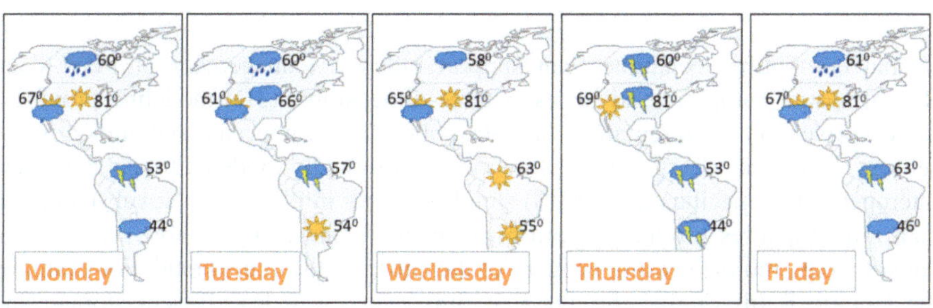

Fig. 3 Weather forecasts showing short-term weather predictions (temperatures in degrees Fahrenheit)

stages of a product's life cycle. Spatial computing is promising in analyzing and understanding FEW life cycles from a data science perspective.

Spatial computing also will have the opportunity to tackle data paucity via remote sensing and volunteered geographic information (VGI), lack of data sharing via the exploration of common spatial data representations across sectors of food (e.g., agriculture), water (e.g., surface water and precipitation maps), and energy (e.g., wind velocity maps), and interchange methods. Spatial database management systems (DBMS) and cloud platforms (e.g., the Google Earth Engine) are also promising in providing support for efficient queries on large global FEW datasets. Another opportunity is to help discover novel, interesting, useful, but potentially hidden FEW interactions at different geographic scales via spatial and spatiotemporal data mining. Finally, spatial decision support systems may identify opportunities to relocate elements of supply chains and redesign landscapes to improve efficiency and reduce unnecessary waste in FEW systems.

2.1 FEW Observations

Current spatial computing techniques, such as the global positioning system (GPS; GPS for US 2015), remote sensing satellites and planes, and ground sensor networks, already are widely used to collect rich data within FEW systems. Such data collections can provide opportunities to leverage rich geocontexts to support a global-scale redundancy of resources. In addition, advances in mobile technology, such as smart phones and location-based social networks, provide tremendous opportunities for collecting FEW data via crowdsourcing, also called VGI. For example, mWater, a mobile application for water quality monitoring, leverages mobile technology and an open data-sharing platform for water safety testing, and allows volunteers to easily find the safest water sources near them.

2.2 FEW Data Management

Spatial computing techniques can support efficient management of FEW data. Recent spatial computing advances in three-dimensional (3D) modeling provide a more convenient representation of data collected from ocean and underground sensors (Heidemann et al. 2012), which was traditionally modeled by open geodata interoperability specification (OGIS) simple features in 2D space. Novel spatial big data infrastructures provide a platform to manage and analyze large-scale spatial datasets (e.g., remote sensing imagery of the entire earth) in the cloud. For example, the Google Earth Engine, for the first time ever, provides efficient storage and computation in the cloud of all kinds of remote sensing imagery of the entire earth surface over several decades. Moreover, a cloud environment nurtures the development of VGI from check-ins, tweets, geotags, and georeports (Shekhar et al.

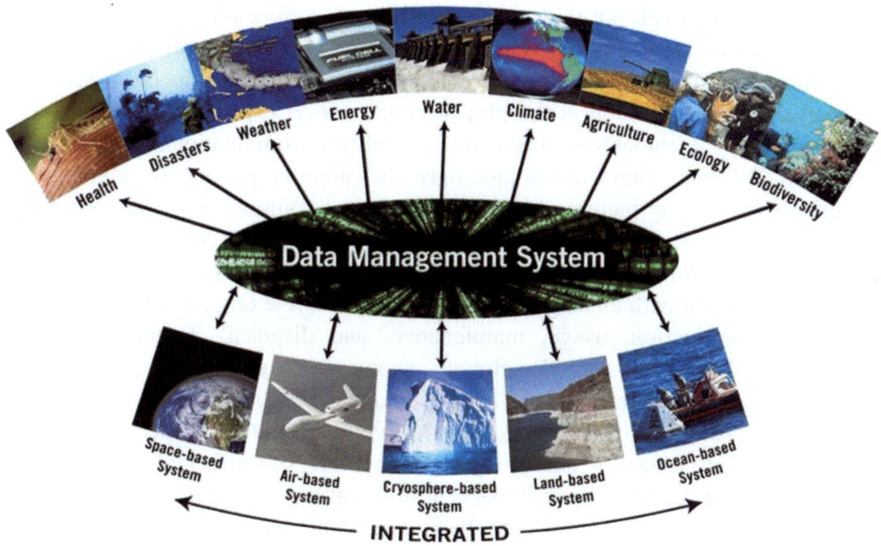

Fig. 4 An illustrative example of FEW data management from GEOSS (GEOSS Portal 2015)

2016). Finally, efficient support for spatial graph queries, such as critical node and path computation, can help enhance the resilience of FEW resources.

However, the FEW datasets are highly heterogeneous because the data collection processes are for various purposes and in various formats and spatiotemporal resolutions, limiting the potential value of the data. Spatial computing can help develop systematic data standards and data-sharing protocols considering various FEW applications as a whole (e.g., the Global Earth Observation System of Systems (Lautenbacher 2006), the Nexus Observatory Platform (UNU-FLORES 2015). Figure 4 illustrates a recent effort, namely, the Global Earth Observations System of Systems (GEOSS) (GEOSS Portal 2015) platform, to create a unified data management system. This system stores and accesses FEW data from heterogeneous sources, including space-, air-, and land-based systems.

2.3 *FEW Data Mining*

Advances in machine learning, spatial statistics, and spatial data mining provide a data science approach for understanding the nexus of FEW resources. Traditionally, FEW systems have been studied via mechanistic process models. These models have many advantages, such as good interpretability and a capability to make future projections, but as observation data are being collected at much higher spatial and temporal resolutions, traditional mechanistic process models may fail to leverage the rich spatiotemporal contextual information that situational assessment of FEW resources require. For example, crop production used to be modeled and studied at a county or state level, but with the availability of high-resolution remote sensing imagery from unmanned aerial vehicles and observations from field sensors a new discipline called precision agriculture (McBratney et al. 2005) has been developed to analyze crop production at the plot or subplot level. Spatial data science techniques will play a crucial role in providing a real-time computational analysis of the rich FEW datasets. These techniques may also identify previously unknown but potentially interesting patterns (e.g., spatiotemporal coupling or telecoupling, spatial hot spots) from FEW data. For example, LCA (UNEP 2016; Rebitzer et al. 2004) is a methodological framework that estimates and assesses the environmental impacts associated with all stages in a product's life cycle (e.g., material extraction, processing, production, usage, maintenance, and disposal). Examples of such environmental impacts include climate change, smog creation, acidification, resource depletion, water use, and land use. Using a LCA approach for the FEW nexus helps provide a more detailed perspective of impacts of a specific change in any of the food, water, or energy systems on other systems, and helps evaluate opportunities for reducing these impacts at different stages of a product's life cycle. Spatial computing is considered to be a promising data analytic technique for understanding a product's life cycle. Moreover, rich FEW data sources may improve the accuracy and timeliness of spatiotemporal predictions. Finally, spatial statistical approaches can help test the significance of these predictions.

2.4 Decision Support

Spatial computing can use FEW data in spatial decision support systems to increase the efficiency and sustainability of FEW resources. Such spatial decision support systems view FEW as a system of systems. A system is an interconnected set of elements for a purpose (Meadows 2008). Figure 5 shows an entity relationship diagram (ER-diagram) representing a system of systems view of the FEW nexus. For example, water is consumed for irrigating crops for food production and the cooling of energy plants, while energy is consumed for producing fertilizer for the process of food production and pumping water from underground for irrigation. In addition, the use of fertilizer in food production pollutes water, and food can be reused to produce energy. The system of systems view helps spatial decision support systems to increase efficient use of FEW resources (Housh et al. 2014). For example, GIS soil maps with nutrient, humidity, and chemical compound details allow users to determine optimal crop type selection in each field and to increase yields by location-aware fertilizer and pesticide use. Another example is hydroelectric production by demand instead of traditional all-time electricity production. Spatial decision support systems are used to determine the peak electricity demand hours, and electricity production is adapted accordingly. Finally, spatial decision support systems provide a unique opportunity to plan supply chain relocation and landscape redesign tasks, depending on the availability of FEW resources in specific locations.

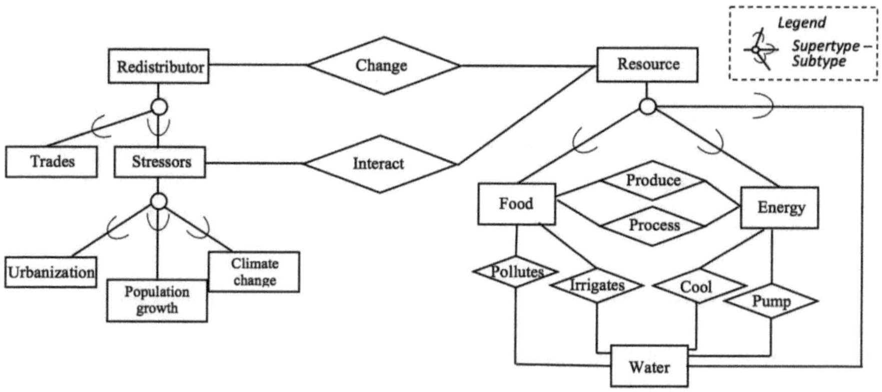

Fig. 5 A FEW system of systems

2.5 FEW Data Visualization

Virtual globes are already widely used to visualize environmental changes in different world regions. These visualizations not only provide historical records of FEW resources but also allow visualization of future scenarios on a global scale. Next-generation virtual globes (e.g., Google Earth, Google Map, Bing maps, and NASA World Wind) will provide a unique opportunity to visualize the interactions and interdependencies of FEW systems across the globe over a long period of time (Google Time Laps). These tools may allow users to have a common operational picture of FEW systems. Such visualization tools not only facilitate current FEW research about the effect of interactions within the FEW nexus but also assist decision-making agencies to demonstrate effectively future effects of their policies.

3 Spatial Computing Challenges

In spite of the potential transformative insights provided by spatial computing about the FEW nexus, significant technical challenges exist due to the unique characteristics of FEW data. This section discusses these challenges and suggests directions for future spatial computing research.

3.1 FEW Observation Challenges

Though recent earth observation platforms (e.g., GEOSS, Earth Observatory) show promise for collecting rich observation data, the observability challenge still exists. For example, collecting data about water quality and quantity is necessary for

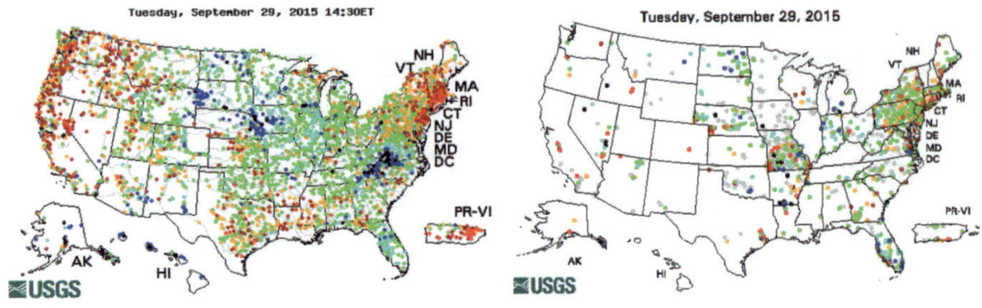

a Surface water **b** Groundwater

Fig. 6 Water source observation locations (USGS 2015)

understanding the water-energy nexus (Healy et al. 2015) and for determining water availability for energy production in the future. To collect water quantity data, a surface water gauging network currently is maintained by the US Geological Survey (USGS 2015) to monitor thousands of sites and groundwater observation wells nationwide. Surface water observation stations (Fig. 6a) have a larger spatial coverage and temporal frequency than groundwater stations (Fig. 6b). In contrast, to collect water quality data, the US Geological Survey operates continuous recorders at about 1,700 sites across the United States. Discrete samples are collected and analyzed by other programs as well. A need exists for remote sensing techniques that allow for the monitoring of water quality and quantity at a large scale at low expense and consuming little time. Moreover, developing new spatial data models for water quality characterization when observation data are missing is necessary.

Another challenge arises from data heterogeneity (e.g., data from different sources and in different spatial and temporal resolutions), which prohibits the integration of FEW observation data across different platforms. Therefore, spatial computing research is needed to design an acceptable standard for FEW data. Finally, data quality or accuracy is an important concern for the data collected from VGI systems.

3.2 FEW Data Management Challenges

Traditionally, FEW data were analyzed by individual disciplines. To facilitate interoperability for the FEW nexus data across disciplines, a comprehensive data management framework needs to be developed. However, such a framework raises new technical challenges. First, FEW data are in different representation models. For example, remote sensing imagery and climate model simulation data are often in raster form, while water census data are in vector form because they are collected at sample wells that are spatial points. Many FEW datasets are collected from 3D

Euclidean space (e.g., ocean data or subsurface data) or spatial network space (e.g., river networks), and often with the additional dimension of time, which requires novel data models and representations. Second, semantic heterogeneity exists in FEW data due to the unique data collection protocols and standards maintained by different agencies. For example, the data schema (e.g., representation, naming) used in water census data in the United States may be quite different from the ones used in China. Third, traditional spatial data management tools were designed to store and manage geometric and raster data. However, some FEW data are from VGI that often contains place-names and prepositions (e.g., near, in, at, along) instead of numerical coordinates (e.g., latitude and longitude). Therefore, a need exists for new methods to interpret these place-names and to clean data errors, evaluate trustworthiness, and avoid bias. Finally, critical node and path computation is very expensive for large FEW datasets, and efficient algorithms are needed to avoid decreases in redundancy.

3.3 FEW Data-Mining Challenges

Spatial data science plays a crucial role in providing computational analysis with rich FEW data. However, significant challenges exist in utilizing these techniques, including the lack of tools for building 3D models and spatiotemporal network models (e.g., anisotropic and asymmetric spatial neighborhood), the inability to simulate the decision-making process of policy makers, and the difficulty of future projections when the assumption of stationarity does not hold.

Data science techniques, for example, have been widely used in land use change modeling, which is an important task in agriculture and water resource management (Geographical Sciences Committee 2014). Compared with other methods, data science approaches can be appropriate for situations where data concerning pattern are available and theory concerning process is scant. However, several challenges exist in utilizing data science techniques for land use change modeling. First, because machine learning and statistical approaches in data science are developed inductively on the basis of the input data, the models are particularly sensitive to an input sample. Second, both machine learning and statistical approaches generally assume stationarity in the relationship between predictor and land change variables (i.e., that the model fitted during the "training" interval can be applied to the subsequent time interval). In cases where such relationships change over time, these approaches cannot be appropriate for the projections of future scenarios. Moreover, machine learning algorithms can easily represent a variety of complex relationships but with a greater risk of overfitting. Finally, interpretation of output can be challenging because many machine learning algorithms are either "black box" or produce a map of "transition potential" for each land transition instead of producing transitional probabilities (NRC 2014).

3.4 FEW Decision Support Challenges

Current spatial decision support systems consider only a single resource (i.e., food, energy, or water). However, a FEW nexus approach requires that all resources be taken into account when decisions are made. In other words, a FEW nexus approach has a system of systems view. For example, in precision agriculture, as illustrated in Fig. 7, detailed observation data about crop fields collected by drones or ground sensors are used to monitor crop health and support decisions about how much water and fertilizer to apply to which plots or subplots to minimize water or energy consumption while maximizing production (NIC 2012). In general, three ways exist to capture the FEW nexus as a system of systems: mechanistic models, an empirical approach, and an optimization approach. FEW decision support systems should support users in the event that FEW resources cause spatial externalities. Also, these systems should take interconnections of FEW resources and the uncertainty of interactions into account for complicated tasks (e.g., supply chain relocation, landscape design, precision agriculture).

Therefore, precision agriculture requires monitoring and predicting FEW resources and taking required precautions. Similarly, FEW decision support systems should be able to monitor and make future projections to prevent FEW resource shortages. Some efforts have already been made to understand the FEW nexus as a system of systems. Currently, the Group on Earth Observations (GEO) (GEOSS Portal 2015), is an intergovernmental organization with 90 members and 67 participating organizations that provides an international framework for collaboration around various societal benefits, including agriculture, as shown in Fig. 8. Due to the international recognition of critical need for improving real-time, reliable, open information about global agricultural production prospects, GEO established the GEO Global Agriculture Monitoring Initiative (GEOGLAM) (Justice 2013), which provides a system of agriculture monitoring systems that uses coordinated, comprehensive, and sustained earth observations to inform decisions and actions in agriculture. The goals of GEOGLAM include supporting,

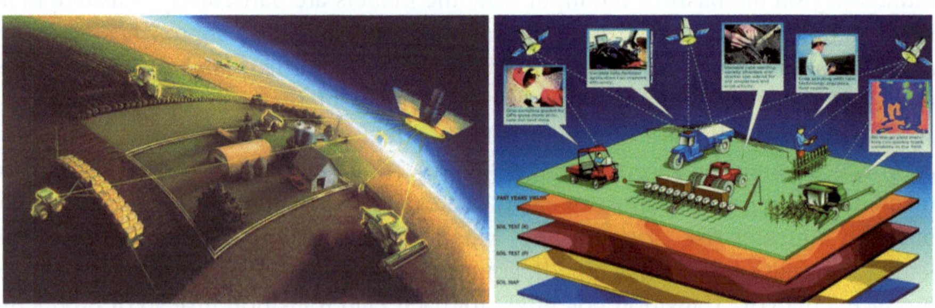

Fig. 7 An illustration of precision agriculture (Plant et al. 2000)

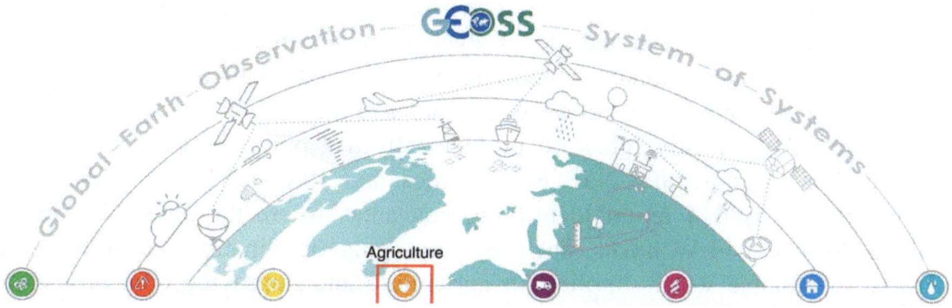

Fig. 8 An illustration of GEO (Justice 2013)

strengthening, and articulating existing efforts through the use of earth observations, developing capacities and awareness at national and global levels, and disseminating information.

3.5 FEW Data Visualization Challenges

Visualizing interactions between resources in the FEW nexus is one of the biggest challenges. First, such interactions happen in different spaces. For example, ocean and underground water data are often presented in a 3D Euclidean space, whereas surface stream flow data are in a 2D spatial network space. Visualizing all these water data in both Euclidean space and spatial network space is nontrivial. Second, previous visualization approaches focus on known information, but the FEW nexus requires more sophisticated techniques to visualize the uncertainty about location, value, recency, and quality of spatiotemporal information. For instance, the lack of site-specific data and the limitations of estimation models result in uncertainty when estimating water resource consumption. To visualize a map of water consumption with uncertainty or to compare two temporal snapshots with uncertain inferred change is nontrivial. Finally, FEW data visualizations should be able to handle unexpected spatial variability, such as the migration of population due to economic reasons.

4 Summary

The FEW nexus framework aims to view these three inextricable resources from a system of systems perspective. By understanding, appreciating, and visualizing the interconnections and interdependencies in FEW resources at local, regional, and global levels, the FEW nexus tries to achieve the goal of reducing unintended resource scarcities. To achieve the goals of resource sustainability and availability, the FEW nexus approach applies the nexus framework at a local level of decision making and also at regional and global levels of policy-making processes.

National and international agencies recently started to focus on solving the sustainability and availability of FEW resources from a FEW nexus perspective. Current initiatives mostly focus on problems from a pure physical science perspective. For example, the NSF Mathematics and Physical Sciences Advisory Committee has a 2014 report identifying key areas where the physical sciences could address the FEW nexus, such as developing desalination technologies to increase sustainable water supplies for agriculture and improving crop protection via biopesticides and genetic techniques. Besides the physical sciences, which no doubt are necessary to solve FEW nexus challenges, spatial computing has the potential to play a critical role in helping domain scientists address the FEW nexus.

In this chapter, we aim to improve the efficiency of FEW nexus thinking from a spatial computing perspective. We envision several key components in the FEW nexus, including data collection, management, mining, and visualization, which may be how spatial computing can help improve our understanding of the FEW nexus. For each component, we also identify the main challenges from a computing perspective.

Acknowledgements This material is based upon work supported by the National Science Foundation under Grant No. 1029711, IIS-1320580, 0940818 and IIS-1218168, U.S. DoD under Grant No. HM1582-08-1-0017, HM0210-13-1-0005, and University of Minnesota via U-Spatial. We would like to thank Kim Koffolt and the members of the University of Minnesota Spatial Computing Research Group for their comments.

References

Andrews-Speed P, Bleischwitz B, Boersma T, Johnson C, Kemp G, VanDeveer S (2014) Want, waste or war? The global resource nexus and the struggle for land, energy, food, water and minerals. Routledge, New York, NY

Bobenrieth E, Wright B (2009) The food price crisis of 2007/2008: evidence and implications. In: Joint meeting of the intergovernmental group on oilseeds, oils and fats (30th session), the intergovernmental group on grains (32nd session), and the intergovernmental group on rice (43rd session), Santiago, Chile, vol 4, no 11, pp 4–6

Flammini A, Puri M, Pluschke L, Dubois D (2014) Walking the nexus talk: assessing the water-energy-food nexus in the context of the sustainable energy for all initiative. Environment and Natural Resources Management. Working Paper, FAO

Geographical Sciences Committee (2014) Advancing land change modeling: opportunities and research requirements. National Academies Press, Washington, DC

GPS4US (GPS for US) (2015) http://www.gps4us.com/news/post/Global-positioning-and-geographic-information-systems-help-create-an-environmentally-friendly-farm-20111228.aspx. Accessed Sept 30, 2015

Group on Earth Observations System of Systems Portal (GEOSS) (2015) http://www.geoportal.org/web/guest/geo_home_stp. Accessed 30 Sept 2015

Healy R, Alley W, Engle M, McMahon P, Bales J (2015) The water-energy nexus: an earth science perspective. US Geological Survey, Reston, VA

Heidemann J, Stojanovic M, Zorzi M (2012) Underwater sensor networks: applications, advances and challenges. Philos Trans R Soc Lond A: Math Phys Eng Sci 370(1958):158–175

Hoekstra A, Hung P (2002) Virtual water trade: a quantification of virtual water flows between nations in relation to international crop trade. Value of water research report series, no 11, IHE Delft

Hoff H (2011) Understanding the nexus. Background paper for the Bonn 2011 Nexus Conference. http://www.sei-international.org/publications?pid=1977. Accessed 30 Sept 2015

Holger H (2011) Understanding the nexus: background paper for the Bonn 2011 nexus conference. http://www.sei-international.org/publications?pid=1977. Accessed 30 Sept 2015

Housh M, Cai X, Ng T, Mclsaac G, Ouyang Y, Khanna M, Sivapalan M, Jain A, Eckhoff S, Gasteyer S, Al-Qadi I (2014) System of systems model for analysis of biofuel development. J Infrastruct Syst 21(3):9385–9404

Justice C (2013) The GEO Global Agricultural Monitoring Initiative (GEOGLAM): overview. http://lcluc.umd.edu/Documents/ScienceTeamMtg/2013_NOV/presentations/GEOGLAM_OverviewTashkent.pdf. Accessed 30 Sept 2015

Lautenbacher C (2006) The global earth observation system of systems: science serving society. Space Policy 22(1):8–11

McBratney A, Whelan B, Ancev T, Bouma J (2005) Future directions of precision agriculture. Precis Agric 6(1):7–23

Meadows D (2008) Thinking in systems: a primer. Chelsea Green Publishing, White River Junction, VT

Miller W (2012) Introducing geodesign: the concept. https://www.esri.com/library/whitepapers/pdfs/introducing-geodesign.pdf. Accessed 30 Sept 2015

Mohtar R, Daher B (2012) Water, energy, and food: the ultimate nexus. In: Encyclopedia of agricultural, food, and biological engineering. CRC Press, Taylor and Francis Group, New York, NY http://wefnexus.tamu.edu/files/2015/01/Mohtar-Daher_Water-Energy-and-Food-The-Ultimate-Nexus.pdf/. Accessed 30 Sept 2015

National Intelligence Council (NIC) (2012) Global Trends 2030: Alternative Worlds. National Intelligence Council, Washington, DC

National Science Foundation (NSF) (2014) Food, energy, and water: transformative research opportunities in the mathematical and physical sciences http://www.nsf.gov/mps/advisory/mpsac_other_reports/nsf_food_security_report.pdf. Accessed 30 Sept 2015

National Science Foundation (NSF) (2015) Dear colleague letter: SEES: interactions of food systems with water and energy systems. http://www.nsf.gov/pubs/2015/nsf15040/nsf15040.jsp. Accessed 30 Sept 2015

NRC (2014) Advancing land change modeling: Opportunities and research. http://nap.edu/catalog.php?record_id=18385

Plant R, Pettygrove G, Reinert W (2000) Precision agriculture can increase profits and limit environmental impacts. Calif Agric 54(4):66–71. doi:10.3733/ca.v054n04p66

Rabalais N, Turner R, Wiseman W Jr (2002) Gulf of Mexico hypoxia, AKA "the dead zone". Annu Rev Ecol Syst 33:235–263

Rebitzer G, Ekvall T, Frischknecht R, Hunkeler D, Norris G, Rydberg T, Schmidt W, Suh S, Weidema B, Pennington D (2004) Life cycle assessment: Part 1: Framework, goal and scope definition, inventory analysis, and applications. Environ Int 30(5):701–720

Scott C, Kurian M, Wescoat J Jr (2015) The water-energy-food nexus: enhancing adaptive capacity to complex global challenges. In: Kurian M, Ardakanian R (eds) Governing the nexus. Springer International Publishing, Berlin, Germany, pp 15–38

Shekhar S, Feiner S, Aref W (2016) Spatial computing: accomplishments, opportunities, and research needs. United Nations Environment Programme. https://nexusobservatory.flores.unu.edu. Accessed 29 Jan 2016

USGS (2015) U.S. Geological Survey Web Site. http://www.usgs.gov. Accessed 30 Sept 2015

UNU-FLORES (2015) Nexus Observatory platform website. https://nexusobservatory.flores.unu.edu
United Nations Environment Programme (UNEP) (2016) Life cycle assessment (2016) http://www.unep.org/resourceefficiency/Consumption/StandardsandLabels/MeasuringSustainability/LifeCycleAssessment/tabid/101348/Default.aspx. Accessed 29 Jan 2016
Webber M (2015) Energy, water and food problems must be solved together. Sci Am 312(2) http://www.scientificamerican.com/article/energy-water-and-food-problems-must-be-solved-together/. Accessed 30 Sept 2015

The Bird's-Eye View from a Worm's-Eye Perspective

C. Dana Tomlin

Abstract Digital cartographic data are generally encoded either as drawings of discrete shapes or as images of continuously varying conditions. These *vector* and *raster* formats differ in terms that are not only operational but also conceptual. To explore the kind of thinking that distinguishes raster processing, several brief examples are presented in the form of classroom brainteasers. These are offered in support of a contention that raster operations are better understood from a local, rather than global, point of view.

Keywords Raster · Map algebra · Cartographic modeling

1 Introduction

> If it ends in "s" without ending in "ness," map it with points, lines, or polygons.
> Otherwise, it is probably better mapped as a field of pixels.

When geographic information systems (GIS) were first introduced, the distinction between *vector*-encoded maps of discrete features and *raster*-encoded maps of continuous fields was largely a matter of practical differences in storage, processing, and presentation capabilities. Though more recent advances on all three of those fronts have since mitigated many of these practical differences, perceptual differences nonetheless remain. In an attempt to reconcile some of the latter, the following pages offer a glimpse of the world as seen through the eyes of one who has long been inclined to view it in terms of images rather than drawings. In Fig. 1 is a straightforward depiction of that distinction.

C. Dana Tomlin (✉)
University of Pennsylvania School of Design, Philadelphia, PA, USA
e-mail: tomlin.dana@verizon.net

D.A. Griffith et al. (eds.), *Advances in Geocomputation*,
Advances in Geographic Information Science,
DOI 10.1007/978-3-319-22786-3_3

Fig. 1 Raster versus vector

Figure 2 makes a similar distinction. It does so, however, in ways that are not only more subtle but also more significant. Here, in the first few lines of what many regard as his greatest poem, William Wordsworth (1815) is masterful in using words to represent phenomena for which the very use of those words can seem antithetical. "Clouds," "hills," "crowds," "bays," "the breeze," and even "the milky

Fig. 2 Fields versus features

way" are all entities whose names imply that each is a discrete object with distinct edges, something comparable to one of the poet's "lakes" or "trees." Clearly, however, they are not. To emphasize this, consider a "simple" question: how many clouds are shown in Fig. 2?

2 Ups and Downs

Or, in terms that are more geographical in nature, how many hills are shown in Fig. 3? The question begs a broader one: what is a hill? And the answer to that question raises even broader implications for geospatial modeling. Geographical space certainly can be characterized by the presence or absence of coherent objects (like hills) that are associated with distinct conditions and that occupy distinct locations. Significantly, however, geographical space also can be characterized in terms of conditions (like topographic elevation) that are not associated with any such object but that instead vary continuously from one location to another. This distinction between discrete features and continuous fields is one whose implications extend well beyond data to the operations by which those data are processed and to the mind-sets with which those operations are employed.

When geographical data are encoded as points, lines, or polygons, locations are recorded among the attributes for each unit of observation. When geographical data are encoded as pixels, however, locations *are* the units of observation for which other attributes are recorded. One of the nice things about the latter is that all manner of geographical phenomena are expressed in a common and uncomplicated format. Even nicer is the fact that, by casting locations (rather than objects) as the elemental units by which data are processed, a number of additional processing

Fig. 3 Topographic elevation

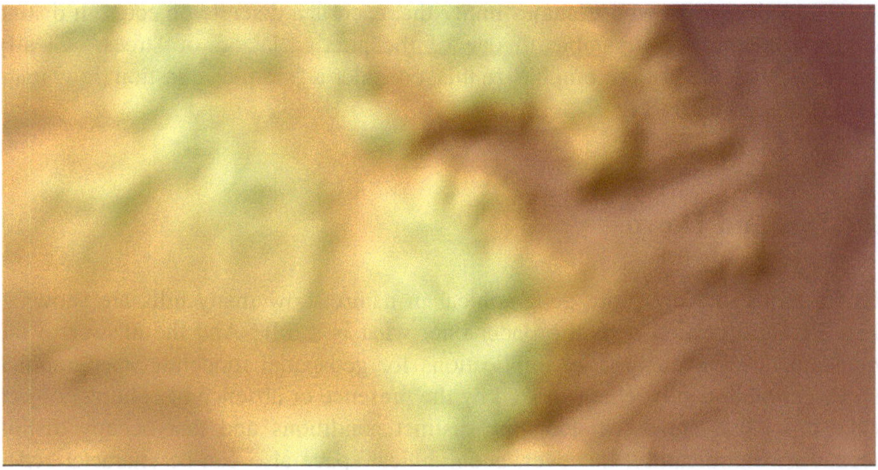

Fig. 4 Topographic elevation smoothed

capabilities can be implemented. Notable among these are operations that model distances, directions, topographic forms, hydrological flows, and other qualities of characterize space itself, rather than objects within it.

To take full advantage of such "map algebraic" capabilities (Tomlin 2012), each operation should be considered not just in terms of its mapwide pattern but also (and more importantly) in terms of its pixel-specific inputs and output. In contrast to (and in concert with) that bird's-eye scan of an entire region, this worm's-eye focus on a typical location often yields greater understanding and control. Consider, for example, the topographic surface depicted in Fig. 4. From the bird's-eye perspective, this might be described as a smoother version of the surface shown in Fig. 3.

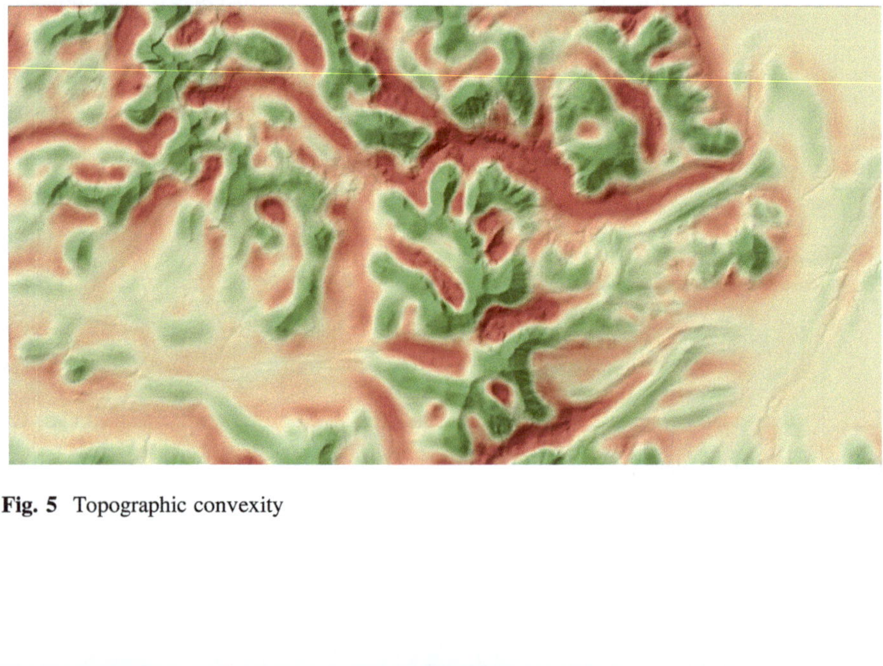

Fig. 5 Topographic convexity

From the worm's-eye perspective, in contrast, it would be seen as a surface on which each pixel's new value is an average of the original values of pixels within its vicinity.

Now consider Fig. 5, where those parts of the same topographic surface that are more convex (hill-like) are shown in darker shades of green, while those that are more concave (valley-like) are shown in darker shades of red. How was this map generated? If the answer is not yet apparent, it is only because you are not thinking like a worm (a worm whose complexion would be green if its own elevation were to exceed the average of its neighbors, and red if it did not).

3 Ins and Outs

Figure 6 presents a similar challenge. Given this map of land (in black) and water (in blue), you can easily see a number of bays. But can your computer see them?

To generate the map shown in Fig. 7 again calls for thinking like a worm. Here, each water pixel has been set to a value indicating the ratio of land to water within that pixel's vicinity. Those that are whiter are those where water is more encompassed by land. In other words, they are more bay-like.

4 Before and After

Now consider the map shown in Fig. 8, where the white star represents a destination and the black dot represents an obstruction. Suppose the lighter shade of gray represents roadways that are more conducive to travel than are nonroad areas depicted in the darker shade of gray.

Fig. 6 Shoreline

Fig. 7 Shoreline concavity

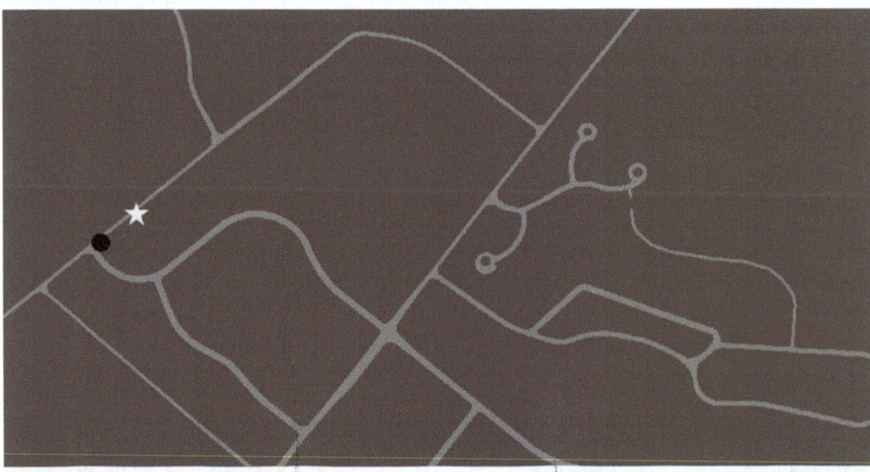

Fig. 8 Travel conditions

Given these conditions, what steps would be required to generate the map presented in Fig. 9? Here, the darker shades of red identify areas where travel time to that destination has been more affected by that obstruction, while the darker shades of green identify areas where the obstruction has had less effect on travel time to the destination.

If you have not yet stopped looking at Fig. 9, do so now. Close your eyes (really) and imagine yourself situated at any given pixel within this image (not unlike a worm). Clearly, whatever impact the obstruction may have on your particular location is a matter of travel time from that location to the designated

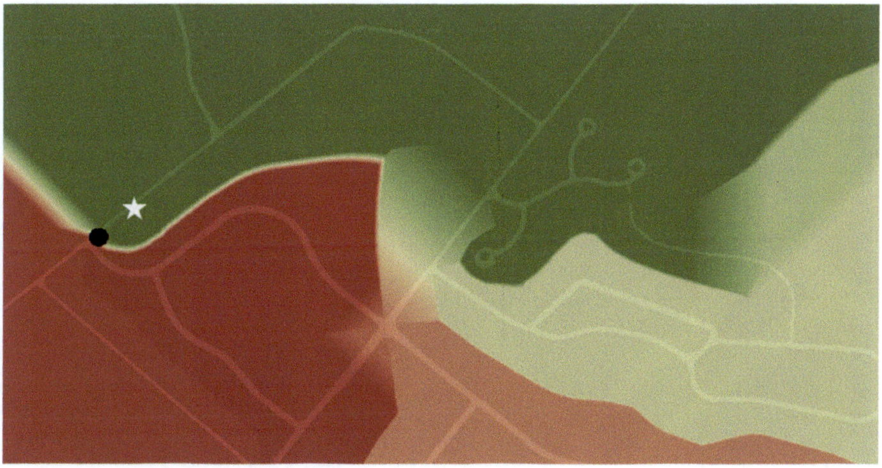

Fig. 9 Increased travel time

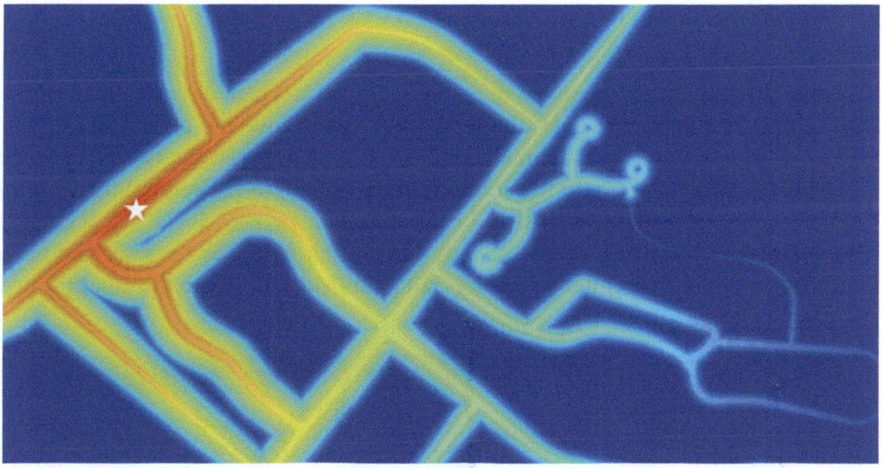

Fig. 10 Unobstructed travel time

destination. To measure that effect of the obstruction, your first step would be to measure your travel time without it, as shown in Fig. 10.

Your next step would be to measure your travel time with the obstruction, as shown in Fig. 11. To generate the map shown in Fig. 9 now requires only that each pixel's postobstruction value (as shown in Fig. 11) be subtracted from its preobstruction value (as shown in Fig. 10).

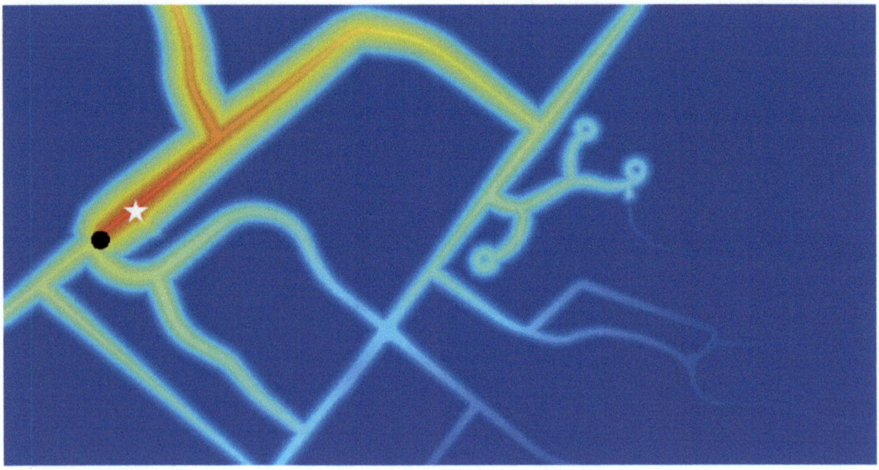

Fig. 11 Obstructed travel time

5 Here and There

A similar problem can be posed in reference to Fig. 12, where darker shades of gray again are used to represent areas through which travel costs are greater. Here, your task is to trace the path of minimum total travel cost between the two white stars.

Figure 13 depicts not only the solution to that problem (in yellow) but also the solution to a more general variation of this problem by showing what amount to Nth-best paths (in varying shades of red). Here, each pixel has been set to a value indicating the total cost of the minimum-cost path from one star to the other through

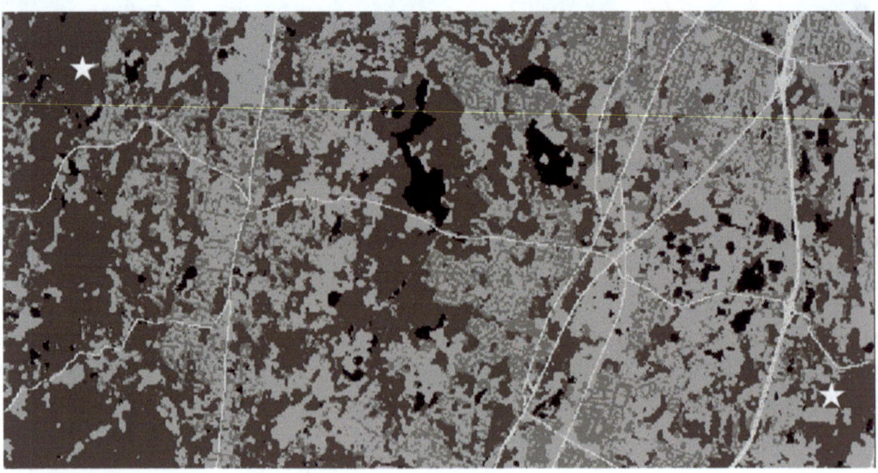

Fig. 12 Incremental travel cost

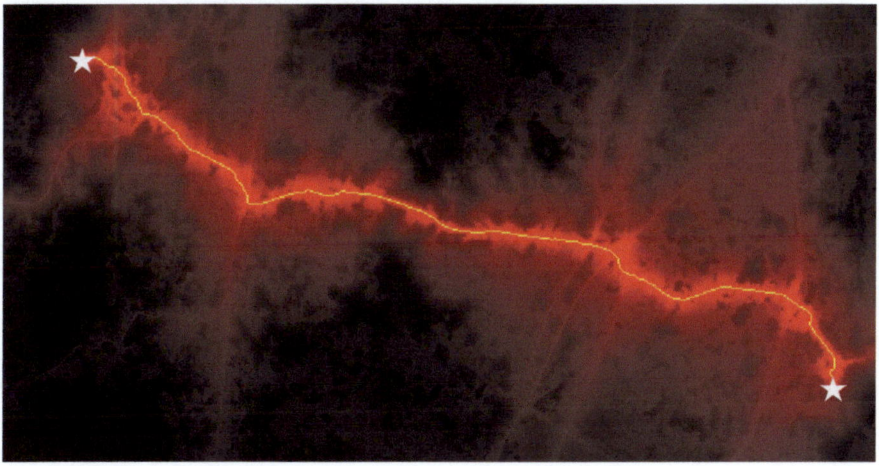

Fig. 13 Cumulative travel cost

that pixel. Lower costs are represented by brighter shades of red. So how was this solution generated? If you select a pixel and think like a worm, you soon realize that not only is this problem similar to the one depicted in Fig. 8 but so is its solution.

6 Corners and Curves

Okay, one final challenge. Among the insular shapes that are shown in Fig. 14, note that some are more rectilinear than others. In fact, these shapes have been colored such that the more rectilinear ones appear in darker shades of green, while those that are less rectilinear appear in darker shades of red. So, how was rectilinearity measured?

As always, we start by thinking like a worm. Each pixel is set to a value indicating the compass direction of the nearest edge of the shape containing that pixel, as illustrated in Fig. 15. Next, each pixel is set to a value indicating the clockwise difference between this compass direction and whichever of north, south, east, or west is closest in the counter-clockwise direction. The standard deviation of these values for any given shape approaches zero when all of that shape's pixels hold directional values that are parallel or perpendicular to one another (regardless of the shape's orientation), and it increases as those directional values vary from this rectilinear alignment.

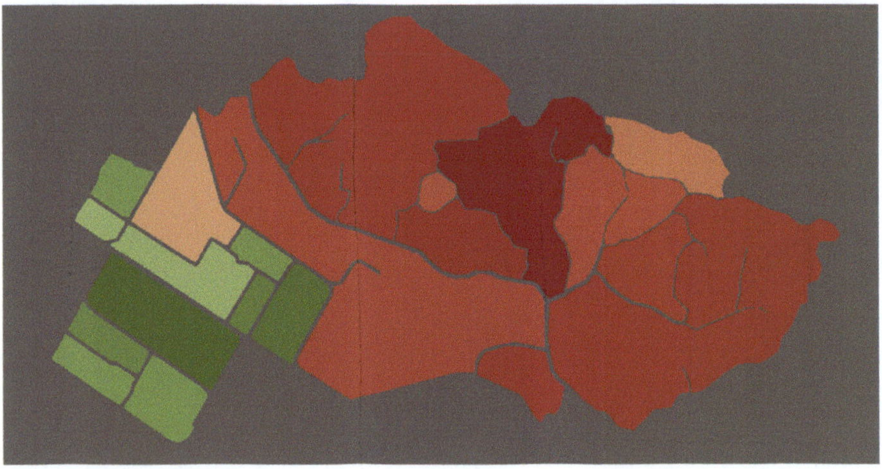

Fig. 14 Rectilinearity

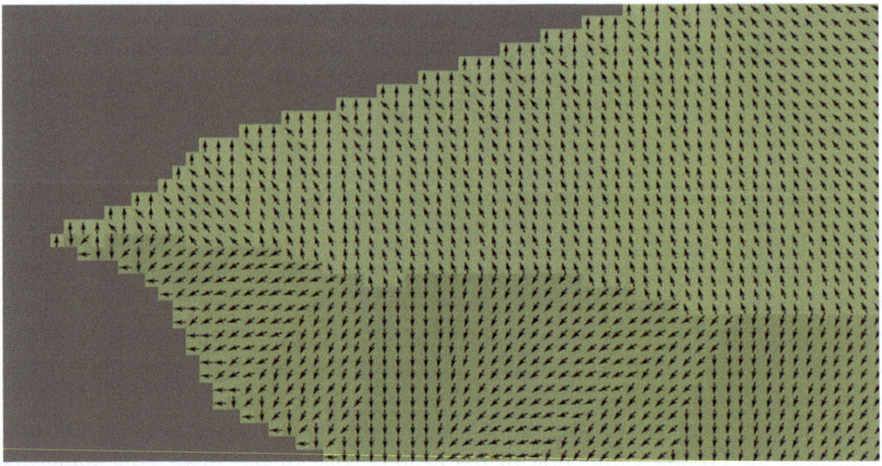

Fig. 15 Edge bearings

7 Conclusion

Given a willingness to think myopically now and then, it is remarkable to see how much can be interpreted by a general-purpose GIS with no more than routine raster processing capabilities. To support that contention, consider the following homework assignment, which refers to the images presented in Fig. 16.

Fig. 16 Recognizing patterns

Please use only the data that have been provided and only ArcGIS to determine

which of three famous painters was responsible for a particular unknown painting, given well known examples of their work. Was it

—J. M. W. Turner (1775–1851) in England,

—Ivan Aivazovsky (1817–1900) in Russia, or

—Zhang Daqian (1899–1993) in China?

This assignment was issued to students who had just been introduced to a standard set of topographic operations several weeks into what was, for most, their first class in GIS. While particular solution strategies varied, well over half of the students were able to successfully use GIS to identify the unknown artist.

The key, of course, is to recognize that black and white are really the same color; they are both just shades of gray.

References

Tomlin D (2012) GIS and cartographic modeling. ESRI Press, Redlands, CA

Wordsworth W (1815) I wandered lonely as a cloud. In: Poems by William Wordsworth, including lyrical ballads, and the miscellaneous pieces of the author, 2 vols. Longman, Hurst, Rees, Orme & Brown, London

Part I
Spatial Data: Construction, Representation, and Visualization

Conference papers in this section focus on spatial data, representation, and visualization, including mapping. One research area concerns compiling data for a fine spatial resolution at which public data are available. Dmowska and Stepinski build a high-resolution population grid for the conterminous USA using dasymetric techniques and block-level population counts. Morton et al. estimate residential energy consumption for a small area using a dasymetric mapping and machine learning approach. Stewart et al. and Sims et al. utilize social media data to construct population dynamics in a small geographic area, such as a building or an athletic sports stadium.

Another group of papers concentrates on geovisualization. Yadav et al. present a prototype system to effectively visualize spatiotemporal data with Google Maps and OpenStreetMap platforms using their application program interfaces (APIs). Pang et al. develop a new visualization system for transit routes by extending the information visualization framework. Bantis et al. analyze spatial-temporal patterns aging people with disabilities using a Bayesian hierarchical model, specifically a Poisson model, and visualize patterns with resulting posterior distributions. The final paper in this section, by Haynes et al., presents Terra Populus, which is a spatial data repository containing heterogeneous data (i.e., vector, raster, and microdata) across the entire globe.

High-Resolution Population Grids for the Entire Conterminous United States

Anna Dmowska and Tomasz F. Stepinski

Abstract To have a more complete awareness of the global environment and how it changes, remotely sensed data pertaining to the physical aspects of the environment need to be complemented by broad-scale demographic data having high spatial resolution. Although such data are available for many parts of the world, there are none available for the United States. Here we report on our ongoing project to develop high-resolution (30–90 m/cell) population/demographic grids for the entire conterminous United States and to bring them into the public domain. Two different, dasymetric modeling-based approaches to disaggregation of block-level census data into a fine grid are described, and resulting maps are compared to existing resources. We also show how to utilize these methods to obtain racial diversity grids for the entire conterminous United States. Our nationwide grids of population and racial diversity can be explored using the online application SocScape at http://sil.uc.edu.

Keywords Population grids · Dasymetric modeling · Demographic data

1 Introduction

Quick and convenient access to high-resolution data for the spatial distribution of population is needed for a wide range of analyses related to resource management, facility allocation, land-use planning, natural hazards, and environmental risk (Dobson et al. 2000; Chen et al. 2004), disaster relief/mitigation (Bhaduri et al. 2002), and socio-environment interactions (Weber and Christophersen 2002). Widely available population and demographic data are a result of aggregating census information into arbitrary areal units to ensure privacy. Consequently, maps showing the

A. Dmowska · T.F. Stepinski (✉)
Space Informatics Lab, Department of Geography, University of Cincinnati, Cincinnati, OH, USA
e-mail: stepintz@uc.edu

A. Dmowska
e-mail: dmowskaa@ucmail.uc.edu

D.A. Griffith et al. (eds.), *Advances in Geocomputation*, Advances in Geographic Information Science,
DOI 10.1007/978-3-319-22786-3_4

spatial distribution of population are in vector form, resulting in an unrealistic level of homogeneity (lack of spatial resolution) in the population distribution. Moreover, these data are inconvenient to use, especially when a study area covers several different administrative regions along which the data are organized (e.g., a metropolitan area that stretches over two or more states).

Given the limitations of areal units-based population data, the focus of investigation has shifted to populations grids that are capable of depicting the population distribution at a higher resolution. These grids also are convenient to use and support means of algorithmic analysis that are not available for vector-based maps. Population grids are obtained using the principle of dasymetric mapping (Wright 1936)—a procedure that subdivides areal units into a regular grid of cells using ancillary information that can serve as a proxy for a more accurate population distribution.

Disaggregation by means of dasymetric modeling is well established and relatively straightforward (Langford and Unwin 1994; Eicher and Brewer 2001; Mennis 2003; Qiu and Cromley 2013), with a majority of applications being applied to small, local study areas for which detailed, local ancillary information—for example, parcels (Jia et al. 2014), buildings (Dong et al. 2010), addresses (Reibel and Bufalino 2005)—is available. However, we are interested in using dasymetric modeling to obtain a continental-scale, high-resolution population grid because it can serve a much bigger number of potential users. Using dasymetric modeling on such a large scale is a technically demanding task due to its needs to handle very large datasets and to develop computationally efficient algorithms. Recognizing a need for broad-coverage population data at a high spatial resolution, population grids have been produced for countries within the European Union, Africa, South America, and Asia (see Table 1).

Until recently, the only public domain gridded population data for the United States (U.S.) were census grids developed by the Socioeconomic Data and Application Center (SEDAC) (http://sedac.ciesin.columbia.edu). SEDAC grids have a number of shortcomings: (1) they have a relatively coarse 1 km resolution (250 m

Table 1 Availability of broad-scale, high-resolution population grids

Project	Region	Resolution	Availability
WorldPop[a]	S. America, Africa, Asia	100 m	http://www.worldpop.org.uk
E.U. pop. grid[b]	Europe	100 m	http://www.eea.europa.eu/
Australian pop. grid	Australia	1000 m	http://www.abs.gov.au
SEDAC-USA	United States	1000/250 m	http://sedac.ciesin.columbia.edu
LandScan-USA[c]	United States	90 m	Not available
SocScape[d]	United States	90/30 m	http://sil.uc.edu/

[a]Tatem et al. (2007), Linard et al. (2012), Gaughan et al. (2013), [b]Gallego (2010), Gallego et al. (2011), [c]Bhaduri et al. (2007), [d]Dmowska and Stepinski (2014)

resolution for selected metropolitan areas), (2) they are a product of simple areal weighting interpolation (Goodchild et al. 1993) rather than disaggregation using dasymetric modeling, and (3) they are available only for 1990 and 2000. The Oak Ridge National Laboratory developed (Bhaduri et al. 2007) the LandScan USA—a 90 m population grid covering the US LandScan uses advanced dasymetric modeling utilizing a multitude of ancillary datasets and comes in two versions, nighttime (density of residential population) and daytime (density of workplace population). However, LandScan 90 is neither in the public domain nor is it commercially available, so it cannot be utilized by the broader scientific community.

Here we report on our ongoing project to bring high-resolution population grids of the entire conterminous United States to the public domain. We have developed two approaches to achieve this goal. Our first approach (hereafter referred to as generation-1 or Gen-1) yields 90 m nationwide population grids for 1990 and 2000. This approach is based on sharpening already existing SEDAC grids using the National Land Cover Dataset (NLCD) (http://www.mrlc.gov/) as ancillary data. Because SEDAC is not producing grids based on the 2010 census, we have decided to change our approach to one that is computationally much more demanding but can be applied to 2010 census data as well as to the data from previous censuses. Our second approach (hereafter referred to as generation-2 or Gen-2) yields 30 m United States-wide population grids for 2010. This approach disaggregates census blocks directly using NLCD 2011 and the newly available, 30 m resolution, United States-wide National Land Use Dataset (NLUD 2010) (Theobald 2014) as ancillary datasets.

For these grids to be easily previewed, we have developed a web-based application called SocScape (Social Landscape) (http://sil.uc.edu). It is designed to facilitate exploration of population density over the entire conterminous United States for 1990, 2000, and 2010. In addition, it can be used to explore the geographical distribution of racial diversity for 1990 and 2000 (the map based on 2010 data is in the development phase). High-resolution racial diversity maps are by-products of our population disaggregation method.

2 Data and Methods

Both the Gen-1 and Gen-2 methods use census data. Gen-2 diaggregates ~11 millions census blocks directly, whereas Gen-1 disaggregates the SEDAC grids. In addition, both methods use the NLCD as ancillary data; but Gen-2 also uses the NLUD 2010 data.

2.1 The Gen-1 Disaggregation Method

Gen-1 calculates 90 m population grids from pre-existing 1 km SEDAC grids. SEDAC grids are a product of simple areal weighting interpolation from 1990 and 2000 census block data; no ancillary data have been used in the process of their creation. We sharpen SEDAC grids from 1,000/250 m/cell to 90 m/cell using land cover (1992 and 2001 editions of the NLCD) as ancillary data. This method was initially selected because it is less computationally demanding than disaggregating directly from census blocks. Computational efficiency comes from working exclusively with grids.

Land cover has been used because it is the only ancillary information that has uniform quality across the entire United States. However, the main editions of 1992 and 2001 NLCD have different land cover legends and cannot be used to produce population grids that can be compared between 1990 and 2000. Because such a comparison is one of our goals, we instead use the NLCD 1992/2001 Retrofit Land Cover Change Product (Fry et al. 2009). This product provides compatible land cover classifications for 1992 and 2001, but at the cost of reducing the number of land cover categories to only eight classes that constitute Level I of the Anderson classification scheme (Anderson et al. 1976). We further reclassify the eight-class land cover data into just three classes: urban, vegetation, and uninhabited area. The Gen-1 dasymetric model uses these three-classes of land cover as its ancillary data.

For details of Gen-1 disaggregation, see Dmowska and Stepinski (2014). Briefly, the population in each SEDAC cell is redistributed to 90 m cells using weights calculated from land cover composition within each 90 m cell and the average population density of the three land cover classes. In metropolitan statistical areas (MSAs), in addition to average population density in land cover classes, we use information about the distribution of population from the finer, 250 m/cell, SEDAC-MSA grids. The population in each 90 m cell is determined by multiplying the population count in a SEDAC cell by a weight specific to a given 90 m cell.

2.2 The Gen-2 Disaggregation Method

Population maps produced by our Gen-1 method offer an efficient means of improving upon the SEDAC grids (see the comparison in Sect. 3). However, because SEDAC has no population grids for 2010, and because SEDAC grids for 1990 and 2000 contain some inconsistencies, we decided to change our disaggregation algorithm to a direct disaggregation from census blocks. This approach requires larger hardware resources and is less computationally efficient, but the outcome justifies the extra effort. In addition, a new ancillary resource—a 30 m/cell land use map (NLUD 2010) over the entire conterminous United States—became available (Theobald 2014) in 2014, and we incorporate it into our Gen-2 method. The Gen-2 method consists of several steps: (1) preprocessing of U.S Census data, (2) preprocessing of ancillary data, (3) sampling population density over different land cover/use classes, and (4) calculating weights and the redistribution of population.

The 2010 United States Census block-level data consist of two components: shapefiles (TIGER/Line Files) with geographical boundaries of ~11 millions blocks, and summary text files that list population data for each block. In the census data preprocessing step, we first join the boundaries shapefile with the data from the summary file to form a vector file. In a key step, which departs from conventional dasymetric modeling, the vector files are rasterized to a 30 m resolution grid. Note that all 30 m cells belonging to a single block initially have the same values. The 30 m resolution was selected because it is the resolution of ancillary datasets; having all datasets as co-registered grids of the same resolution improves the computational efficiency of the disaggregation algorithm.

The major problem with using land cover (NLCD 2011) as ancillary information is that it may not differentiate correctly between inhabited and uninhabited impervious areas. We use land use data (NLUD 2010) to make this differentiation. In the ancillary data preprocessing step, we combine information from NLCD 2011 and NLUD 2010 to define six land cover/use classes: developed open space, developed low intensity, developed medium intensity, developed high intensity, vegetation, and uninhabited. Following Mennis and Hultgren (2006), representative population density for each of the six land cover/use classes is established using a set of blocks (selected from the entire United States) having relatively homogeneous land cover (90 % for developed classes, and 95 % for vegetation classes). Within each block class, weights are calculated using class abundances in the block and their representative population densities (Mennis 2003). Finally, the population in each (rasterized) block is redistributed to its constituent cells using the weights. Note that once the weights are calculated, they can be used not only for disaggregation of total population, but also for disaggregation of segments of populations defined by any attribute for which block-level data are available. In particular, these procedure can be used to disaggregate race-specific populations. This procedure preserves the block-level value of a ratio between different races because no race-specific ancillary information is used.

The Gen-2 algorithm redistributes ~11 millions census blocks over 8 billion grid cells. The output grid file size is 139 GB. To keep computational cost under control, calculations were performed for each state separately, and results were joined into a single map for the entire United States. These calculations were done using a computer with an Intel 3.4 GHz, 4-cores processor and 16 GB of memory running the Linux operating system. All calculations were performed using Python scripts written for GRASS GIS 7.0 software. The total time of Gen-2 calculations was 66 h.

3 Results

The best way to examine population density grids generated by our Gen-1 and Gen-2 algorithms is to explore them using the SocScape web application. SocScape is a computerized map application that allows a population density map to be overlaid on a base layer of either a street map or an aerial image. It works on desktops, laptops, and mobile devices. It supports downloading of data in either GeoTiFF or PNG formats. Note, however, that only classified data (as seen in the application) can be downloaded. The numerical data (people counts per cell) are too large to be distributed via SocScape, but are available upon request from the authors. Here we use three examples to demonstrate how our grids compare with other available population density resources. The first example compares different population density maps/grids in an urban setting (Cincinnati, Ohio). The second example compares different population density maps/grids in a rural setting (Somerset, Ohio). Finally, the third example shows how our method can be used to produce high-resolution racial diversity grids.

Figure 1 compares different population density grids in a portion of the Cincinnati (Ohio) metropolitan region. This site captures a region around the Ohio River, with Cincinnati located north of the river and Kentucky located south of the river. The industrial transportation corridor (that includes railroad tracks and Interstate 71/75) runs through the middle of the site, from the Ohio River northward. To the west of this corridor are residential neighborhoods, and to the east is the downtown area. For reference, Fig. 1a shows a satellite image of the site, and Fig. 1b shows a land cover map (NLCD 2011) of the site. Five population density maps are compared: census block-based (Fig. 1c), SEDAC 1 km grid (Fig. 1d), SEDAC 250 m grid (available for the Cincinnati MSA) (Fig. 1e), our Gen-1 90 m grid (Fig. 1f), and our Gen-2 30 m grid (Fig. 1g). All grids have the same legends, and the land cover legend is also shown for reference.

Because sizes of census blocks are small in heavily populated areas, a block-based map (Fig. 1c) offers more actual details in the downtown area than the other maps do, except for our Gen-2 grid (Fig. 1g). Outside the downtown, where blocks are large and they may include parks and other sparsely inhabited areas, the block-based map still offers advantages over the SEDAC 1 km grid but not over the other grids. The SEDAC 1 km grid (Fig. 1d) captures only the most basic features of the population distribution. The presence of a river and of an industrial transportation corridor cannot be deduced from this map. The SEDAC 250 m grid (Fig. 1e) shows some additional details. The river and the industrial transportation corridor are delineated; however, parks are not distinguished from built-up areas. Our Gen-1 grid (Fig. 1f) offers a significant improvement over the SEDAC 250 m grid. The river and the industrial transportation corridor are well delineated and parks and green spaces are distinguished from built-up areas. Comparing the Gen-1 grid with the block-based map, however, reveals that some industrial, uninhabited areas are shown as inhabited because some NLCD classes contain residential and non-residential buildings. Also, the resolution of our Gen-1 map in the downtown area is not as good as it is in the block-based map. Based on a comparison with a satellite image (Fig. 1a) and a land cover map (Fig. 1b), as well as our familiarity with the site, we conclude that our Gen-2 grid (Fig. 1g) offers the best depiction of population density in this site. Utilizing land use ancillary data results in a proper delineation of uninhabited areas and the parks, and forested areas also are correctly shown as having low population density. Finally, the resolution in the downtown area is as good as, or better than, it is in the block-based map.

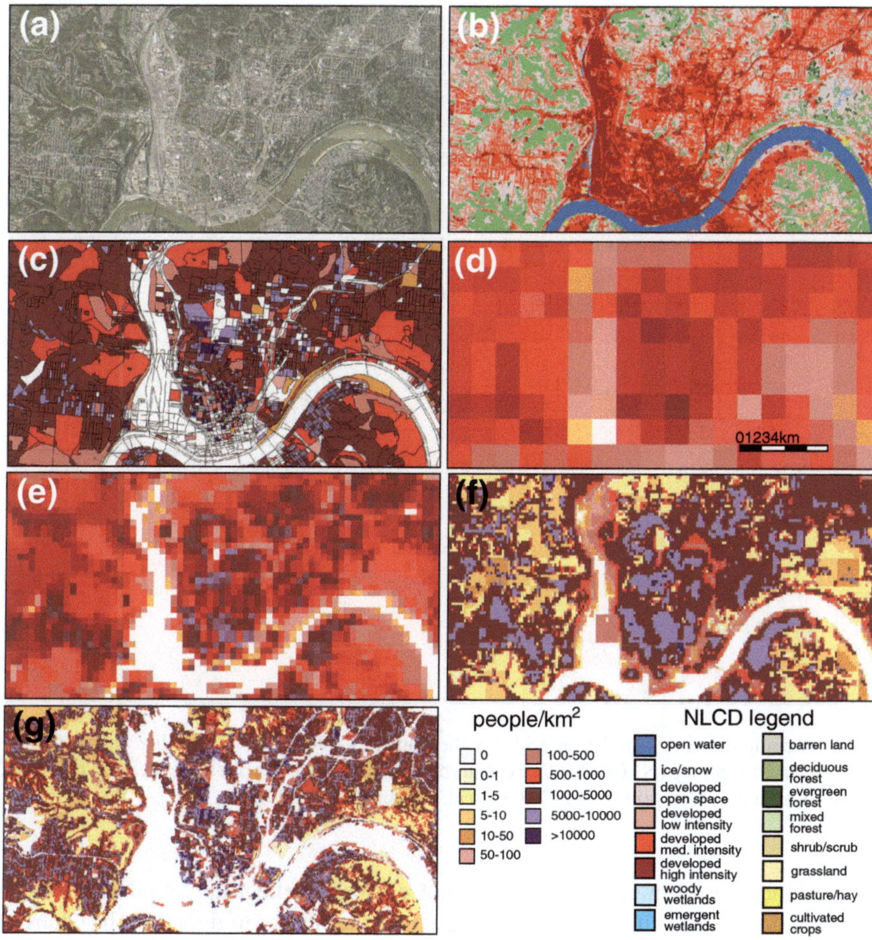

Fig. 1 A comparison of population grids for the Cincinnati (Ohio) site. **a** Satellite image (Google Maps), **b** land cover map (NLCD 2011), **c** census block-based map of population density, **d** SEDAC 1 km grid, **e** SEDAC 250 m grid, **f** Gen-1 90 m grid, **g** Gen-2 30 m grid

Figure 2 compares different population density grids in a rural site located around the village of Somerset (Ohio). For reference, Fig. 2a shows a satellite image of the site, and Fig. 2b shows a land cover map (NLCD 2011) of the site. The site consists mostly of agricultural land. Apart from the village, the population is concentrated in farmhouses located predominantly along roads. Because the site is sparsely populated, census blocks are relatively large. Four population density maps are compared: census block-based (Fig. 2c), SEDAC 1 km grid (Fig. 2d), Gen-1 90 m grid (Fig. 2e), and Gen-2 30 m grid (Fig. 2f). All grids have the same legend (see Fig. 1 for the legend). Note that, unlike the Cincinnati site, the SEDAC 250 m grid is not available for this area.

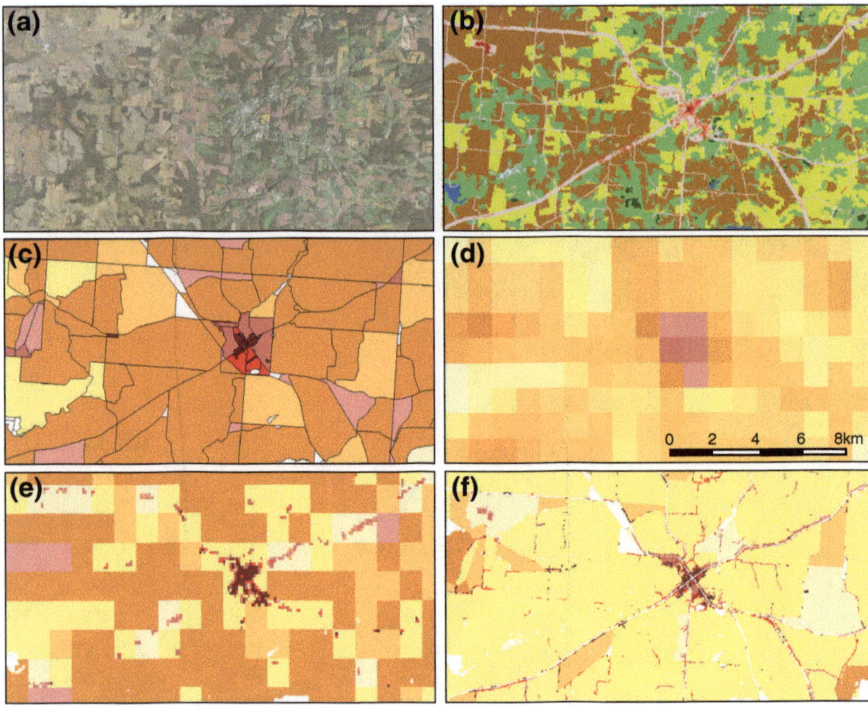

Fig. 2 A comparison of population grids for the Somerset (Ohio) site. **a** Satellite image (Google Maps), **b** land cover map (NLCD 2011), **c** census blocks-based map of population density, **d** SEDAC 1 km grid, **e** Gen-1 90 m grid, **f** Gen-2 30 m grid. For legends, see Fig. 1

The block-based map (Fig. 2c) does not correctly reflect an actual distribution of population in this site. The color assigned to the blocks to depict their values of population density reflects an average density over each entire block, whereas most of the block area is uninhabited or very sparsely populated because it is covered by crops and pastures. The SEDAC 1 km grid (Fig. 2d) offers a fair approximation to an overall distribution of population, but without any details for the village of Somerset (population 1,418). The Gen-1 grid (Fig. 2e) was able to resolve the village, but its population density over farmland is too high (although this value is still low, given that the area is rural). The Gen-2 grid (Fig. 2f) recognizes individual farm houses, and thus lowers the population density of farmland as people are assigned to very small regions at the locations of farm houses. Additional information from the land use data delineates uninhabited land such as the state forest (the light yellow area at the right edge of the region).

Figure 3 shows maps of racial diversity for an urban site covering Cincinnati, Ohio (Fig. 3a and b) and a rural site around Fresno, California (Fig. 3c and d). Our racial diversity maps (Fig. 3a and c) are 90 m grids produced using the Gen-1 disaggregation method on 2000 census data. The total population is segmented with respect

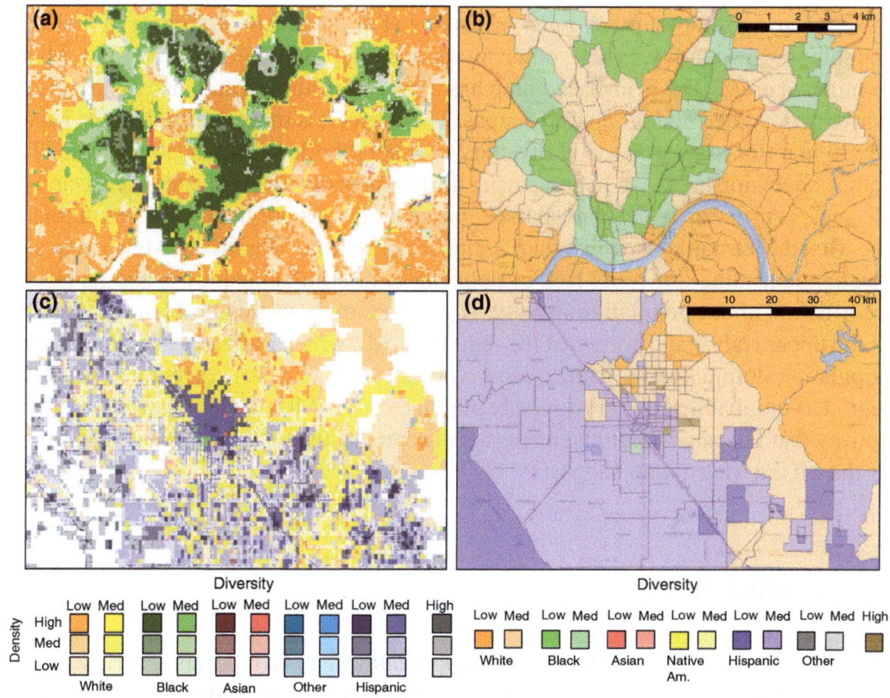

Fig. 3 A comparison of racial diversity maps for sites in urban and rural settings. **a** Racial diversity, 90 m grid for the Cincinnati (Ohio) area, **b** Mixed Metro census tracts-based map for the Cincinnati area, **c** Racial diversity, 90 m grid for the Fresno (California) area, **d** Mixed Metro census tracts-based map for the Fresno area

to race/ethnicity into the following groups: whites, blacks, Asians, Hispanics, and others. Each group is disaggregated separately in a way to preserves the value of total population in each 90 m cell. Using values of population densities for all race groups, we classify grid cells into 33 categories, taking into account diversity level (low, medium or high), dominant race, and population density (low, medium, and high). Uninhabited areas are grouped into the separate 34th category. Details of this classification are given in Dmowska and Stepinski (2014). The two diversity grids shown in Fig. 3 are parts of the U.S.-wide racial diversity grid that can be explored in the SocScape web application. For comparison, we also show (Fig. 3b and d) racial diversity maps of the same sites that are available from http://mixedmetro.com/. The Mixed Metro maps are not grids; instead, they are based on census tracts. They are produced using a classification of population into 13 diversity/race categories. Their classification is similar to ours, but does not include population density, and uses division into six instead of five racial groups. The legends of the two classifications have been constructed to correspond to each other as much as possible.

Comparing the two maps in an urban site (Fig. 3a and b), we observe that they roughly correspond to each other in delineating white-dominated and black-dominated areas. The grid-based map has a better resolution, and also distinguishes between inhabited and uninhabited areas. Therefore, it is more useful than the Mixed Metro map for assessing racial diversity of a neighborhood at the street scale. The same conclusion holds when comparing the two maps in the rural site (Fig. 3c and d). Because census tracts in more sparsely populated areas are larger, the resolution of the Mixed Metro map is worse than in the urban setting. The Mixed Metro map also can be misleading due to its lack of delineation of uninhabited areas and its lack of information about population density. A good example of this problem is observed in the upper right-hand corner of the map (Fig. 3d), which indicates a large area with population dominated by whites. However, our map (Fig. 3c) correctly depicts this region as uninhabited or very sparsely populated. The Mixed Metro map also fails to convey that Hispanic-dominated areas are urban clusters that are not as wide-spread as the Mixed Metro map suggests.

4 Conclusions

Our project to develop high-resolution population and demographic grids for the entire United States has resulted in 90 m population density grids for the years 1990 and 2000, and a 30 m grid for 2010. It has also resulted in 90 m racial diversity grids for 1990 and 2000. These grids are available for exploration and downloading from the SocScape web application at http://sil.uc.edu.

The Gen-2 method, used to calculate a 2010 edition of the population grid, cannot be fully applied to 1990 and 2000 data because the land use data (Theobald 2014) pertain only to 2010. However, the Gen-2 method can recalculate population grids for 1990 and 2000, but with land cover as the only ancillary data. The Gen-2 method is a significant improvement over our older Gen-1 method (Dmowska and Stepinski 2014) because it only uses original census data, which gives us a full control over the quality of its results. It is difficult-to-impossible to perform a formal assessment of our grids, accuracy. Such an assessment requires comparison with sub-block resolution data, such as, for example, parcel data. This is not feasible for the entire United States but can be performed with very small regions for which parcel data have been utilized to calculate population density. Using results available in the literature, we conclude that our 30 m population grid agrees well with parcel-derived population density maps for a small area in Alachua County, Florida (Jia et al. 2014).

Grids of demographic variables other than population density can be calculated using weights established by the population model. Examples of such variables (available at the census block level) are race, age, and income. No ancillary data specific to race, age, or income exist that would allow directly disaggregating these variables within a block; but we can disaggregate them according to the population model. By narrowing the locations where people live within a block, we increase the spatial resolution of these variables, although we would not be able to account for variation of, for example, racial diversity within a block. Nevertheless, as illustrated by Fig. 3, using grids of demographic variables (like racial diversity) instead of maps based on census areal units results in a much better depiction of the actual spatial distribution of these variables.

Acknowledgements This work was supported by the University of Cincinnati Space Exploration Institute.

References

Anderson JR, Hardy EE, Roach JT, Witmer RE (1976) A land use and land cover classification system for use with remote sensor data. Tech rep, Geological Survey Professional Paper 964

Bhaduri B, Bright E, Coleman P, Dobson J (2002) LandScan: locating people is what matters. Geoinformatics 5(2):34–37

Bhaduri B, Bright E, Coleman P, Urban ML (2007) LandScan USA: a high-resolution geospatial and temporal modeling approach for population distribution and dynamics. GeoJournal 69(1–2):103–117

Chen K, McAneney J, Blong R, Leigh R, Hunter L, Magill C (2004) Defining area at risk and its effect in catastrophe loss estimation: a dasymetric mapping approach. Appl Geogr 24(2):97–117

Dmowska A, Stepinski TF (2014) High resolution dasymetric model of US demographics with application to spatial distribution of racial diversity. Appl Geogr 53:417–426

Dobson JE, Bright EA, Coleman PR, Durfee RC, Worley BA (2000) LandScan: a global population database for estimating populations at risk. Photogram Eng Remote Sens 66(7):849–857

Dong P, Sathya R, Nepali A (2010) Evaluation of small-area population estimation using LiDAR, Landsat TM and parcel data. Int J Remote Sens 31(2):5571–5586

Eicher CL, Brewer CA (2001) Dasymetric mapping and areal interpolation: implementation and evaluation. Cartogr Geogr Inf Sci 28:125–138

Fry JA, Coan MJ, Homer CG, Meyer DK, Wickham JF (2009) Completion of the National Land Cover Database (NLCD) 1992–2001 land cover change retrofit product. Tech rep, U.S. Geological Survey Open-File Report 2008–1379

Gallego FJ (2010) A population density grid of the European Union. Popul Environ 31(6):460–473

Gallego FJ, Batista F, Rocha C, Mubareka S (2011) Disaggregating population density of the European Union with CORINE land cover. Int J Geogr Inf Sci 25(12):2051–2069

Gaughan AE, Stevens FR, Linard C, Jia P, Tatem AJ (2013) High resolution population distribution maps for Southeast Asia in 2010 and 2015. PLoS One 8(2):e55,882

Goodchild M, Anselin L, Deichmann U (1993) A framework for the areal interpolation of socioeconomic data. Environ Plann A 25:383–397

Jia P, Qiu Y, Gaughan AE (2014) A fine-scale spatial population distribution on the high-resolution gridded population surface and application in Alachua County, Florida. Appl Geogr 50:99–107

Langford M, Unwin D (1994) Generating and mapping population density surfaces within a geographical information system. Cartogr J 31(1):21–26

Linard C, Gilbert M, Snow RW, Noor AM, Tatem AJ (2012) Population distribution, settlement patterns and accessibility across Africa in 2010. PLoS One 7(2):e31,743

Mennis J (2003) Generating surface models of population using dasymetric mapping. Prof Geogr 55(1):31–42

Mennis J, Hultgren T (2006) Intelligent dasymetric mapping and its application to areal interpolation. Cartogr Geogr Inf Sci 33(3):179–194

Qiu F, Cromley R (2013) Areal interpolation and dasymetric modeling. Geogr Anal 45(3):213–215

Reibel M, Bufalino ME (2005) Street-weighted interpolation techniques for demographic count estimation in incompatible zone systems. Environ Plann A 37(1):127–139

Tatem AJ, Noor AM, vonHagen C, DiGregorio A, Hay SI (2007) High resolution population maps for low income nations: combining land cover and census in East Africa. PLoS One 2(12):e1298
Theobald DM (2014) Development and applications of a comprehensive land use classification and map for the US. PloS One 9(4):e94,628
Weber N, Christophersen T (2002) The influence of non-governmental organisations on the creation of Natura 2000 during the European Policy process. For Policy Econ 4(1):1–12
Wright J (1936) A method of mapping densities of population: with Cape Cod as an example. Geogr Rev 26(1):103–110

A Hybrid Dasymetric and Machine Learning Approach to High-Resolution Residential Electricity Consumption Modeling

April Morton, Nicholas Nagle, Jesse Piburn, Robert N. Stewart and Ryan McManamay

Abstract As urban areas continue to grow and evolve in a world of increasing environmental awareness, the need for detailed information regarding residential energy consumption patterns has become increasingly important. Though current modeling efforts mark significant progress in the effort to better understand the spatial distribution of energy consumption, the majority of techniques are highly dependent on region-specific data sources and often require building- or dwelling-level details that are not publicly available for many regions in the United States. Furthermore, many existing methods do not account for errors in input data sources and may not accurately reflect inherent uncertainties in model outputs. We propose an alternative and more general hybrid approach to high-resolution residential electricity consumption modeling by merging a dasymetric model with a complementary machine learning algorithm. The method's flexible data requirement and statistical framework ensure that the model both is applicable to a wide range of regions and considers errors in input data sources.

Keywords Energy modeling · Dasymetric modeling · Machine learning

A. Morton (✉) · N. Nagle · J. Piburn · R.N. Stewart · R. McManamay
Oak Ridge National Laboratory, 1 Bethel Valley Road, Oak Ridge, TN 37831, USA
e-mail: mortonam@ornl.gov

N. Nagle
e-mail: naglenn@ornl.gov

J. Piburn
e-mail: piburnjo@ornl.gov

R.N. Stewart
e-mail: stewartrn@ornl.gov

R. McManamay
e-mail: mcmanamayra@ornl.gov

D.A. Griffith et al. (eds.), *Advances in Geocomputation*,
Advances in Geographic Information Science,
DOI 10.1007/978-3-319-22786-3_5

1 Introduction

As urban areas continue to grow and evolve in a world of increasing environmental awareness, the need for detailed information regarding energy consumption patterns has become increasingly important. Because the residential sector alone accounts for approximately 30 % of all energy consumption worldwide (Swan and Ugursal 2009), a detailed spatial understanding of residential energy consumption, in particular, is important for supporting efforts to promote conservation, technology implementation, and other necessary changes in the existing energy infrastructure.

Because most open access energy data are coarse, they prevent finer resolution spatial analyses that are needed to make meaningful decisions. Consequently, several authors have proposed a variety of statistical- and engineering-based techniques for estimating residential energy use at a more detailed level (Swan and Ugursal 2009). Although these methods mark significant progress in the energy modeling field, the majority are highly dependent on region-specific data sources and often require building- or dwelling-level details that are not publicly available for many regions in the United States (US). Furthermore, many of these methods do not account for errors in input data sources and may not accurately reflect inherent uncertainties in model outputs.

In light of these limitations, we propose an alternative and more general hybrid approach to high-resolution residential electricity consumption modeling by merging the dasymetric model proposed by Nagle et al. (2014) with a complementary machine learning algorithm. Rather than basing the model on sparsely available high-resolution data sources, we choose to disaggregate publicly available datasets into higher resolution target regions. The flexible data requirement, along with the model's statistical framework, ensures that the model both is applicable to a wide range of regions and considers errors in input data sources.

2 Related Work

The residential energy modeling and dasymetric modeling fields have been widely studied by several researchers. Residential energy modeling typically involves collecting data for specific households and then using these data to learn or predict household-level energy consumption, often via a machine learning algorithm (Swan and Ugursal 2009). In the energy modeling discipline, a variety of machine learning algorithms, such as neural networks and support vector machines, have been trained using data related to household attributes, such as demographic, economic, and appliance use-related variables, to predict energy consumption (Swan and Ugursal 2009). Aydinalp et al. (2002) develop a consumption model by training a neural network using detailed household billing data, whereas Parti and Parti (1980) applied linear regression to similar household electrical billing data and a detailed survey of appliance use. In addition, Kadian et al. (2007) utilized appliance distributions and microlevel data sources to develop an engineering-based energy consumption model for the residential sector of Delhi, while Saidur et al. (2007) created a residential energy model for Malaysia based on variance distribution estimates of appliance ownership, appliance power rating and efficiency, and appliance use. Although these methods mark significant progress in the energy modeling field, they are highly dependent on region-specific data sources and often require building- or dwelling-level details (e.g., household billing data, demographics, and appliance ownership information) that are not publicly available for many regions.

Dasymetric modeling includes a suite of techniques to more precisely depict the spatial distribution of population within a spatially aggregate region (Slocum el al. 2009) and is closely related to the field of areal interpolation, a discipline concerned with the procedures for transferring attribute data from one partitioning constituting a set of source units to a set of target units (Qiu and Cromley 2013). Mennis and Hultgren (2006) suggest the earliest reference to dasymetric mapping occurred in 1922, when Russian cartographer Semenov Tian-Shansky (McMaster and McMaster 2003) produced a population map of European Russia. Since then, several other traditional approaches, including the binary method (Eicher and Brewer 2001), the class percent and limiting variable methods (Wright 1936; McCleary 1969; Eicher and Brewer 2001), and the expectation maximization approach (Dempster et al. 1977) have been developed. More recently, Langford (2013) explored the accuracy of areal interpolation in providing population estimates for target zones much smaller than those of the finest geographic resolution census division, and Nagle et al. (2014) presented the penalized maximum entropy dasymetric model (PMEDM), a dasymetric technique unique in its ability to consider explicitly the uncertainty often present in population data and ancillary data layers.

3 Methodology

To create a flexible high-resolution electricity modeling framework that does not depend on sparse high-resolution data, we borrow techniques from the field of dasymetric modeling and machine learning. More specifically, rather than start with household-level data that may or may not be available at our desired spatial resolution, we begin by disaggregating publicly available coarse (and often uncertain) household counts into our desired smaller regions and then use a complementary machine learning algorithm to estimate electricity consumption for each of the disaggregated households. We then sum over the estimated consumption values for households placed within specific target regions to produce aggregate estimates at the census block group level. Although many dasymetric techniques exist, we use the dasymetric approach proposed by Nagle et al. (2014) because of its ability to consider the uncertainty typically present in population data and ancillary data layers. In the following section, we discuss the hybrid dasymetric and machine learning approach to residential electricity consumption modeling in greater detail.

More formally, suppose we have access to a source sample containing n of N households from region s partitioned into $t \in \{1, \ldots, T\}$ target regions, and assume p_{it} is the unknown probability that a household "like" household i is in target region t. In addition, assume each survey response contains a variable or vector of variable values c_i that can be used by some machine learning function $f(c_i)$ to estimate household electricity consumption.

Our first goal is to estimate the expected number of households $w_{it} = Np_{it}$ that are like household i in target region t by using complementary ancillary data and a dasymetric model. Suppose we are given this ancillary information in the form of household counts with specific characteristics for nested geographies within s. More specifically, let $\widehat{pop}_{ka}$ represent an uncertain estimate of the number of housing units with characteristic k in subregion a (i.e., the number of five-bedroom housing units in a specific tract in Tennessee), where e_{ka} is the positive or negative error that is the difference between the estimated and true housing unit count, and σ_{ka}^2 is the variance of the error. Given the preceding constraints, as well as prior probabilities q_{it} for all p_{it}, we determine the number of households $w_{it} = Np_{it}$ that are like household i in target region t, as well as each of the errors e_{ka}, by solving the optimization problem

$$\text{MIN: } n \sum_{it} p_{it} \log \frac{p_{it}}{q_{it}} + \sum_{it} \frac{e_{ka}^2}{2\sigma_{ka}^2} \tag{1}$$

subject to the relaxed pycnophylactic constraints:

$$\sum_i \sum_{t \subseteq a} Np_{it} \cdot I_k(w_{it}) = \widehat{pop}_{ka} + e_{ka} \text{for each constraint } k \text{ and sub-region } a, \tag{2}$$

where

$$I_k(i) = \begin{cases} 1 & \text{If household i has characteristic } k \\ 0 & \text{Otherwise} \end{cases}. \tag{3}$$

The term $\sum_{it} p_{it} \log \frac{p_{it}}{q_{it}}$ in the objective function is equal to the Kullback–Leibler divergence of q from p. Thus, we can interpret our objective function as choosing the p that minimizes the information lost when using the prior distribution q to approximate p and choosing the error terms that minimize the sum of the ratio of squared error terms over variance terms.

Once the preceding estimates w_{it} have been determined, the final aggregate electricity consumption for target region t can be estimated with the function

$$C_t = \sum_t Np_{it} \cdot f(c_i). \tag{4}$$

4 Application and Results

To illustrate the utility of the proposed hybrid model, we estimate the aggregate electricity consumption at the census block group level for Anderson, Union, and Knox counties of the Knoxville metropolitan statistical area.

4.1 Datasets

We obtain the source population from the 2008–2012 household-level Public Use Microdata Sample (PUMS) of the American Community Survey (ACS) (US Census Bureau 2012a), which includes detailed demographic and household characteristics, as well as a variable denoting whether the average monthly electricity cost is provided and, if so, its value in dollars per household for a 5 % sample of households chosen from course geographic units called public use microdata areas (PUMAs) (US Census Bureau 2012a). To determine the constraints and variances, we use the 2008–2012 ACS summary tables (US Census Bureau 2012b), which contain both census tract and block group level average totals and their corresponding 90 % margins of error (MOEs) (US Census Bureau 2009).

The weights w_{is} provided for each household by the PUMS survey represent the number of households that are like household i in PUMA s. We assume each unique household has the same probability of belonging to each target region and thus let $q_{it} = \frac{w_{is}}{T \cdot \sum_{r=1}^{n} w_{rs}}$. We select the census tract and block group level ACS summary table totals summarized in Table 1 as our constraints $\widehat{pop}_{ka}$ and their corresponding 90 % MOEs to derive our error variances σ_{ka}^2. Note that one–four-person household counts were not included in this study because of additional data-processing requirements. However, they can be included in future studies and would likely improve the results. When the variable denoting the average monthly electricity cost is provided, we set c_i equal to it and then compute the average monthly consumption with the function $f(c_i) = c_i/r_i$, where $r_i = 0.097$ is the average rate per kilowatt-hour (kWh) reported by the Knoxville Utilities Board (2012). When the average rate is not provided, we use Breiman's (2001) random forest regression, a popular technique in the machine learning community because of both its ability to handle categorical variables and its robustness against overfitting (Liaw and Wiener 2002) to estimate c_i using the same constraint variables for training. Once the missing c_i s are learned, we again use the function $f(c_i) = c_i/r_i$ to estimate the average monthly consumption in kWh. In this particular application of the model, machine learning is used to learn the average monthly electricity cost c_i only when it is missing from a microdata sample. However, in applications using other household datasets, the average monthly electricity cost may not be provided at all. In such cases, a more sophisticated machine learning algorithm would likely be helpful in estimating parameters for a more complex function $f(c_i)$, where c_i likely represents a vector of several variables available in the microdata sample.

Table 1 Census tract and block group level constraints

Constraint	Tract	Block group
0-person households	X	X
1-person households	X	
2-person households	X	
3-person households	X	
4 or more-person households	X	
0-bedroom households	X	X
1-bedroom households	X	X
2-bedroom households	X	X
3-bedroom households	X	X
4-bedroom households	X	X
5 or more-bedroom households	X	X
Houses built 2010 or later	X`	X
Houses built 2000–2009	X	X
Houses built 1990–1999	X	X
Houses built 1980–1989	X	X
Houses built 1970–1979	X	X
Houses built 1960–1969	X	X
Houses built 1950–1959	X	X
Houses built 1940–1949	X	X
Houses built 1939 or earlier	X	X
Housing units	X	X

4.2 Results and Discussion

Figure 1 shows the normalized estimated average census block group level electricity distribution in kWh per m^2 for Anderson, Union, and Knox counties. As expected, the census block groups closer to the downtown Knoxville area and the University of Tennessee, Knoxville, have a much higher normalized average residential electricity consumption estimate than the more rural areas lying outside the major cities. In addition, most of the uninhabited areas, such as Blue Ridge State Park and the forested areas to the west of Rocky Top, have very low normalized average residential electricity consumption estimates.

Although validation of the final results by comparing the actual and predicted aggregate electricity among census block groups would be very useful, doing so is extremely difficult because of privacy issues, publicly unavailable data, and the number of utility providers that typically cover a study area, each with different data-sharing policies and regulations. For example, the three counties covered in this study area contain 11 utility companies, 8 of which are classified as municipal and 3 as cooperative. Upon contacting these companies, most were not willing to share any data, primarily due to privacy issues and/or resource issues (paying or finding time for employees to process data).

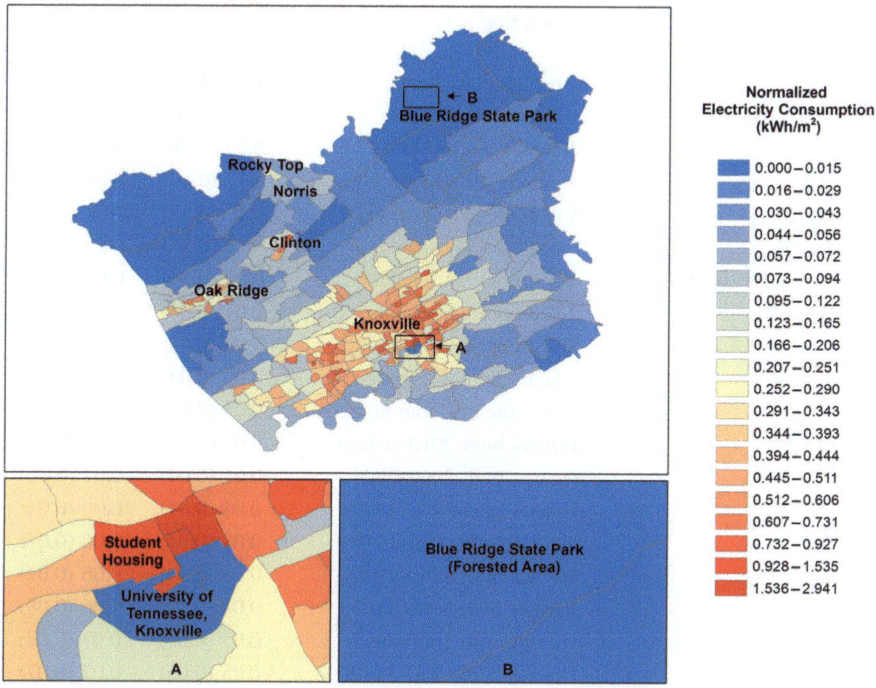

Fig. 1 Normalized census block group level residential electricity consumption estimates (kWh/m^2) for Anderson, Union, and Knox counties

Despite the challenges in evaluating final results, however, evaluation of the reasonableness of subresults coming from the two most influential components of the model is possible: the disaggregation of households and the predicted monthly electricity consumption for different household types. Table 2 shows the mean relative error (MRE) and the standard deviation of the relative error (SRE), both rounded to two digits, between the nonzero original constraint totals at the tract and census block group levels, and the new totals after the households have been disaggregated. The counts match most closely for total housing units, with a 0.01 MRE and 0.02 SRE for both census tracts and block groups. This outcome is reasonable because the variance σ_{ka}^2 of this category is much smaller than the variance of other categories, thus penalizing a large e_{ka} when optimizing the objective function in expression (1). All constraints for the year a house was built, and number of one–five-bedroom households, follow closely with MREs and SREs less than 0.08. The number of one-four-person households have higher MREs, varying between 0.13 and 0.17, followed by the number of zero-bedroom houses

Table 2 MREs between the original housing counts and disaggregated households along with the SREs

Constraint	Tract MRE (SRE)	Block group MRE (SRE)
0-person households	**2.31 (1.41)**	**0.76 (0.15)**
1-person households	0.17 (0.10)	Not selected
2-person households	0.15 (0.08)	Not selected
3-person households	0.17 (0.10)	Not selected
4 or more-person households	0.13 (0.07)	Not selected
0-bedroom households	0.20 (0.31)	0.23 (0.30)
1-bedroom households	0.06 (0.06)	0.07 (0.07)
2-bedroom households	0.04 (0.04)	0.06 (0.08)
3-bedroom households	0.02 (0.03)	0.03 (0.04)
4-bedroom households	0.03 (0.03)	0.05 (0.04)
5 or more-bedroom households	0.05 (0.04)	0.05 (0.05)
Houses built 2010 or later	0.08 (0.07)	0.07 (0.07)
Houses built 2000–2009	0.04 (0.04)	0.07 (0.07)
Houses built 1990–1999	0.04 (0.04)	0.06 (0.06)
Houses built 1980–1989	0.03 (0.04)	0.06 (0.07)
Houses built 1970–1979	0.04 (0.04)	0.06 (0.07)
Houses built 1960–1969	0.04 (0.04)	0.06 (0.08)
Houses built 1950–1959	0.05 (0.06)	0.07 (0.08)
Houses built 1940–1949	0.05 (0.05)	0.07 (0.07)
Houses built 1939 or earlier	0.07 (0.07)	0.07 (0.07)
Housing units	0.01 (0.02)*	0.01 (0.02)*

Bold numbers indicate the highest MRE in the census tract and block group categories, whereas * indicates the lowest MREs in each category

with an MRE of 0.20 and 0.23 for census tracts and block groups, respectively. The number of zero-person households have the highest MRE, for both tract and block groups, at 2.31 and 0.76, respectively. Although higher MREs are not ideal, they are consistent with what we would expect given the model, because the variances for these categories are much larger than the variances for other categories, thus leaving room for large errors when optimizing the objective function in expression (1). Overall, despite higher MREs and SREs for specific categories, these results are promising, because the differences between constraints and disaggregated values are not unreasonably large, and the results are easy to explain with respect to interaction between the objective function defined in expression (1) and the variances derived from the MOEs provided in the ACS summary tables.

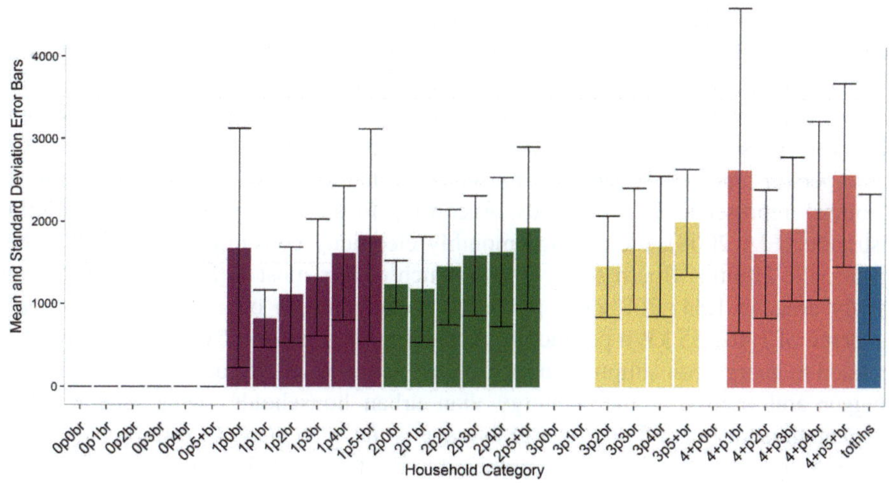

Fig. 2 Bar chart of mean monthly electricity consumption estimates for each household category, together with standard deviation error bars

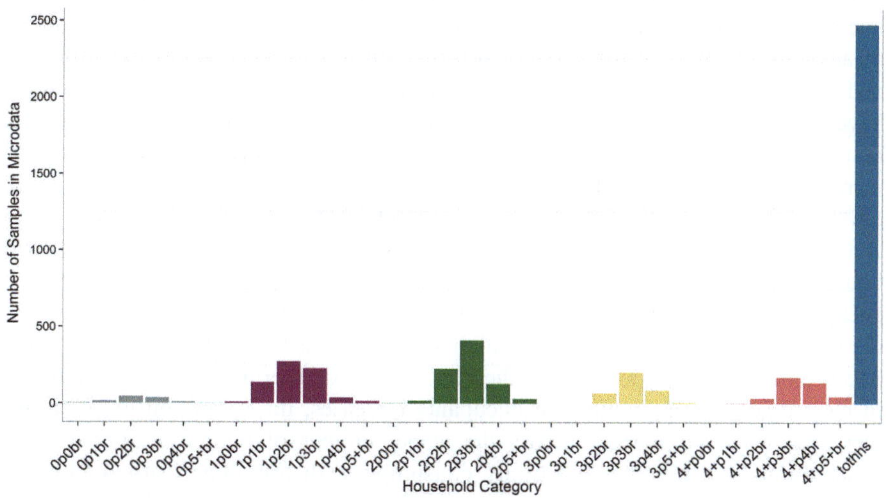

Fig. 3 Bar chart of the number of samples in the microdata by household category

The household monthly electricity consumption distributions computed for each of the microdata samples through the function $f(c_i)$ also greatly influence the final aggregate results. Thus, another way of analyzing and validating the model is by evaluating the reasonableness of the means and standard deviations of the monthly electricity consumption distributions over different category types. Figure 2 shows the mean monthly electricity consumption and standard deviation error bars calculated from the microdata for 30 different households, where a household category is defined by the number of bedrooms and number of occupants. Figure 3 represents the number of microdata samples in each category. Note that "0p0br" denotes the zero-person and zero-bedroom household category, "0b1br" denotes the zero-person and one-bedroom household category, and so on. Furthermore, "tothhs" represents all households. According to the Energy Information Administration (EIA 2008), the average monthly electricity consumption in 2008 in the Knoxville Utilities Board service area, which covers most of Knox County in our study area, is 1,343.708 kWh per month, compared to our overall study area's average of 1,467.55 kWh per month. According to Muratori (2014) in a policy brief related to energy consumption in rural areas, rural households are, on average, 30 % bigger and use 10 % more energy than urban households. Thus, one possible explanation for our inflated overall average is the percentage of rural versus urban areas covered by our study area compared to the Knoxville Utilities Board's service area; our study area has a much greater percentage of rural area than the Knoxville Utilities Board's service area. Furthermore, the data show that all zero-person households have a mean electricity consumption of 0 kWh, which is consistent with the fact that unoccupied households typically do not consume electricity. In addition, with the exception of zero-bedroom households, four or more-person and one-bedroom households, and the three categories missing samples, each of the means within a category of a fixed number of occupants increases as the number of bedrooms increases. The zero-bedroom households that do not follow this trend all have much lower sample counts than the other categories, indicating that their mean is not representative of the true population and would likely be corrected with an increase in samples. In addition, the errors introduced from these categories do not have as large of an effect on the model output as the other errors, because studio apartments are not as common in the Knoxville area (representing only 1 % of households in the ACS summary tables), and households with three or more occupants and only one bedroom are rare. Thus, though we are not as confident in the estimates for these categories, their effect on the final output is likely small because of their small presence in the population. Another trend we observe is that, in most cases, as the number of occupants increases, the mean consumption also increases when comparing across the same number of bedrooms. Given these observations, the monthly electricity consumption distributions learned from our function $f(c_i)$ seem to be within a reasonable range for the Knoxville area.

5 Conclusion

In this chapter, we present a novel hybrid dasymetric and machine learning approach to high-resolution residential electricity consumption modeling and demonstrate the utility of the method by using it to estimate and analyze aggregate census block group level residential consumption within a growing urban area. The model overcomes existing limitations by requiring only commonly available data and by providing a well-defined method for handling uncertain input data sources. Furthermore, the dasymetric framework provides new opportunities for other scientific areas that would benefit from finer resolution spatial analyses using existing open access information.

Acknowledgements This manuscript has been authored by employees of UT-Battelle, LLC, under contract DE-AC05-00OR22725 with the US Department of Energy. Accordingly, the United States Government retains, and the publisher, by accepting the article for publication, acknowledges that the United States Government retains, a nonexclusive, paid-up, irrevocable, worldwide license to publish or reproduce the published form of this manuscript, or allow others to do so, for United States Government purposes.

References

Aydinalp M, Ugursal V, Fung A (2002) Modeling of the appliance, lighting, and space-cooling energy consumptions in the residential sector using neural networks. Appl Energy 71(2): 87–110

Breiman L (2001) Random forests. Machine learning 45(1):5–32

Dempster A, Laird N, Rubin D (1977) Maximum likelihood from incomplete data via the EM algorithm. J R Stat Soc Ser B (Methodological) 39(1):1–38

Eicher C, Brewer C (2001) Dasymetric mapping and areal interpolation: implementation and evaluation. Cartogr Geogr Inf Sci 28(2):125–138

Energy Information Administration (2008) Form eia-826 database monthly electric utility sales and revenue data. Energy Information Administration, Washington, DC

Fabrikant S (2003) Commentary on 'A history of twentieth-century American academic cartography' by Robert McMaster and Susanna McMaster. Cartogr Geogr Inf Sci 30:81–84

Kadian R, Dahiya R, Garg H (2007) Energy-related emissions and mitigation opportunities from the household sector in Delhi. Energy Policy 35(12):6195–6211

Knoxville Utilities Board (2012) Building for the future. Available via Knoxville Utilities Board. http://www.kub.org. Accessed 6 Feb 2015

Langford M (2013) An evaluation of small area population estimation techniques using open access ancillary data. Geogr Anal 45(3):324–344

Liaw A, Wiener M (2002) Classification and regression by randomForest. R News 2(3):18–22

McCleary G (1969) The dasymetric method in thematic cartography. Dissertation, University of Wisconsin, Madison

Mennis J, Hultgren T (2006) Intelligent dasymetric mapping and its application to areal interpolation. Cartogr Geogr Inf Sci 33(3):179–194

Muratori M (2014) Rural energy use and the challenges for energy conservation and efficiency. Our energy future: socioeconomic implications and policy options for rural America. National Agricultural and Rural Development Policy Center, New York City, pp 147–162

Nagle N, Buttenfield B, Leyk S, Spielman S (2014) Dasymetric modeling and uncertainty. Ann Assoc Am Geogr 104(1):80–95

Parti M, Parti C (1980) The total and appliance specific conditional demand for electricity in the household sector. Bell J Econ 11(1):309–321

Qiu F, Cromley R (2013) Areal interpolation and dasymetric modeling. Geogr Anal 45(3): 213–215

Saidur R, Masjuki H, Jamaluddin M (2007) An application of energy and energy analysis in residential sector of Malaysia. Energy Policy 35(2):1050–1063

Slocum T, McMaster R, Kessler F, Howard H (2009) Thematic cartography and geovisualization. Pearson Prentice Hall, Upper Saddle River

Swan L, Ugursal V (2009) Modeling of end-use energy consumption in the residential sector: a review of modeling techniques. Renew Sustain Energy Rev 13(8):1819–1835

US Census Bureau (2009) A compass for understanding and using American Community Survey data: what researchers need to know. U.S. Census Bureau, Washington, DC

US Census Bureau (2012a) 2008–2012, American Community Survey microdata. http://factfinder2.census.gov. Accessed 6 Jan 2015

US Census Bureau (2012b) 2008–2012, American Community Survey summary tables. http://factfinder2.census.gov. Accessed 6 Jan 2015

Wright J (1936) A method of mapping densities of population: with Cape Cod as an example. Geogr Rev 26(1):103–110

Can Social Media Play a Role in the Development of Building Occupancy Curves?

Robert Stewart, Jesse Piburn, Eric Weber, Marie Urban, April Morton, Gautam Thakur and Budhendra Bhaduri

Abstract The demand for building occupancy estimation continues to grow in a wide array of application domains, such as population distribution modeling, green building technologies, public safety, and natural hazards loss analytics. While much has been gained in using survey diaries, sensor technologies, and dasymetric modeling, the volume and velocity of social media data provide a unique opportunity to measure and model occupancy patterns with unprecedented temporal and spatial resolution. If successful, patterns or occupancy curves could describe the fluctuations in population across a 24 h period for a single building or a class of building types. Although social media hold great promise in responding to this need, a number of challenges exist regarding representativeness and fitness for purpose that, left unconsidered, could lead to erroneous conclusions about true building occupancy. As a mode of discussion, this chapter presents an explicit social media model that assists in delineating and articulating the specific challenges and limitations of using social media. It concludes by proposing a research agenda for further work and engagement in this domain.

R. Stewart (✉) · J. Piburn · E. Weber · M. Urban · A. Morton · G. Thakur · B. Bhaduri
Oak Ridge National Laboratory, 1 Bethel Valley Road, Oak Ridge, TN 37831, USA
e-mail: stewartrn@ornl.gov

J. Piburn
e-mail: piburnjo@ornl.gov

E. Weber
e-mail: weberem@ornl.gov

M. Urban
e-mail: urbanml@ornl.gov

A. Morton
e-mail: mortonam@ornl.gov

G. Thakur
e-mail: thakurg@ornl.gov

B. Bhaduri
e-mail: bhaduribl@ornl.gov

D.A. Griffith et al. (eds.), *Advances in Geocomputation*,
Advances in Geographic Information Science,
DOI 10.1007/978-3-319-22786-3_6

Keywords Social media • Building occupancy • Population • Small area estimation

1 Introduction

Building occupancy estimation is critical to a wide array of application domains, including population distribution modeling, green building technologies, public safety, and natural hazards loss analytics. Researchers and practitioners in natural hazard domains estimate human fatalities by considering human occupancy patterns in combination with building construction practices and geophysical models (Jaiswal et al. 2011). Research in green building technologies includes studies on the relationship between occupancy and energy consumption in residential and commercial facilities. Detecting real-time occupancy levels with motion detectors, infrared devices, and wireless fidelity (Wi-Fi) connectivity can lead to improved energy efficiency by automatically adjusting climate controls, lighting, and information technology (IT) infrastructure (e.g., Hong and Lin 2013; Martani et al. 2012; Melfi et al. 2011; Meyn et al. 2009). Public health scientists concerned with the effects of exposure to indoor pollutants have conducted surveys to understand building occupancy (e.g., Klepeis et al. 2001). Small area estimation estimates populations for day and night at a 3 arc-sec (~90 m^2) spatial resolution by considering a wide array of ancillary data using areal interpolation and dasymetric modeling techniques (e.g., Leyk et al. 2013; Langford 2013; Qiu and Cromley 2013; Zandbergen and Ignizio 2010), including information about facilities, transportation, and businesses (e.g., Bhaduri et al. 2007). The Population Density Tables (PDT) project has responded to these building level occupancy demands by estimating ranges for average day and night population density for over 50 building types (Stewart et al. 2016).

Human activity and building occupancy dynamics are well connected, and scholars have studied this connection in a variety of important ways. Geographers, for example, have studied individual movement patterns using travel diaries (e.g., Axhausen et al. 2000), time use surveys, social media, and mobile phone location data (e.g., Noulas et al. 2012; Gonalez et al. 2008; Schlich and Axhausen 2003). To move to an even finer temporal resolution, modelers need occupancy signatures that characterize population density over smaller time intervals within specific time frames. Some facilities (e.g., theaters) may be able to monitor occupancy (e.g., ticket sales) in near-real time. Research in using building sensor technologies as a means for detecting occupancy continues to grow (e.g., Melfi et al. 2011) as well. Unfortunately, paying for these data over many facilities is expensive and the data only apply to a small subset of facility types.

How viable is social media data in developing occupancy signatures? Specifically, given aggregate count data for a specified time frame (e.g., daily visitors), can social media further disaggregate this into smaller time steps (e.g., hourly visitors)? We motivate the problem by proposing an occupancy model that explicitly situates

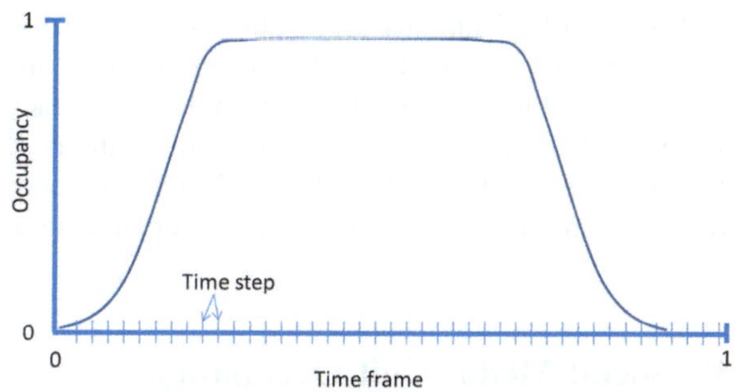

Fig. 1 A unit occupancy signature can be scaled to a specific facility and time frame

social media at the center of estimation. We conclude with a research agenda based on the needs of applying this model in practice.

2 Unit Occupancy

To begin, we establish a time frame and temporal resolution (time step). Examples include time frames of monthly, daily, or hourly operation. The time step is smaller and is the desired temporal resolution. Because a selected time frame may vary by institution (e.g., hours of operation), we normalize the time frame and time step onto the unit domain [0, 1] without loss of generality. Because different institutions vary by popularity, we normalize the amplitude onto [0, 1] as well. This produces a *unit occupancy model* that can be used to disaggregate time frame data to the finer scale given by the selected temporal time step. Figure 1 shows an example.

The model for occupancy at time t, O_t, is structured in terms of normalized visitor[1] arrival times a_i and visit duration v_i of the ith visitor, as in the following equation.

$$O_t \sim K \sum_{i=1}^{V} f(a_i, v_i, t), \tag{1}$$

where $f(a_i, v_i, t) = \begin{cases} 1 & \text{if } t \in [a_i, a_i + v_i] \\ 0 & \text{otherwise} \end{cases}$

Here, K normalizes maximum occupancy to one, and V is the number of arrivals occurring within a time frame. The idea is simply that the proportion of people present at any given moment is the sum of people arriving before and leaving after that moment. Therefore, the focus is on developing arrival and visit duration models. Scaling the model to any particular setting amounts to scaling the number of arrivals during a time frame (*V*).

[1]The term "visitor" means, more broadly, the occupant, and includes visitors, employees, and so forth.

The cost of continuous observation (e.g., video) makes estimation of arrival and visit times over repeated time frames intractable. In lieu of this, we focus on more readily available social media as an indicator of facility population dynamics. We begin by defining the term "social media author" (SMA) as any individual producing social media content that indicates presence at an institution under study. We further our inquiry by writing the occupancy model for the SMA subpopulation.

3 Social Media Unit Occupancy

Given a social media filter applied to the social media stream(s), a set of SMAs and posts indicating presence at a facility is identified. Let $SMA \equiv \{sma_1, \ldots, sma_Q\}$ represent the set of SMAs, and $e_i \equiv \{e_i^1, \ldots, e_i^N\}$ represent the set of relevant posts for the ith SMA. If $e_i^{range} = e_i^N - e_i^1$ indicates the minimum visit duration, we model the full visit duration of the ith SMA as the conditional distribution $v \sim \varphi(\cdot\,|e_i^{range})$, where the visit duration is modeled as a random variable parameterized by external, ancillary visitor data (e.g., Stewart et al. 2016; Morton 2013). The arrival time of an individual SMA is modeled as

$$a_i \sim \lambda\left(a_i^{\min}, e_i^1 | \delta\right), \tag{2}$$

where $a_i^{min} = e_i^1 - (v_i - e_i^{range})$, with a_i^{min} and e_i^1 being, respectively, the earliest and latest arrival time[2] possible for the ith SMA, and δ parameterizing the model λ for individual arrival times (e.g., λ could be a uniform distribution). Given the set of *sample* arrivals a_i from one or more institutions, we fit the following continuous *population* arrival model:

$$p(a) = \theta(\pi), \tag{3}$$

where the probability of an arrival a is given by the probability distribution θ with model parameters π.

Occupancy curve estimation is carried out by drawing realizations from the population arrival model (Eq. 3), and then conditional realizations are drawn from the duration model, giving pairs $(a_k, v_k|a_k)$. Calculation of O_t for SMAs is according to Eq. 1 and is straightforward. Inferring occupancy to population assumes that SMA arrivals and visit durations are not fundamentally different than those of the population. If this critical assumption is valid, then SMA occupancy dynamics is an estimate of population dynamics.

---

[2]Assuming the first post e_i^1 occurred while at the facility, then this is the latest arrival time possible for the ith visitor. Such an assumption could be confirmed for georeferenced data.

4 Results

We apply the model to High Museum of Art Facebook check-ins collected for 15 min intervals from September 6, 2013 to January 8, 2014, using the Facebook Graph application programming interface (API). The time frame is daily hours of operation, and the time step is 1 min intervals. We adopt a uniform visit duration model U[1,3][3] for large museums (Stewart et al. 2016) with individual check-in times coincident with arrival times. Figure 2a shows the fitted population arrival model, and Fig. 2b the resulting SMA occupancy model with 95 % confidence bands.

The results suggest that SMAs arrive uniformly throughout the day such that the population increases gradually, reaching peak occupancy about two-thirds the way between opening and closing hours. With over 400,000 annual visitors visiting across 311 days of operation (http://www.high.org), the museum averages 1,280 visitors per day. We simulate 1,280 visitors per day using the unit occupancy to disaggregate people counts throughout the day. The shape is the same, but the amplitude now reflects total visitor counts. For example, peak occupancy occurs near 2 P.M. and averages 380–424 people.

5 A Model-Based Research Agenda

The model structure is valuable in that it supports articulation and discussion of specific research needs in responsibly applying the model across multiple facilities.

- Which facilities will generate a viable amount of social media data, and which will not? For example, libraries are unlikely to generate as much SMA content as museums do.
- Are SMA arrivals and visit times substantially different than for non-SMAs? Is there anything different about SMAs that would cause them to arrive and remain in patterns very different than those for the rest of the population?
- The previous two points suggest great value in furthering knowledge about the specific relationship between demographics and facility popularity. Could separate surveys of SMAs and non-SMAs support this inquiry?
- Can we develop adequate filters to identify spatially unreferenced SMAs, and how much error might be incurred if we can? Georeferenced data are, at best, only a small percentage of the social media volume and may create problems for narrowing filters.

---

[3]A uniform distribution means that visitors remain at the museum for some period of time between 1 and 3 h.

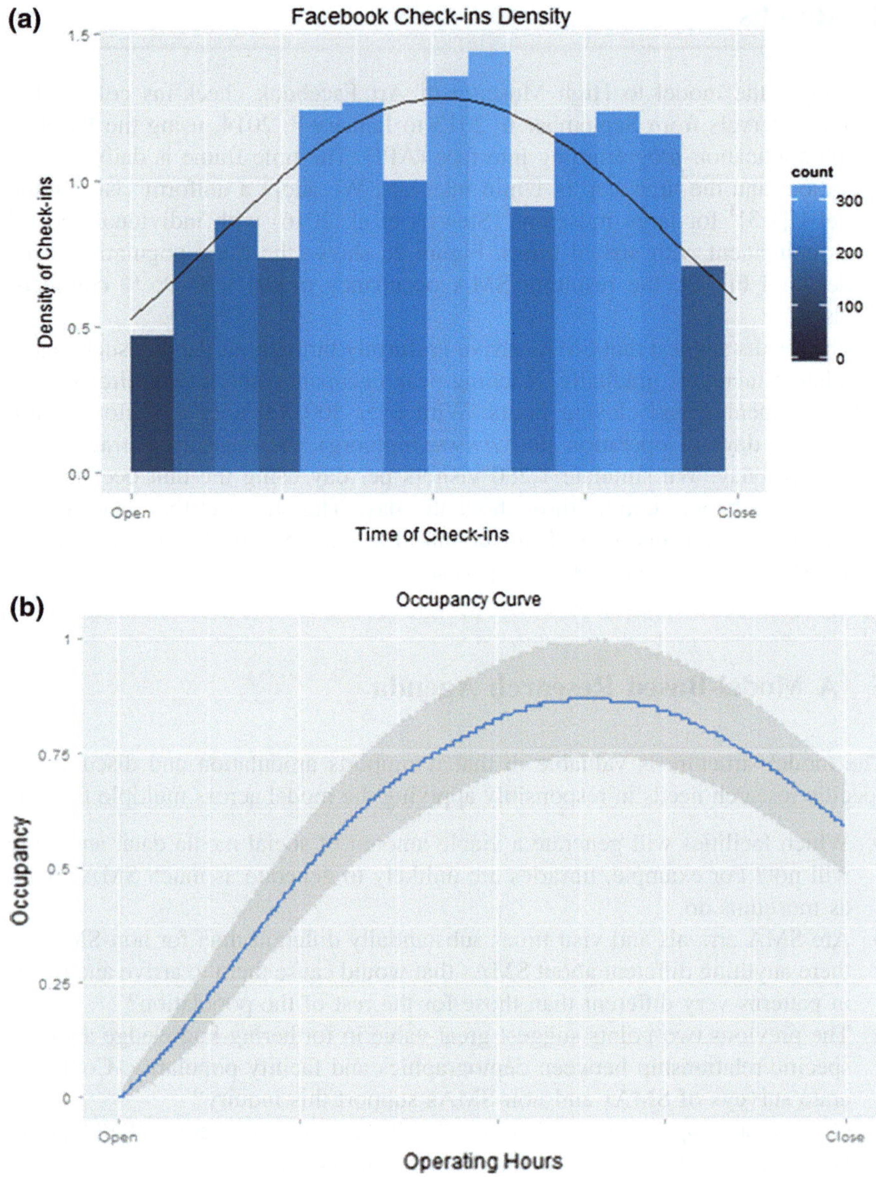

Fig. 2 **a** Arrival times data and model, **b** unit occupancy (*blue line*) with uncertainty (*gray bands*)

- What are cost-effective means of validating these model results? Validating the model will require some collaboration/direct observation for at least some time steps.

The first two questions are about data sufficiency and are not unique to this problem. The third question points to the novelty and benefit of this approach. Here the question moves from how well SMA counts represent population counts to how well SMA arrivals represent population arrivals. The SMA subpopulation for one facility may be 1 % and another 30 %, but if SMA arrival and visit duration curves separately are representative, then unit occupancy can theoretically disaggregate time frame population totals into smaller time steps. Validation could be addressed for some facilities through collection of open source time step data (e.g., museum monthly totals for validating annual-to-month disaggregation) or through limited real-time observations. More research and development is required in each of these areas.

We present here a novel model for leveraging and illuminating the challenges of social media data in estimating occupancy dynamics. The model presented here formally encapsulates social media contributions and focuses data quality concerns explicitly on when SMAs arrive and depart, and narrowly focuses the goals in subsequent research efforts.

Acknowledgment This manuscript has been authored by employees of UT-Battelle, LLC, under contract DE-AC05-00OR22725 with the US Department of Energy. Accordingly, the United States Government retains, and the publisher, by accepting the article for publication, acknowledges that the United States Government retains, a nonexclusive, paid-up, irrevocable, worldwide license to publish or reproduce the published form of this manuscript or allow others to do so, for United States Government purposes.

References

Axhausen K, Zimmermann A, Schönfelder S, Rindsfüser G, Haupt T (2000) Observing the rhythms of daily life: a six-week travel diary. Transportation 29(2):95–124

Bhaduri B, Bright E, Coleman P, Urban M (2007) LandScan USA: a high-resolution geospatial and temporal modeling approach for population distribution and dynamics. GeoJournal 69 (1):103–117

Gonzalez MC, Hidalgo CA, Barabasi A-L (2008) Understanding individual human mobility patterns. Nature 453(7196):779–782

Hong T, Lin H-W (2013) Occupant behavior: impact on energy use of private offices. Berkeley National Laboratory and the Green Energy and Environment Laboratories, Industrial Technology Research Institute, Taiwan, ROC

Jaiswal K, Wald D, Earle P, Porter K, Herne M (2011) Earthquake casualty models within the USGS Prompt Assessment of Global Earthquakes for Response (PAGER) system. In: Spence R, So E, Scawthorn C (eds) Human casualties in earthquakes. Springer, New York, pp 83–94

Klepeis N, Nelson W, Ott W, Robinson J, Tsang A, Switzer P, Behar J, Hern S, Engelmann W (2001) The National Human Activity Pattern Survey (NHAPS): a resource for assessing exposure to environmental pollutants. J Expo Anal Environ Epidemiol 11:231–252

Langford M (2013) An evaluation of small area population estimation techniques using open access ancillary data. Geogr Anal 45:324–344. doi:10.1111/gean.1201

Leyk S, Nagle NN, Buttenfield BP (2013) Maximum entropy dasymetric modeling for demographic small area estimation. Geogr Anal 45:285–306. doi:10.1111/gean.12011

Martani C, Lee D, Robinson P, Britter R, Ratti C (2012) ENERNET: studying the dynamic relationship between building occupancy and energy consumption. Energy Build 47:584–591

Melfi R, Rosenblum B, Nordman B, Christensen K (2011) Measuring building occupancy using existing network infrastructure. In: Proceedings of the 2011 international green computing conference and workshops, IEEE Computer Society, Orlando, FL, July 25–28, 2011

Meyn S, Surana A, Lin Y, Oggianu S, Narayanan S, Frewen T (2009) A sensor-utility-network method for estimation of occupancy distribution in buildings. In: Joint 48th IEEE conference on decision and control and 28th Chinese control conference, Shanghai, China, December 16–18, 2009

Morton A (2013) A process model for capturing museum population dynamics. Master's thesis, Department of Mathematics, California State Polytechnic University

Noulas A, Scellato S, Lambiotte R, Pontil M, Mascolo C (2012) A tale of many cities: universal patterns in human urban mobility. PLoS ONE 7(5):e37027. doi:10.1371/journal.pone.0037027

Qiu F, Cromley R (2013) Areal interpolation and dasymetric modeling. Geogr Anal 45:213–215. doi:10.1111/gean.12016

Schlich R, Axhausen K (2003) Habitual travel behaviour: Evidence from a six-week travel diary. Transportation 30(1):13–36

Stewart RN, Urban M, Morton A, Duchscherer S (2016) A Bayesian machine learning model for estimating building occupancy from open source data. Nat Hazards. doi:10.1007/s11069-016-2164-9

Zandbergen P, Ignizio D (2010) Comparison of dasymetric mapping techniques for small-area population estimates. Cartogr Geogr Inf Sci 37(3):199–214

Application of Social Media Data to High-Resolution Mapping of a Special Event Population

Kelly M. Sims, Eric M. Weber, Budhendra L. Bhaduri, Gautam S. Thakur and David R. Resseguie

Abstract Society's increasing participation in social media provides access to new sources of near-real-time data that reflect our activities in space and in time. The ability for users to capture and express their geolocations through their phones' global positioning system (GPS), or through a particular location's hashtag or Facebook page, provides an opportunity for modeling spatiotemporal population dynamics. One illustrative application is the modeling of dynamic populations associated with special events such as sporting events. To demonstrate, Twitter posts and Facebook check-ins were collected across a 24 h period for several football game days at the University of Tennessee, Knoxville, during the 2013 season. Population distributions for game hours and nongame hours of a typical game day were modeled at a high spatial resolution using the spatiotemporal distributions of the social media data.

Keywords Social media · Population · Population modeling

K.M. Sims · E.M. Weber · B.L. Bhaduri (✉) · G.S. Thakur · D.R. Resseguie
Oak Ridge National Laboratory, PO Box 2008 MS6017, Oak Ridge, TN 37831, USA
e-mail: bhaduribl@ornl.gov

K.M. Sims
e-mail: simskm@ornl.gov

E.M. Weber
e-mail: weberem@ornl.gov

G.S. Thakur
e-mail: thakurg@ornl.gov

D.R. Resseguie
e-mail: resseguiedr@ornl.gov

D.A. Griffith et al. (eds.), *Advances in Geocomputation*,
Advances in Geographic Information Science,
DOI 10.1007/978-3-319-22786-3_7

1 Introduction

Modeling population distributions at a high spatial and temporal resolution requires accounting for the dynamic nature of human populations. Models and representations of population that rely on census counts necessarily miss these dynamics. However, some recent efforts incorporate daytime or diurnal distributions (Bhaduri et al. 2007; Kobayashi et al. 2011) and episodic or tourist populations (Jochem et al. 2012; Charles-Edwards and Bell 2012) in order to better capture population dynamics. All of the efforts to extend population modeling beyond static representations of nighttime populations contribute to the overall goal of achieving what Martin et al. (2015) call "a full representation of population time," which is important for emergency preparedness and response, risk modeling, and many other applications. To continue these efforts, publicly available data feeds from social media offer an additional opportunity to improve population models at high resolutions (Bukhari et al. 2012; Birkin and Malleson 2013).

As of late 2015, the world's two leading social media platforms are Twitter, with 320 million monthly active users, and Facebook, with 1.5 billion monthly active users. Only a portion of social media data has associated location information, however. For example, although an estimated 500 million tweets were sent per day as of October 2013 (Twitter, Inc. 2013), estimates of the portion of tweets that are geolocated have ranged from 0.47 % (Cheng 2010) to 3.17 % (Morstatter et al. 2013).

A natural application of the geolocated subset of social media data is the modeling of episodic populations associated with special events having high attendance and a significant presence on social media; in particular, this study focuses on game-day college football fans at the University of Tennessee (UT), Knoxville. Using tweets (from Twitter) and check-ins (from Facebook), this research integrates this new form of data in a high-resolution dasymetric population distribution model.

2 Methods and Results

The area within a 1.5 mile radius around the UT football stadium was chosen for this study, and the population associated with football game days was modeled. Geolocated tweets and check-ins were collected for the 24 h period surrounding the scheduled kickoff for each home game in 2013. Seven terms associated with the university were used to filter tweets from Twitter's streaming application programming interface (API) (Table 1). A cumulative count of Facebook check-ins was captured every 30 min for 95 establishments associated with game-day activities (e.g., restaurants and tailgating locations).

Two scenarios were modeled: (1) a "nongame-hours" scenario and (2) a "game-hours" scenario. Each model outputs a population estimate for each cell in a

Table 1 The total seasonal count of game-day geocoded tweets associated with the University of Tennessee, 2013

UT term/phrase	Number of geocoded tweets
"Tennessee"	11,582
"Vols" (abbr. of "Volunteers," the team nickname)	7,837
"GBO" (abbr. of "Go Big Orange," a common chant)	2,495
"VFL" (abbr. of "Vol For Life," a common slogan)	1,612
"Neyland" (name of football stadium)	1,144
"Football time in Tennessee" (common slogan)	1,135
"Big orange" (a common alias for the team)	418

raster grid with three arc-second resolution (~90 m). The 2012 version of the LandScan USA (Bhaduri et al. 2007) gridded nighttime dataset was used as a baseline population distribution to which the new modeled distributions could be added to create the final output grids. The LandScan USA nighttime distribution better represents the study area on a Saturday (all game days were Saturdays) than the daytime dataset, because the daytime dataset assumes a weekday distribution of workers and students, which is very different from what would be expected in the area on a weekend.

A consistent measure of social media activity was required for each raster cell. First, the tweets and check-ins (collectively referred to as "posts" hereafter) were divided into two sets based on their timestamps. Posts were considered for the game-hours scenario if they occurred less than 2 h prior to kickoff or less than 3 h after kickoff. All posts outside of those hours were considered for the nongame-hours scenario. A count of tweets and a count of check-ins were computed for each raster cell for each of the two scenarios, resulting in four raw count rasters.

Because of the limited amount of geolocated posts and the spatial errors in the associated location information, some locations that attract event populations may have no representation in the social media data. To overcome this limitation, kernel density estimation with a radius of two grid cells was performed on each of the four raw count rasters to estimate tweet and check-in densities across the grid. Then, the densities were scaled so that each value represents posts per cell and can be interpreted as an interpolated count.

For each scenario, a linear relationship between social media activity and event population is assumed. For each raster cell i, the special event population (y) is described by

$$y_i = \beta_T w_i, \quad (1)$$

where β_T is a linear coefficient specific to the scenario T, and w_i is the number of posts at cell i. The β coefficient is the number of fans represented by each geolocated post. If available, an observed value or estimate of the total population in the study area associated with a scenario can be used to estimate β_T:

$$\beta_T = E_T / \sum_{i=1}^{n} w_i, \tag{2}$$

where E_T is the estimated total special event population for scenario T. The study area total for the game-hours and nongame-hours scenarios can be estimated by summing two separate components of the special event population: A, the ticketed fans, which is estimated by averaging the recorded attendance for each game, and α, all the other (nonticketed) people in the area specifically for the event:

$$E_T = \lambda_T (A + \alpha), \tag{3}$$

where λ is a parameter representing the estimated portion of the peak game-day population. Ultimately, the final population estimate for each cell i is the sum of the baseline LandScan USA population (L_i) and the special event population (y_i):

$$P_i = L_i + y_i. \tag{4}$$

In many special event situations, data often will not be available to support precise estimates of the parameters, α and λ, in Eq. (3). But event officials often have expert knowledge and are privy to information that allows reasonable estimates of these parameters. Ultimately, a software solution aimed at event officials allows such knowledge to be incorporated in the parameterization. Figures 1 and 2 show example representations of the nongame-hours and game-hours scenarios, respectively, using rough estimates of these parameters. The estimate for α is 30,000, which is meant to include nonticketed fans, city security, local business workers, and stadium staff, security, and teams. For the game-hours scenario, λ was set at one, assuming the population peaks during the game. The nongame-hours scenario is meant to represent a moment approximately 3 h before kickoff, for which λ equals two-thirds.

Large populations can be seen in and near the stadium in both Figs. 1 and 2 but with much greater concentration in Fig. 2. Figure 1 shows greater concentrations in areas on and near campus that are popular for pregame tailgate parties, as well as along Cumberland Avenue and in the downtown area, where restaurants, bars, and shops are concentrated. Different parameterizations (for α and λ) lead to different absolute population values; the overall pattern and the ratios among the values, however, remain the same (because the spatial distribution is based on only the social media data).

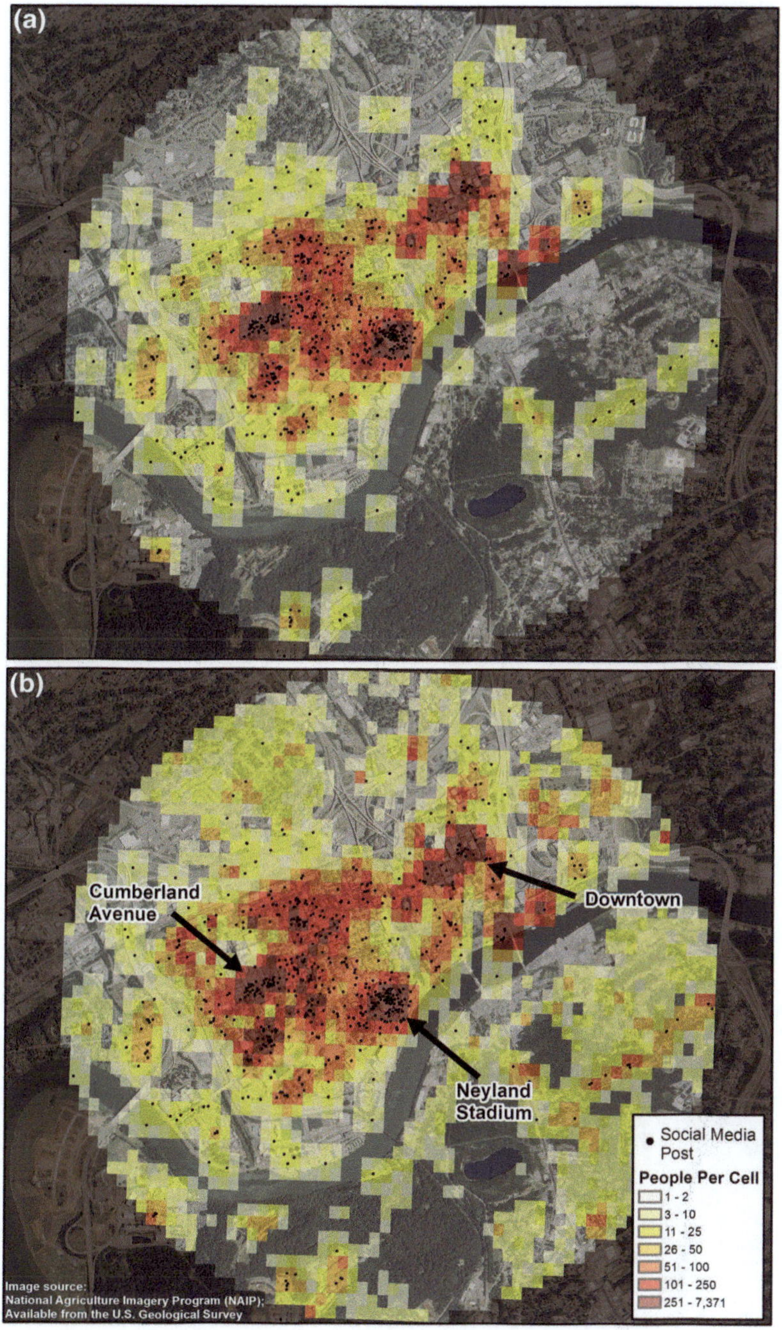

Fig. 1 An example game-day population distribution around the University of Tennessee, Knoxville, during nongame-hours: **a** modeled event population, **b** combined population (baseline + event population)

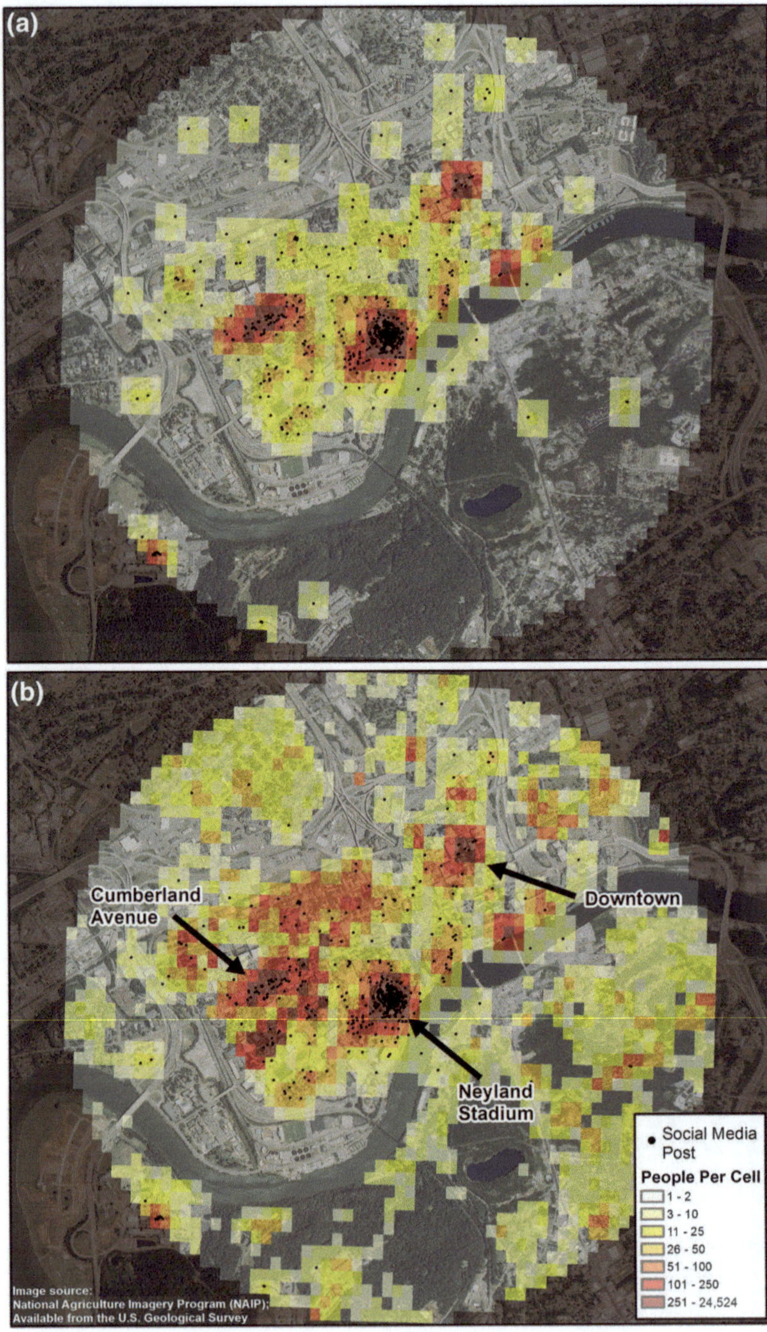

Fig. 2 An example game-day population distribution around the University of Tennessee, Knoxville, during game-hours: **a** modeled event population, **b** combined population (baseline + event population)

3 Discussion

The population distributions in Figs. 1 and 2 generally correspond with the spatial patterns familiar to the authors and therefore suggest social media having a positive relationship with this particular special event population. However, greater quantities and greater spatial precision of observations of the population would be needed to test the assumed linear fit (or to fit alternative models). Greater quantities of data also would allow more refined temporal resolution, rather than having to aggregate data from across several football games into two general scenarios, as was done here.

How to estimate the total population present for an event also deserves future research. In the college football scenario, attendance data are helpful for estimating a portion of the total population, but the α parameter (nonticketed event population) from Eq. (3) is more difficult to estimate. A scenario including ancillary data about the counts of subpopulations, such as security personnel and event staff, would have greater certainty. The data available to an analyst, and the analyst's familiarity with an event, play an important role in establishing reliable estimates of this parameter.

The filtering of the social media data is also a crucial step that relies on knowledge about an event and its location. The authors were able to eliminate a large subset of irrelevant social media data through search-term filtering of Twitter data and identification of relevant event-related establishments from the Facebook data. Of course, other events could prove more challenging because relevant search terms or relevant establishments could be unknown. Again, the development of a software solution that allows analysts with detailed knowledge about an event and location of interest to easily implement a model like the one demonstrated here would be a reasonable next step in expanding this methodology to other events.

Acknowledgements This manuscript has been authored by employees of UT-Battelle, LLC, under contract DE-AC05-00OR22725 with the US Department of Energy. Accordingly, the United States Government retains, and the publisher, by accepting the article for publication, acknowledges that the United States Government retains a nonexclusive, paid-up, irrevocable, worldwide license to publish or reproduce the published form of this manuscript, or allow others to do so, for United States Government purposes.

References

Bhaduri B, Bright E, Coleman P, Urban ML (2007) LandScan USA: a high-resolution geospatial and temporal modeling approach for population distribution and dynamics. GeoJournal 69(1):103–117. doi:10.1007/s10708-007-9105-9

Birkin M, Malleson N (2013) The spatial analysis of short-term population movements with social media data. In: Proceedings of the 12th international conference on geocomputation, LIESMARS, Wuhan University, Wuhan, China, May 23–25, 2013

Bukhari I, Wojtalewicz C, Vorvoreanu M, Dietz JE (2012) Social media use for large event management: the application of social media analytic tools for the Super Bowl XLVI. In: Homeland Security (HST), 2012 IEEE international conference on technologies for homeland security, Westin Hotel, Waltham, MA, November 13–15, 2012

Charles-Edwards E, Bell M (2012) Estimating the service population of a large metropolitan university campus. Appl Spat Anal Policy 6(3):209–228. doi:10.1007/s12061-012-9079-y

Cheng Z, Caverlee J, Lee K (2010) You are where you tweet: a content-based approach to geo-locating twitter users. In: Proceedings of the 19th ACM international conference on information and knowledge management, Toronto, Canada, October 26–30, 2010

Jochem WC, Sims KM, Bright EA, Urban ML, Rose AN, Coleman PR, Bhaduri BL (2012) Estimating traveler populations at airport and cruise terminals for population distribution and dynamics. Nat Hazards 68:1325–1342. doi:10.1007/s11069-012-0441-9

Kobayashi T, Medina RM, Cova TJ (2011) Visualizing diurnal population change in urban areas for emergency management. Prof Geogr 63(1):113–130. doi:10.1080/00330124.2010.533565

Martin D, Cockings S, Leung S (2015) Developing a flexible framework for spatiotemporal population modeling. Ann Assoc Am Geogr 105(4):754–772. doi:10.1080/00045608.2015.1022089

Morstatter F, Pfeffer J, Liu H, Carley KM (2013) Is the sample good enough? Comparing data from Twitter's streaming API with Twitter's firehose. In: Kiciman E, Ellison NB, Hogan B, Resnick P, Soboroff I (eds) Proceedings of the seventh international conference on weblogs and social media, Cambridge, MA, July 8–11, 2013, Palo Alto, CA. The AAAI Press, pp 400–408

Twitter, Inc (2013) The SEC commission, Form S-1. http://www.sec.gov/Archives/edgar/data/1418091/000119312513390321/d564001ds1.htm. Accessed 10 Dec 2013

Animating Maps: Visual Analytics Meets GeoWeb 2.0

Piyush Yadav, Shailesh Deshpande and Raja Sengupta

Abstract Improved visualization techniques for spatiotemporal data have potential to reveal interesting insights from GIS data. Thus, visual analytics has captured the attention of GIS researchers in the recent past. Furthermore, developments in free tools, such as Google Map API and OpenStreetMap, provide easy access to geospatial data that can be leveraged by visual analytics. In this chapter, we propose such a system that utilizes free GIS APIs to visualize spatiotemporal data effectively. Our application allows a user to create a time slide bar control to connect time and position of various GIS objects on a map and to display them in animated mode. This control can handle vector and raster data with equal ease. The resulting effective visualization makes it is very easy to understand some complex spatiotemporal patterns, as exemplified in this chapter.

Keywords Visual analytics · Google map API · Spatiotemporal data · Geospatial data

1 Introduction

Spatiotemporal information plays a vital role in understanding many complex geographic processes. However, visualization techniques tend to neglect temporal properties of data and thus undermine their importance (Andrienko et al. 2003). Recent research has focused on animated visualization techniques to augment

P. Yadav · S. Deshpande
Tata Research Development and Design Centre, Pune, India
e-mail: piyush.yadav1@tcs.com

S. Deshpande
e-mail: shailesh.deshpande@tcs.com

R. Sengupta (✉)
Department of Geography and School of Environment,
McGill University, Montreal, Canada
e-mail: raja.sengupta@mcgill.ca

D.A. Griffith et al. (eds.), *Advances in Geocomputation*,
Advances in Geographic Information Science,
DOI 10.1007/978-3-319-22786-3_8

meaningful insights from complex and large spatiotemporal datasets (Demšar and Virrantaus 2010). Furthermore, the arrival of GeoWeb 2.0 attracted considerable citizen participation in the process of map development (Haklay et al. 2008). Individuals new to geospatial technology such as geographic information systems (GIS) can easily access and share spatial information with such softwares as Google Maps and OpenStreetMap (OSM). Combining visualizations of spatiotemporal data with a GeoWeb 2.0 interface provides a sophisticated mechanism to understand dynamic processes for a variety of geographic phenomenon and provides novice users with a powerful tool for analysis.

2 Literature Review

Dynamic visualization of information has been a key agenda for geovisualization research in the recent past (MacEachren and Jan-Kraak 2001). Several scholars specifically suggest mechanisms by which animated maps can be effective means of visual communication. Harrower (2003) suggests four challenges for animated maps: disappearance, attention, complexity, and confidence. Andrienko et al. (2007, 2010) propose the field of visual analytics specifically for the visualization of spatially explicit events that are temporally interconnected; such events, when synchronized and displayed on a map, potentially can reveal interesting facts. Furthermore, multicomponent maps have been touted as appropriate mechanisms to view complex spatiotemporal phenomenon (Opach et al. 2013). However, Shipley et al. (2013) caution that human beings rarely observe spatiotemporal objects in isolation and that animation methods that do so are doomed to failure. Harrower (2007) also cautions against cognitive overload caused by animations and suggests that the major limitations are not hardware or software, but rather the visual and cognitive loads an animated map places on a map reader. We remain cognizant of these issues when developing our animation methodology.

The preliminary research summarized in this chapter extends the notion of visual analytics by developing a fully customizable, bar-type time legend (Harrower and Fabrikant 2008) for various map application programming interfaces (APIs) like Google Maps and OSM. This extension gives users the ability to represent spatiotemporal information by associating various GIS objects on a map and then synchronizing them with a time legend to visualize changing patterns. For example, the changing migration pattern of animals (e.g., migratory birds) is a complex spatiotemporal phenomenon and one for which data can be downloaded freely from websites like movebank.org (Sarkar et al. 2015). Observing the data using just static maps (e.g., with the attribute information stored in tables and/or graphs) or in the absence of the available web mapping platforms does not provide complete insight into the dynamics of the migration process. Integrating available space-time data about migration for display on an animated visualization system against the backdrop of Google Maps helps to provide a better understanding of this phenomenon to interested researchers. Such visualization shows a potential to generate new insights quickly, which would otherwise be very difficult to generate in other data formats. Because an online map API itself contains a lot of preexisting information, a new form of visual analytics can be achieved by combining this information with GeoWeb 2.0 platforms like Google Maps and OSM.

3 System Design

For this chapter, various tools for drawing and animating spatiotemporal information with Google Maps and OSM were developed using their APIs, ones that can handle both vector and raster data. Specifically, we attempted to construct a system that facilitates the creation of specific animations by users, either by providing a digitized input to an application or by connecting it to a spatiotemporal database. These two application modes are explained next.

3.1 Mode of Operation

In the automatic mode, users can connect to an external spatiotemporal database to upload information from the database onto our system, whereas in the user-customization mode, the user can digitize new vector geographic primitives (points, lines, polygons) on-screen.

3.1.1 Automatic Mode

In this mode, users can import data by connecting to an external spatiotemporal database (Fig. 1). Initially, a user provides the credential of a database with which a connection has to be made. The server handles the connection process by executing various queries to fetch spatiotemporal information from the database. This mode handles both vector and raster data. Vector objects, such as line, circle, and polyline, can be drawn using the map API; consequently, they are imported from the external spatiotemporal database, converted, and then stored in the vector table of our application's database. Each object occupies one tuple/row of our database and is stored as either a point, line, or polygon object. In contrast, raster data are stored in our application's internal database in two ways (depending on the nature of the external spatiotemporal database). If the imported raster consists of a series of raster images, then this series is directly stored in the raster table of our application's internal database. If the database consists of spatial information in the form of pixel values, then a Python process is applied to convert these pixels to images, which then are stored in our application's raster database. All data are stored with their temporal information.

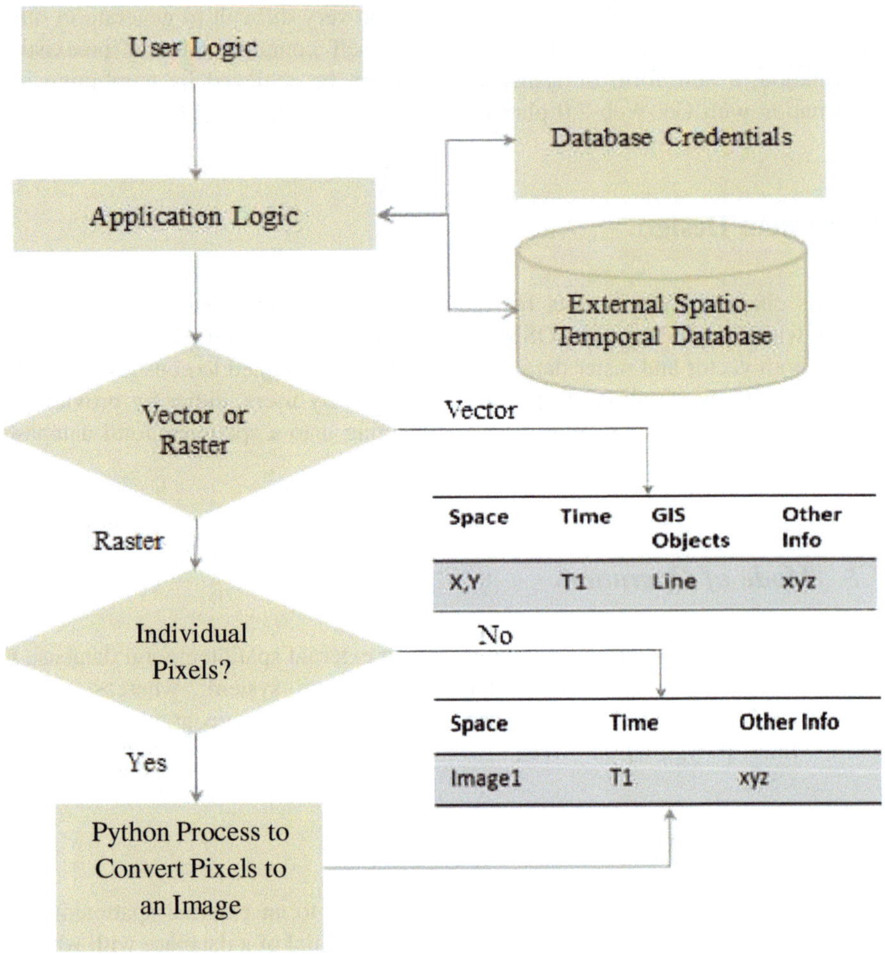

Fig. 1 Data fetching process from an external data source

Figure 2 portrays the process of visualization of stored data. For vector data, the system creates spatial primitives, such as point, line, and polyline, for visualization automatically using map APIs. A user can specify additional graphical properties, such as size, shape, and color, for the attributes as and when required. The application automatically creates controls and integrates the data with a map and slider, creating a dynamic slider-map visualization. Similarly, slider controls are associated with each raster image according to its timestamp. Thus, with each hover on a slider, a query is executed and a raster image gets updated. The time lag between changing images has been kept to a minimum, giving the impression of variations happening in the same image.

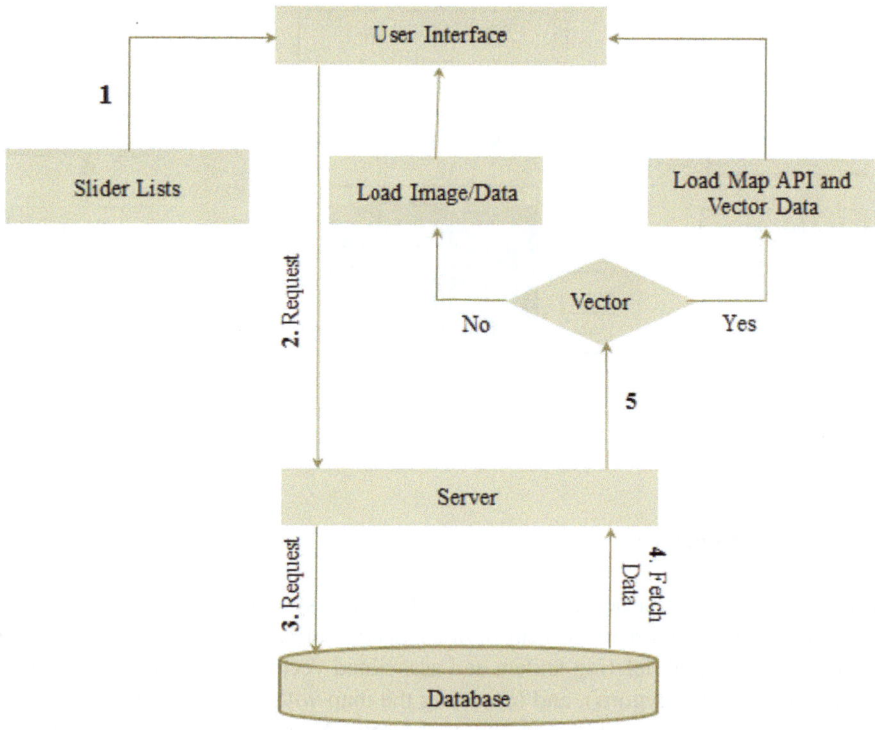

Fig. 2 Data retrieval process from an application database for a slider

3.1.2 User Customization Mode

In this mode, users can perform on-screen digitizing to add a geographic feature (e.g., line, polygon, circle, marker, polyline) onto a map and associate temporal information (i.e., a timeline) with it. Furthermore, different graphical properties can be assigned to a vector layer according to the importance of events and geographical features; for example, users can assign different colors or add a marker to the features as required by their cartographic needs. Thus, a newly digitized map can be saved for further analysis at a later stage.

3.2 System Architecture

The developed application is broadly divided into four parts (Fig. 3).

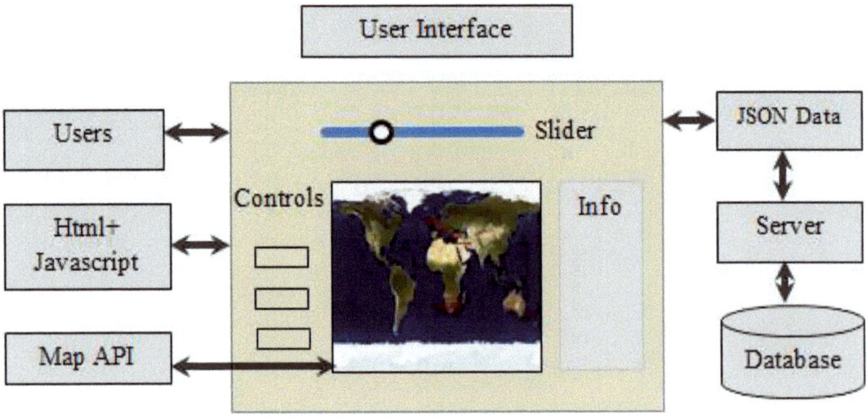

Fig. 3 Architecture

3.2.1 User Interface

HTML and JavaScript act as the front ends to the system (Fig. 4). The interface provides various data management options: specifically, it allows for (a) importing data, (b) adding and deleting sliders and associated vector geographic objects (e.g., points, lines, and polygons), and (c) saving the map with sliders. Users can create a customized slider-map by using these controls. Note that raster data currently cannot be user generated (i.e., they can only be imported from an external source). Maps can be saved with sliders and can be accessed by clicking on the saved link provided in the interface. The database connectivity for an external spatiotemporal database is done by providing various credentials that are shared with the server hosting the spatiotemporal database.

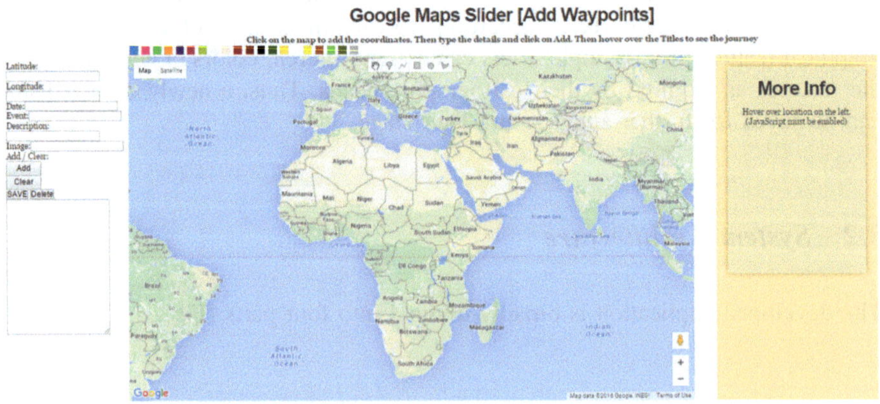

Fig. 4 The user interface for data management

3.2.2 Server

An Apache web server is used for running the application. This server, which handles the JavaScript object notation (JSON) data, is responsible for the following tasks: store and retrieve spatiotemporal information from a database following a "request response" call. As shown in Fig. 1, a "request server" (application logic) initiates a connection to an external spatiotemporal database, fetches the data, and stores them in the application database. Vector information is stored as (x, y) coordinates of the various geographic primitives; that is, point, line, and polygon. Each geographic primitive occupies one line of the data table. Additionally, various Python processes have been written to handle raster data. These processes convert the pixel information into images and associate the temporal information fetched from the external spatiotemporal database. Single images are stored in their entirety, whereas individual pixels are combined into images (albeit while retaining their temporal stamp) and then stored as multiple images.

3.2.3 Map API

The application presently uses Google Maps JavaScript API (2014) v3 and OSM to create events on a map. Various features of the map API (e.g., marker, polyline, circle, marker) have been used to create overlays on maps.

3.2.4 Database

An structured query language (SQL) database in Apache stores the data for the customized maps. The customized data are stored in a tag-based format (just like XML), and the tag-based data are converted to JSON while interacting with the map API. The database handles vector and raster data differently. The vector table handles various GIS objects such as line and circle, associating its temporal and other information in different columns. Similarly, the raster table handles the raster images with its relevant temporal information.

4 Results

We demonstrate the utility of the developed application discussed in this chapter with two working examples. In particular, we emphasize that the "order" and "rate of change (D)" (two visual properties that describe the presentation of information in sequence, and the speed at which this information is presented) are useful means for visualizing information (Kobben and Yaman 1995). The first example shows Vasco Da Gama's 1497 voyage to India from Portugal. Each segment of the voyage is recorded as a separate element with a timestamp, which then is displayed on the bar-type legend (Fig. 5). Users can choose the display color for each line segment,

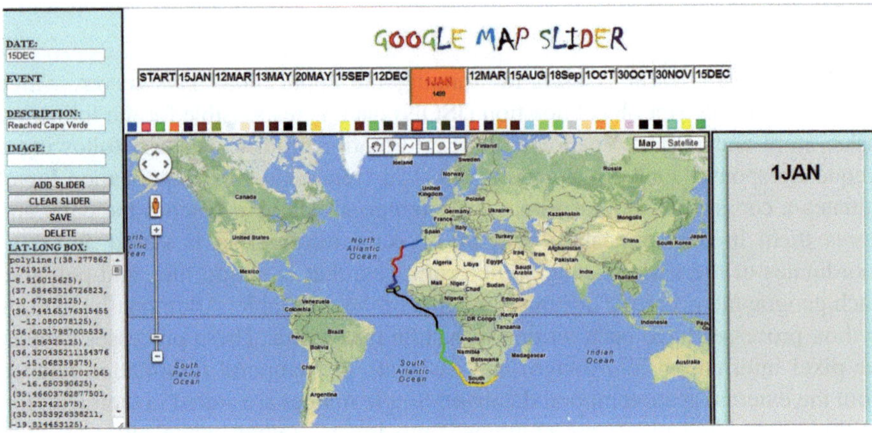

Fig. 5 Visualizing the voyage of Vasco Da Gama

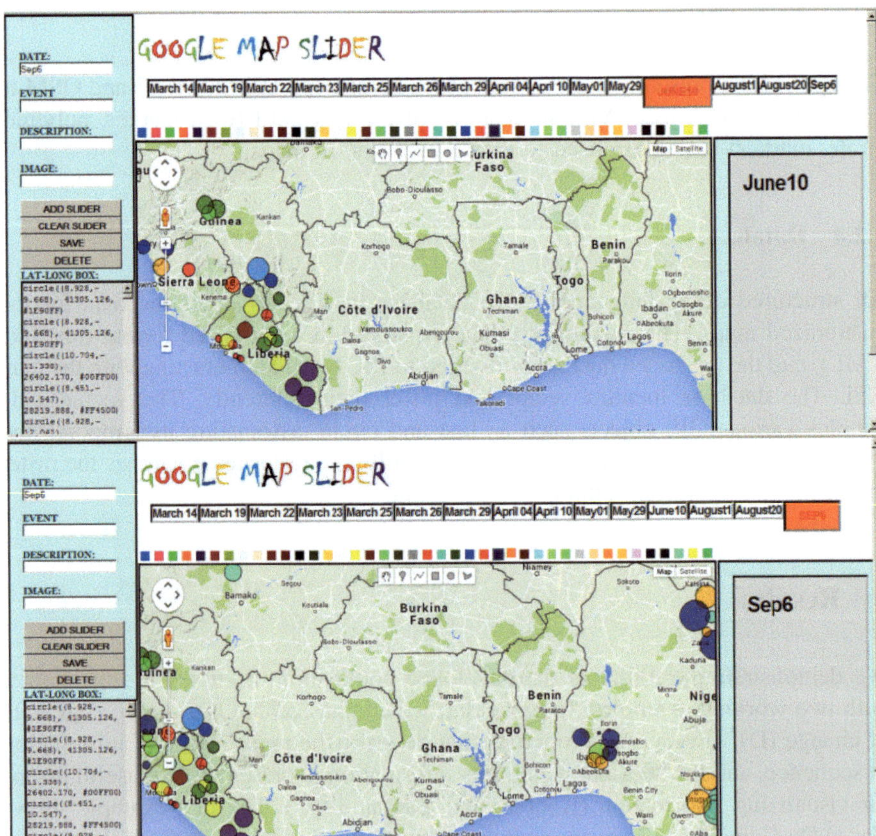

Fig. 6 Visualizing the spread of the Ebola Virus in Africa

and the same color is used during visualization of the information at a later stage. This can be used for visual analytics by picking a particular date (e.g., January 1) and visualizing the information in as a spatiotemporal sequence of this voyage. In the future, one can envision adding other objects (such as storms) that could have affected the voyage itself.

The second example (Fig. 6) visualizes how the Ebola virus spread across various African countries in 2014. The initial spread of Ebola started from three coastal African countries (Guinea, Liberia, and Sierra Leone) but with time also spread to other nations (e.g., Nigeria and Mali). Thus, these types of visual analytics can show the spatial spread of diseases with respect to time.

5 Conclusion

In this chapter, we present a prototype application with a spatiotemporal bar-type slider we developed and tested for commonly used GeoWeb 2.0 applications like Google Maps and OSM. This application provides users with significant control for developing their own slider-map view with various customization options. Users also can save a map and use it for future reference. The application addresses a classic spatiotemporal visualization problem of information overload by allowing users simple control of their animated maps. It also attempts to provide a more visually stimulating experience in which the user can view graphical representations of various events through a navigational slider while controlling the relevancy of information via "order" and "rate of change (D)." Furthermore, different graphical properties can be assigned to digitized objects by a nonexpert user, allowing their importance to be reflected on-screen. In doing so, the application melds the power of GeoWeb 2.0 with visual analytics for the benefit of enhanced spatiotemporal data visualization by a novice user.

References

Andrienko G, Andrienko N, Demšar U (2010) Space, time and visual analytics. Int J Geogr Inf Sci 24(10):1577–1600

Andrienko N, Andrienko G, Gatalsky P (2003) Exploratory spatio-temporal visualization: an analytical review. J Vis Lang Comput 14(6):503–541

Andrienko G, Andrienko N, Jankowski P (2007) Geovisual analytics for spatial decision support: setting the research agenda. Int J Geogr Inf Sci 21(8):839–857

Demšar U, Virrantaus K (2010) Space–time density of trajectories: exploring spatio-temporal patterns in movement data. Int J Geogr Inf Sci 24(10):1527–1542

Google Map JavaScript API (2014) https://developers.google.com/maps/web/. Accessed 10 Mar 2016

Haklay M, Singleton A, Parker C (2008) Web mapping 2.0: the neogeography of the GeoWeb. Geogr Compass 2(6):2011–2039

Harrower M (2003) Tips for designing effective animated maps. Cartogr Perspect 44:63–65

Harrower M (2007) The cognitive limits of animated maps. Cartographica Int J Geogr Inf Geovis 42(4):349–357

Harrower M, Fabrikant S (2008) The role of map animation for geographic visualization. In: Dodge M, McDerby M, Turner M (eds) Geographic visualization: concepts, tools and applications. Wiley, Chichester, pp 49–65

Kobben B, Yaman M (1995) Evaluating dynamic visual variables. In: Teaching animated cartography. Seminar at the Escuela Universitaria de Ingeniera Tecnica Topografica, Madrid, Spain, Aug 30–Sept 1, 1995. http://cartography.geo.uu.nl/ica/Madrid/kobben.html. Accessed 10 Mar 2016

MacEachren A, Jan-Kraak M (2001) Research challenges in geovisualization. Cartogr Geogr Inf Sci 28(1):3–12

Opach T, Gołębiowska I, Fabrikant SI (2013) How do people view multi-component animated maps? Cartogr J 51(4):330–342

Sarkar D, Chapman C, Griffin L, Sengupta R (2015) Analyzing animal movement characteristics from location data. Trans GIS 19(4):516–534

Shipley T, Fabrikant SI, Lautenschiitz AK (2013) Creating perceptually salient animated displays of spatiotemporal coordination. In: Raubal M, Mark DM, Frank AU (eds) Cognitive and linguistic aspects of geographic space: new perspectives on geographic information research. Springer, Berlin, pp 259–270

Atvis: A New Transit Visualization System

Jiaxuan Pang, Charles Tian, Yan Huang, Bill Buckles and Arash Mirzaei

Abstract This chapter presents Atvis, a system to visualize transit data. Atvis is capable of representing bus route and traffic information at an abstract level. Compared with the traditional map-based geographical information system, Atvis focuses more on ridership, visualizing traffic data at or between stops, and interstop relationships, disregarding information that may not be essential to a user. This chapter describes the design and implementation of the Atvis visualization system and discusses how it can be used to observe transit patterns.

Keywords Transit · Visualization · Bus · Route · Traffic · Public transportation · Abstract data

1 Introduction

Due to urbanization and an increase in private vehicles, most cities are becoming extremely crowded. Moreover, with rising awareness of environmental conservation, people are looking for more economical and eco-friendly ways to travel. As of 2011, there were 7,100 public transportation providers in the United States

J. Pang (✉) · Y. Huang · B. Buckles
University of North Texas, Denton, USA
e-mail: tjpjx2012@gmail.com

Y. Huang
e-mail: huangyan@unt.edu

B. Buckles
e-mail: bbucskles@unt.edu

C. Tian
University of Texas at Austin, Austin, USA
e-mail: c.y.t@me.com

A. Mirzaei
North Central Texas Council of Governments, Arlington, USA
e-mail: amirzaei@nctcog.org

D.A. Griffith et al. (eds.), *Advances in Geocomputation*,
Advances in Geographic Information Science,
DOI 10.1007/978-3-319-22786-3_9

(American Public Transportation Association 2013). These providers range from single-vehicle response servers to large multimodal systems. The largest among them, Metropolitan Transportation Authority's (MTA's) New York City Transit, provided 3.3 billion trips covering 12.2 billion miles in 2011. Cumulatively, bus transportation accounts for more than one-half of the 10.3 billion total trips taken, and 36 % of passenger miles covered for the year, making it the most popular form of public transportation. As a result, optimization of bus routes and stop allocations is crucial in helping public transportation providers meet the needs of people in rapidly developing cities.

Visualizing traffic and route data allows urban planners to observe bus systems and to draw conclusions. Deciding upon a suitable visualization model is one of the greatest challenges in developing an effective visualization. A variety of such models for map and route data exist, ranging from highly abstract (e.g., one-dimensional route maps Wolff (2007)) to highly representational (e.g., 3D models Kwan (2002)). Representational visualizations provide more detail and fidelity but do not necessarily offer greater usability. Agrawala (2002) states that "usability peaks near the abstract end of the spectrum." Nevertheless, many existing visualization methods for bus route data involve some geographical component (Sadahiro et al. 2015; Krause et al. 2012).

Although such map-based visualizations are widely used, alternative visualization methods for discrete data, such as public transit routes, should be considered. First, map-based geographic visualization systems often struggle with the issue of readability, especially when large-scale data are presented (Mashima et al. 2011). Second, the discrete nature of bus routes and stops marginalizes the importance of geographical features. Oftentimes, the unnecessary geographical detail of such systems can distract decision makers from observing individual and systematic traffic patterns. A well-designed bus visualization system must have the appropriate level of abstraction to provide maximal usability.

This chapter presents Atvis, an abstract system for visualizing public transit information. Atvis proposes a system of arc and bar diagram representations of bus route data on a one-dimensional line map. Arc diagrams commonly have been used to visualize "the connection between different parts of a set of objects" (Xiao and Chun 2009, pp. 183–191) for data in graph theory (Saaty 1964), e-mail threads (Kerr 2003), and musical sequences (Wattenberg 2002). We extend such existing implementations of arc diagrams by manipulating arc thickness and height to represent bus ridership levels and distance between bus stops. Additionally, we associate a bar diagram with the arc diagram, which provides more detailed information regarding the boarding/alighting of bus passengers. Atvis visualizes key traffic statistics, allowing decision makers to observe the flow of passengers along a given transit route.

2 Atvis Visualization Model

Atvis provides a more efficient way to visualize spatial and temporal transit data. With the combined use of arc and bar diagrams, decision makers can not only observe the passenger flow between stops but also focus on the number of passengers at each individual stop.

2.1 Goals and Objectives

The primary objective of Atvis is to provide decision makers with an intuitive representation of bus data in order to explore spatial and temporal relationships and to observe systematic traffic patterns. Specifically, one can observe visually

1. distance between stops;
2. passenger allocation throughout a system for any time interval; and
3. boarding and alighting at each stop, individually and cumulatively, at any time interval.

We first explored traditional visualization methods using a map-based data presentation. However, this adds a high level of geographical detail, creating more distraction than information. Figure 1 summarizes our process in transitioning from a geographic information system (GIS)-based model to an abstract data visualization.

2.2 Atvis Model Design

Atvis allows users to view and analyze trends and information for customized bus routes and datasets. Our system offers three major advantages:

1. multidimensional data profiles in a two-dimensional format;
2. seamless integration of spatial and temporal information; and
3. a combination of local and global information in one view.

Figure 2 depicts the Atvis visualization model, using sample data for two routes (represented in blue and gray) and consisting of 20 stops. The endpoints of each arc in an arc diagram represent stops along a visualized route. These stops are arranged according to the route's stop order. For multiple routes, a single linear order is unlikely to be consistent with all routes' stop orders. For example, we observe in Fig. 2 that stops 1, 5, 10, 15, and 20 are shared by both routes, while the rest are specific to a single route. In these latter cases, we can manipulate the suborder of stops between shared stops to represent accurately each route's stop order. The height of each arc represents the distance between two stops, while the thickness

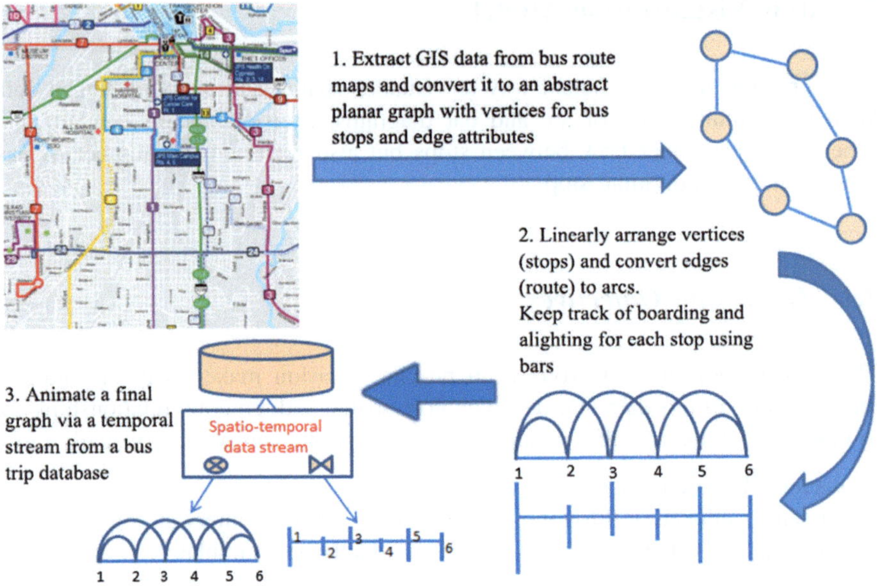

Fig. 1 Geographical to abstract visualization

represents ridership data. The bar diagram consists of bars that correspond to stops in an arc diagram. The component of each bar above the horizontal axis represents the number of boarded passengers, while the component below this axis represents the number of alighted passengers. The combined height of these two components measures the total passenger flow at each stop. Table 1 summarizes how data are represented through the arc and bar diagrams.

3 An Atvis Visualization Demonstration Program

Based on the Atvis model, we created a demonstration program to present the system's representation of actual transit data. Our demonstration contains both a backend component,[1] which allows users to define routes and input ridership data, and a frontend component,[2] which visualizes these data.

---

[1]http://hpproliant.cse.unt.edu/nctcog/backend/.

[2]http://hpproliant.cse.unt.edu/nctcog/.

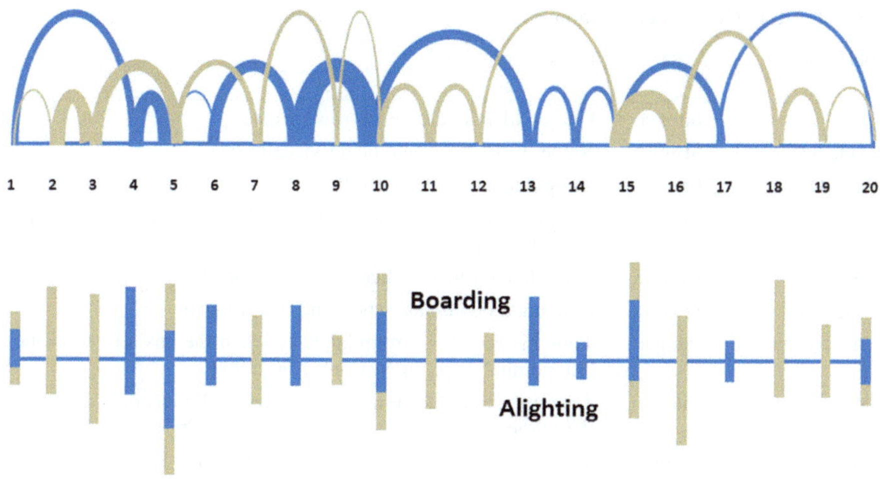

Fig. 2 The Atvis visualization model

Table 1 The Atvis visualization model

Arc legend	Description of arc legend	Bar legend	Description of bar legend
Integer	Bus stop order	Integer	Bus stop order
Arc color	Bus route	Bar color	Bus route
Arc thickness	Number of passengers on the bus	Bar above axis	Number of passengers boarding
Arc height	Distance between stops	Bar below axis	Number of passengers alighting

3.1 Data Description

The bus dataset we used for visualization was provided by the North Central Texas Council of Governments (NCTCOG). It contains 1,827 records from November 30, 1999, detailing bus trips on Route 2 in Fort Worth, Texas. Each record represents a single stop made on a trip along the route. Table 2 offers a detailed description of the NCTCOG bus data, and Fig. 3 is a map of Route 2 provided by NCTCOG.

3.2 Backend System

The demo backend system allows users to manage, define, and upload bus routes and corresponding datasets to be visualized. Figure 4 describes the flow of a route and trip data submission. Users must provide the name of each stop, as well as the distance in miles between stops. Figure 5 is a screenshot of the route creation system.

Table 2 NCTCOG data description

Field name	Data type	Field description
ID	Integer	Unique identifier for each row of data
Route_name	String	Name of the bus route being visualized
Trip_ID	Integer	Unique identifier for each bus trip
Stoporder	Integer	Unique identifier for each stop, based on the order of all stops along the route
Latitude	Double	Latitude coordinate where the bus stopped
Longitude	Double	Longitude coordinate where the bus stopped
Arrival_time	Integer	Time in seconds from midnight to when the bus arrived at the "Stoporder" stop, on November 30, 1999
Departure_time	Integer	Time in seconds from midnight to when the bus departed from the "Stoporder" stop, on November 30, 1999
Dwell_time	Integer	Time in seconds that the bus dwelled at the "Stoporder" stop
Board	Integer	Number of passengers who boarded the bus at the "Stoporder" stop
Alight	Integer	Number of passengers who alighted the bus at the "Stoporder" stop

Datasets are uploaded to a server as a plain text file. Users must also indicate the route with which their dataset corresponds. Figure 6 offers an example of how a dataset is properly formatted. The format must comply with the following requirements:

1. Stop names for each data entry match stop names for the corresponding route.
2. Data entries are separated by a line break.
3. Items in each data entry are separated by a vertical bar.
4. Items in each data entry are in the following order: Trip ID, Stop Name, Time, #Board, #Alight.
5. Trip ID, Time, #Board, and #Alight must be integers.
6. Time must be in seconds.

3.3 The Frontend System

Our frontend system can visualize multiple routes and datasets that are defined in our backend system. Thus, we offer a menu that allows users to select the data they wish to display, as shown in Fig. 7. The system progresses through time in accumulative or real-time fashion and can be slowed down, sped up, paused, and resumed by a user. Changes in passenger flow are reflected when the system clock reaches the time that they occurred. Users can specify start and end times for data

Fig. 3 Map of route 2

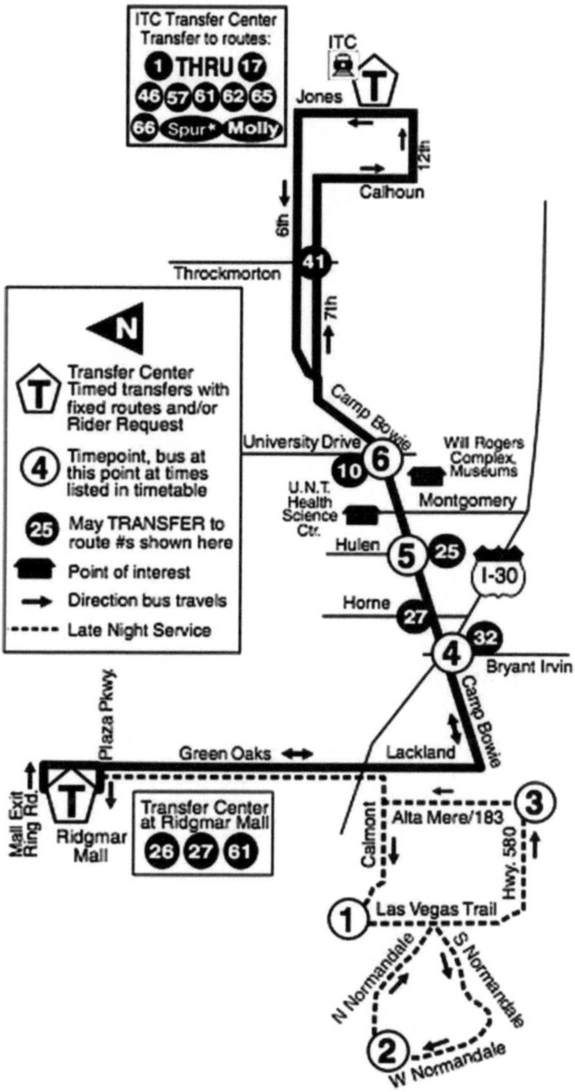

visualization, allowing for a more straightforward analysis of bus traffic flow based on the time of day.

Figure 8 is a screenshot of the Atvis visualization demonstration system. Our original model and demonstration differed in a few key ways:

1. Our final implementation allows for the visualization of only one route and dataset at a time.
2. We add interfaces that allow users to control simulation timing and visualization methodologies (described in greater detail in Sect. 3.4).

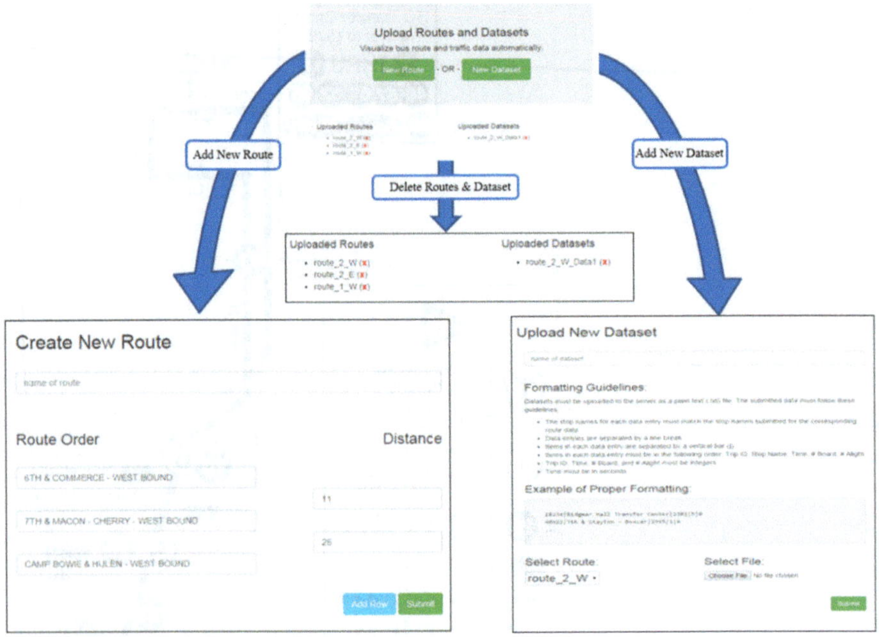

Fig. 4 Atvis backend data management system

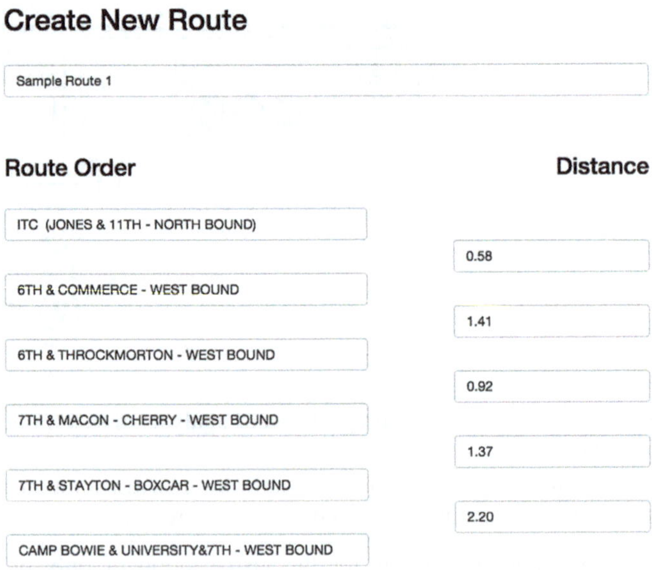

Fig. 5 The backend route creation system

```
18234|Ridgmar Mall Transfer Center|2381|5|0
48922|7th & Stayton - Boxcar|2395|2|4
...
```

Fig. 6 An example of a proper dataset format

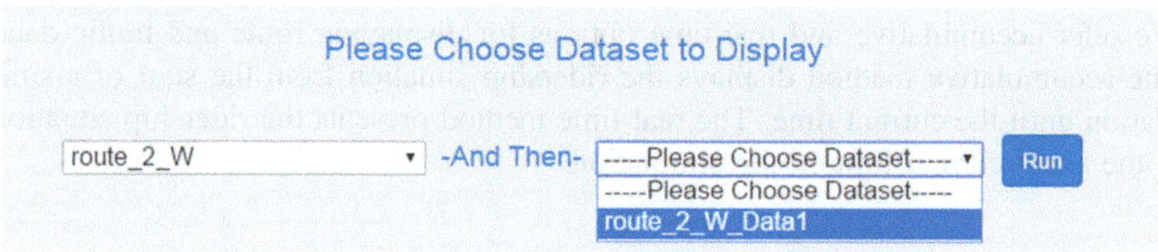

Fig. 7 The frontend data selection

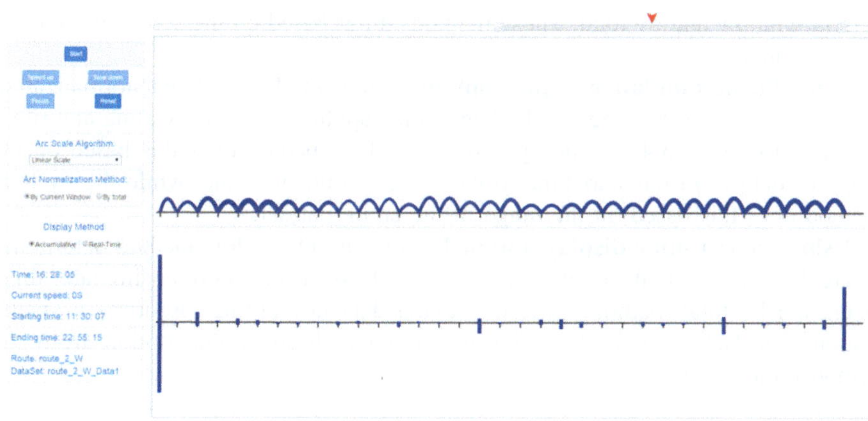

Fig. 8 The frontend data visualization system

3. We add a timeline that allows users to specify the starting/ending time of a simulation and to view current simulator time.
4. Users can mouse over both the arc and bar diagrams to view more detailed ridership information at each stop.
5. Users can zoom in and out of the arc and bar diagrams to view ridership and passenger flow statistics in greater or lesser detail.

3.4 *Visualization Methodologies*

The Atvis demo supports a variety of visualization options that display different types of information using different algorithms and calculation methods.

3.4.1 The Display Method

We offer accumulative and real-time options for displaying route and traffic data. The accumulative method displays the ridership situation from the start of a simulation until the current time. The real-time method presents the ridership situation at the exact current time in the simulation.

3.4.2 The Arc Normalization Method

Atvis normalizes ridership values as arc thickness using two different methods: by current window and by total. These methods differ based on the scope of data that are presented.

Using the accumulative display option Both methods calculate a normalization factor to use in determining arc thickness, then applied to the arc scaling algorithm (described in Sect. 3.4.3). The by current window method does this based on the range of data from the start time to the current simulator time, while the by total method does this based on the range of the entire dataset.

Using the real-time display option The by current window method determines an arc thickness normalization factor based on the data at the current simulator time. Using the by total method while displaying data in real-time produces a visualization that lacks detail. As a result, we have disabled this option from the demonstration system.

3.4.3 The Arc Scaling Algorithm

The arc scaling algorithm controls the method in which ridership values are presented in an arc graph. Our demonstration system offers three different algorithms for scaling data in this view:

1. Linear scale—the base scaling function utilized by the following two scaling algorithms. A scale is determined for a dataset by linearly distributing thickness values throughout the range of data.
2. Square root scale—takes the square root of all ridership values and then applies the linear scale to them.
3. Relative magnitude scale—takes the square root of the largest ridership value and then uses the linear scale to scale the rest of the values relative to this value.

4 Discussion/Conclusion

Atvis allows a transit planner to observe bus data visually, to pinpoint individual stops and transitions, and to formulate research hypotheses. The following is a list of questions that can be posted to or answered by the Atvis system:

1. Which stop has the highest boarding/alighting level at 8:11 a.m.?
2. How many people ride a bus between stop 2 and stop 3 from 9 a.m. to 10 a.m.?
3. Which two bus stops have the most ridership on Saturday?
4. Does the distance between two stops have any correlation with their ridership levels during rush hours?
5. Do any stops have abnormal boarding and alighting ratios for a given time period?

Atvis excels at visualizing transit data at an abstract level. Our focus on ridership and interstop relationships creates a visualization that emphasizes key bus/route statistics and removes the distraction of unwanted geographical data, allowing for more direct observation of transit patterns. Our innovative use of the arc diagram in conjunction with a bidirectional bar graph offers a holistic representation of transit data in a single view.

Our demonstration system is not currently implemented fully for multiple routes and datasets. Future implementation of this functionality will be useful to transit planners. A simple extension is to add the relationship between trips headed toward opposite bounds but along the same geographical route. A more general extension is to add a multiroute using suborder and multi-dataset visualization to allow users to compare ridership and passenger flow through different bus systems at the same simulator time.

Our system can also be improved by expanding data presentation to allow zooming in on individual bus trips. Doing so allows Atvis to help analysts not only to determine trends in bus data for the overall system but also to itemize ridership and passenger flow information for each trip. This feature would allow users to analyze data from a different perspective while maintaining our visualization's other advantages.

References

Agrawala M (2002) Visualizing route maps. Dissertation, Stanford University

American Public Transportation Association (2013) Public transportation fact book, 64th edn. American Public Transportation Association, Washington, DC

Kerr B (2003) Thread arcs: An email thread visualization. In: Munzner T, North S (eds) INFOVIS 2003: IEEE symposium on information visualization. IEEE Computer Society, Seattle, WA, pp 211–218

Krause J, Spicker M, Schäfer M, Strobelt H, Wörteler L, Zhang L (2012) Interactive visualization for real-time public transport journey planning. In: Proceedings of SIGRAD on the interactive visual analysis of data. Linnaeus University, Växjö, Sweden, 2012

Kwan M (2002) Feminist visualization: re-envisioning GIS as a method in feminist geographic research. Ann Assoc Am Geogr 94(4):645–661

Mashima D, Kobourov S, Hu Y (2011) Visualizing dynamic data with maps. In: 2011 IEEE Pacific visualization symposium, Hong Kong, 2011, pp 155–162. doi:10.1109/PACIFICVIS.2011.5742385

Saaty TL (1964) The minimum number of intersections in complete graphs. Proc Natl Acad Sci 52 (3):688–690

Sadahiro Y, Tanabe T, Pierre M, Fujii K (2015) Computer-aided design of bus route maps. doi:10.1080/15230406.2015.1077162

Wattenberg M (2002) Arc diagrams: visualizing structure in strings. In: INFOVIS 2002: IEEE symposium on information visualization, 2002, pp 110–116. doi:10.1109/INFVIS.2002.1173155

Wolff A (2007) Drawing subway maps: a survey. Informatik—Forschung und entwicklung 22 (1):23–44

Xiao N, Chun Y (2009) Visualizing migration flows using Kriskograms. doi:10.1559/152304009788188763

Mapping Spatiotemporal Patterns of Disabled People: The Case of the St. Jude's Storm Emergency

Thanos Bantis, James Haworth, Catherine Holloway and John Twigg

Abstract Emergency management can greatly benefit from an understanding of the spatiotemporal distribution of individual population groups because it optimizes the allocation of resources and personnel needed in case of an emergency caused by a disaster. In practice, however, vulnerable population groups, such as people with disability, tend to be overlooked by emergency officials. Tasks such as identifying people who are vulnerable in an emergency generally are approached statically using census data, without taking into account the spatiotemporal dynamics of disabled people's concentrations as observed in large metropolitan areas such as London, United Kingdom. Transport data gathered by automatic fare collection methods combined with accessibility covariates have the potential of being a good source for describing the distribution of this concentration. As a case study, data from the peak of the St. Jude's Day storm in London on October 28, 2013, were used to model the within-day fluctuation of disabled people, employing discrete spatiotemporal variation methods. Specifically, Poisson spatiotemporal generalized linear models were built within a hierarchical framework, ranging from simple to more complex ones, taking into account spatiotemporal interactions that emerge between space-time units. The performance of the resulting models in terms of their ability to explain the effects of the covariates, as well as predict future disabled peoples counts, were compared relative to each other using the deviance

T. Bantis (✉)
Center for Urban Sustainability and Resilience, University College London, London, UK
e-mail: thanos.bantis.13@ucl.ac.uk

J. Haworth
SpaceTimeLab for Big Data Analytics, University College London, London, UK
e-mail: j.haworth@ucl.ac.uk

C. Holloway · J. Twigg
Department of Civil Environmental and Geomatics Engineering, University College London, London, UK
e-mail: c.holloway@ucl.ac.uk

J. Twigg
e-mail: j.twigg@ucl.ac.uk

D.A. Griffith et al. (eds.), *Advances in Geocomputation*,
Advances in Geographic Information Science,
DOI 10.1007/978-3-319-22786-3_10

information criterion and posterior predictive check criterion. Analysis of results revealed a distinct spatiotemporal pattern of disabled transport users that potentially could be used by emergency planners to inform their decisions.

Keywords Disability • Emergencies • Bayesian hierarchical models • Poisson generalized linear model • Spatiotemporal

1 Introduction

Emergency management in the context of disasters is a complicated issue requiring both detailed planning and a high level of alert from both emergency managers and citizens themselves. It cannot be approached in a generalized way because different population groups, having different needs and capabilities, are affected to varying degrees. This is especially true for people with a disability because they tend to be overlooked by emergency officials (Kailes and Enders 2007; Twigg et al. 2011; McGuire et al. 2007; Rooney and White 2007).

The value of self-preparedness has been documented in numerous emergency preparedness guidelines for people with disabilities. As an example, Kailes (2002) elaborates on the importance of being self-reliant by integrating emergency planning as part of everyday routine activities. Beyond that, some authors comment about the very important role that disabled people have to play in emergency management as active participants in decision making (Wisner 2002; Abbott and Porter 2013).

Emergency officials should include all population groups when preparing emergency plans, and people with disabilities are no exception. However, the way in which the preparation, delivery, and impact of different approaches on disaster risk reduction has not been well documented (Alexander 2011; Smith et al. 2012). In a literature review, Smith et al. (2012) comment that one of the most frequently reported reasons of inequality in disaster situations for disabled people is the lack of knowledge regarding their needs (more specifically, lack of knowledge about the number of people with disabilities and where they live), together with inaccurate information related to disability groups within the population.

Currently, common practice in identifying the location and number of disabled people involved in an emergency relies either on census data or the creation and maintenance of lists of partners or contact numbers that could be used during the information-gathering phase (Norwood 2011; HMGovernment 2008). This latter information is difficult to keep up to date and largely is based only on home address (Metz et al. 2002; Morrow 1999). The constant movement of people in a large metropolitan area such as London makes the determination of these locations and numbers even more difficult because individuals may be on the move throughout the day, quite often away from the boundaries of the local authorities in which they are registered. As such, a clear need exists to investigate data sources that are able to locate disabled people in a dynamic way.

One such data source is the automatic fare collection (AFC) network, of which London's Oyster card system is an example. Although AFCs cannot be used directly to identify the location of individuals, they have the potential of providing near-real-time information about the spatiotemporal distribution of disabled people, as well as their approximate locations and mobility as far as public transport use (Lathia et al. 2012).

In this study, Oyster card data are used to model the within-day fluctuation of disabled users during an emergency. The hourly fluctuations during the St. Jude's Day storm of October 28, 2013 are used because the storm hit hardest on this day in southeast England, causing widespread disruption (AIR Worldwide 2013). A Bayesian hierarchical modeling structure is used to formulate a Poisson generalized linear model (GLM) while paying special attention to interaction between different space-time units. The study of the model's results can benefit emergency managers in a number of different ways. First, by identifying the potential location and numbers of disabled people involved, emergency managers can plan more efficiently the allocation of personnel and resources needed during an emergency. Second, using the predictive potential of the model, better strategic decisions can be made for future emergencies. Third, looking at the interaction between spatial and temporal structures can help reveal gaps in the current understanding of disabled people's mobility patterns. Finally, the conditional approach followed imposes a smoothing effect across the crude observations. As such, the results can be more easily interpreted by emergency managers. Section 2 furnishes a description of the data and the study area. Section 3 introduces the model specification, while the results are presented in Sect. 4. Section 5 discusses the results in the context of emergency response. Section 6 addresses limitations and future directions.

2 Data

This section introduces the datasets that are used for modeling purposes.

2.1 *The Oyster Card System*

Transport for London's (TfL) automated fare collection system uses radio-frequency identification (RFID) stamped cards (called Oyster cards) as a unified transportation ticketing system for many public means of travel. This includes the Underground, National Rail, other rail services, and buses. Although TfL has recently combined the use of contactless payment cards with Oyster cards, only the Oyster card option was available at the time of the data collection. Information related to individual trips is captured each time an Oyster card is used. A brief description of Oyster card dataset characteristics is given in Table 1 (this

Table 1 Relevant Oyster card fields

Field	Description
CARDTYPEKEY	Contains population groups as determined by the different fare types (adults, children, students, elderly people, and disabled people)
PRESTIGEID	Contains unique pseudo ID (in the sense that the data remain anonymous) for each record, generated by TfL
STATIONOFFIRSTENTRYKEY ROUTEID	Contains entry/exit information in case of travel by rail, and bus route number in case of travel by bus
TIMEOFFIRSTENTRY	Contains time/day information

tabulation resulted from personal communication in 2014 with Jonathan Reades, who furnished a document titled "TfL Data Source Documentation").

The population group attribute contained within the CARDTYPEKEY field is important within the scope of this study because it distinguishes disabled people from nondisabled people. This feature, combined with information about travel preferences, has the potential to reveal mobility patterns at a high temporal resolution that could be used to provide an image of the spatiotemporal activity patterns of disabled individuals.

This dataset has the following limitations, among others:

1. It does not provide information about specific disability types.
2. For bus trips, only boarding information is stored.

The first limitation means that inferences have to be made for the full range of disability types as specified and assessed by the relevant organizations (London Councils 2014). Because buses are the predominant means of public transportation for disabled people (Transport for London 2012), for reasons of consistency, only boarding information for rail travel is considered in the analysis.

The Oyster card dataset obtained contains every single journey made using Oyster cards for a period of eight weeks, from the end of October to the middle of December in 2014. This amounts to more than 100 million records, most of which refer to journeys made using the Underground.

2.2 A Case Study

London's borough of Croydon was chosen for analysis because it has a representative percentage of disabled population with respect to its general population, compared to the other London boroughs (London's Poverty Profile 2014). Figure 1 shows Croydon's location along with the distribution of public transport means.

Oyster card records with a disabled pass attribute were used to represent disabled people boarding a bus/rail service while keeping the total number of passengers as the exposure count. The resulting observations represent aggregated counts at each

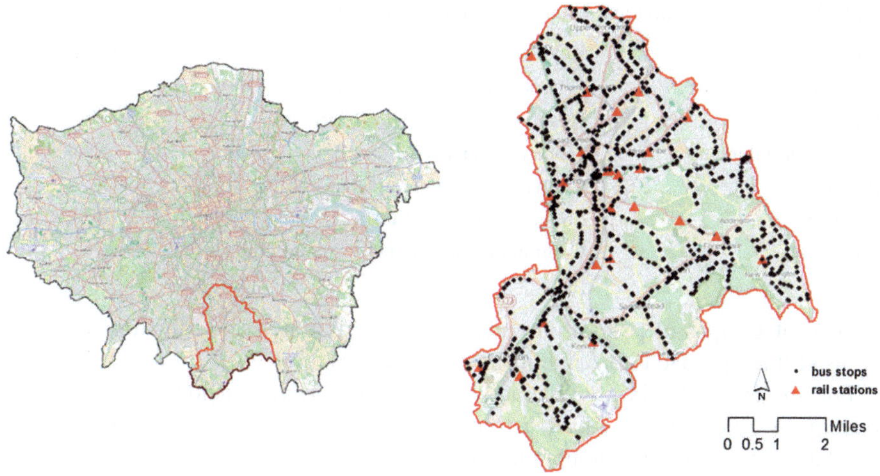

Fig. 1 The case study area

bus stop/rail station at the moment of Oyster validation. For the temporal domain, October 28, 2013 (St. Jude's Day) was used, because this was the peak of the storm in London. The day was discretized into 16 approximately hourly intervals during public transport operational times, excluding night buses; for example, 04:30–01:30 the next day.

2.3 Choosing Covariates

This section presents a description of the covariates used to link the Oyster card counts with elements of the build environment/services. We hypothesize that after accounting for variation in the distribution of disabled people in regular conditions, as expressed by the opportunities/destinations and public transport accessibility level (PTAL) covariates described in the following sections, the remaining variation captured could be attributed to the emergent behavior caused by the storm.

2.3.1 Opportunities/Destinations

An important influence on the number of people visiting a particular destination is the total number of opportunities/potential destinations within a geographic area. This concept is closely related to the notion of accessibility of a particular area as defined by the total number of locations at which an activity can be found within a prespecified spatial extent. The larger the number of opportunities within reach of a particular areal unit, the larger the accessibility of that unit. Other more

Table 2 POI classification and counts according to categories

Category	Keywords	Count
Education	School, university, education	2006
Medical	Hospital, health, center, NHS	379
Public entertainment	Library, theater, cinema	58
Religious	Church, synagogue, mosque	1554
Shopping	Shop, shopping center	1733
Social cubs	Community, social center, social club	701

sophisticated types of accessibility measures exist (Church and Marston 2003), but we used this approach here due to its simplicity.

For this research, six possible activity categories were considered (representing locations where general social activities take place): education, medical, public entertainment, religious, shopping, and social clubs.

The range of possible activities was assessed with points of interest (POIs) within each areal unit, using the OpenStreetMap (Openstreetmap.org 2014) POI database. A simple categorization between POI categories and travel preference categories was employed using keywords appearing in the official name of a POI. Table 2 summarizes the total number of POIs and the keywords used.

2.3.2 PTAL

The PTAL level is a measure of accessibility of areas to public transport. It was first designed by the borough of Hammersmith and Fulham in 1992, and after a series of reviews and tests, the borough-led PTAL development group agreed that it is the most appropriate method for calculating accessibility of areas to public transport in London (Transport for London 2010). Its calculation takes into account different elements of public transport accessibility, such as walking time from any geographical point to public transport access points, reliability and frequency of public transport means, average waiting time, and public transport means density.

The PTAL index has weaknesses, such as its failure to include the effects of overcrowding, congestion, ability to board in general, and the difficulties of interchanging between different travel means. For disabled people, these are important aspects because they are among the most influential factors depriving them of access to public transport services (Jolly et al. 2006).

The calculation of a PTAL index is relatively easy and straightforward. The reader is referred to Transport for London (2010) for details. The final product is a map with contour lines representing the transitions among six levels of accessibility, ranging from low to high (1–6). Levels 1 and 6 are further subdivided into two for further clarity. Figure 2 shows that PTAL essentially quantifies the density of public transport per unit area, with the higher values corresponding to Croydon city center.

Fig. 2 Croydon PTAL levels

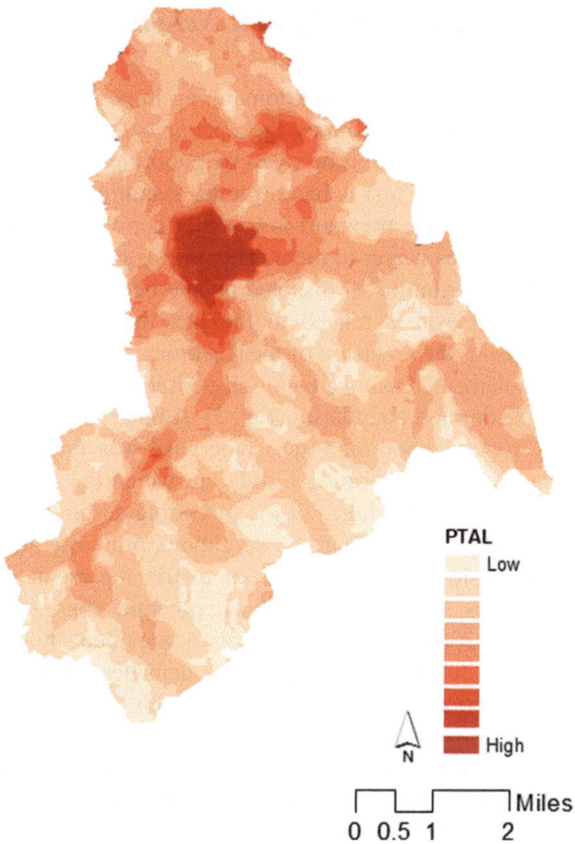

3 Methods

In line with the nature of the Oyster card data, discrete spatial variation methods are used to model the spatiotemporal distribution of aggregated disabled users' counts. However, before that, there is a need to define the spatial and temporal domain appropriately.

3.1 Data Preparation

The spatial domain was defined by a combination of a buffer distance around the bus stops and train stations and a Voronoi tessellation. The recommended walking distance for disabled people without a rest (May et al. 1991) is 150 m. For this research, a buffer distance of 200 m was chosen to encourage the creation of a spatially contiguous area, as far as this is possible.

3.2 Defining the Spatial Neighborhood

After defining the spatial domain, an important step is to define the neighborhood structure that determines the spatial connectivity and hence the way in which information is shared between neighboring areas. This can benefit the modeling in two ways. The first is based on the assumption that the number of disabled people boarding public transport is spatially correlated because people quite often tend to use the stops nearer to them. This translates into the assumption of continuity in the spatial distribution of disabled people between adjacent public transport stops. The second reason is related to the fact that such a structure benefits inference by increasing the precision of local estimates. In this study, the neighborhood structure is implemented using a contiguity criterion that is realized when adjacent areal units share common boundaries (Fig. 3). A binary weight matrix representing a weight of one for neighbors and zero otherwise was implemented; at this stage, no information exists supporting a different weighting scheme.

However, the choice of the preceding described neighborhood structure resulted in some areal units with no neighbors, primarily along the boundaries of the study area. Besides being an issue of matrix representation for the entries having zero neighbors, such artifacts can have potential consequences for inference (Bivand and Portnov 2004). In this study, this was evident by the slow convergence and wider credible intervals of the posterior densities for those areas. Currently, there is no best practice about how to proceed with these cases, because including them in an analysis results in undefined entries in the spatial weight matrix (e.g., in the case of a row standardized matrix), while omitting them results in a reduced sample size (Bivand and Portnov 2004). A common practice (Banerjee et al. 2004; Bivand et al. 2008) is to include a vector of zero length in the weight matrix, accepting the fact that for these units imposing spatial smoothing falls outside the model's assumptions. This strategy was adopted in this study.

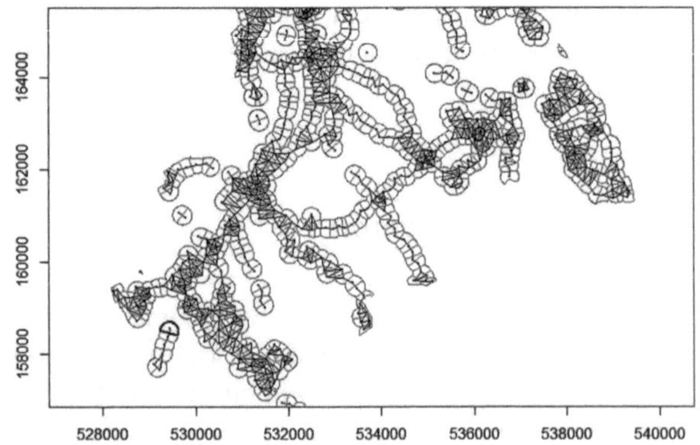

Fig. 3 Neighborhood structure (detail from Croydon borough)

3.3 Modeling

We begin discussion of the modeling procedure with the assumption that there exists a process that generates a relatively high number of disabled people counts for a few, typically more accessible, stops, and relatively low counts for the majority of the remaining stops. A Poisson process was chosen as the base model. To include covariate information as well as spatial, temporal, and spatiotemporal interaction terms, the rate parameter of the Poisson process is modeled in a GLM specification. Finally, the entire modeling procedure is conceptualized within a Bayesian hierarchical modeling framework that allows for increased flexibility by placing prior densities on the unknown parameters. It also allows for better interpretability of the results by providing posterior densities and uncertainty intervals that often resonate better with one's everyday interpretations (Willink and Lira 2005). This modeling approach is most commonly encountered in disease mapping, where the relevant counts are reported disease incidents per areal unit (Gelfand et al. 2010).

To summarize, the modeling was done using Poisson spatiotemporal GLM built within a Bayesian hierarchical modeling framework, ranging from simple to more complex ones, while taking into account the spatiotemporal interactions that emerge between the space-time units.

The final model is

$$Y_{it}|\lambda_{it} \sim Poi(\lambda_{it}), \quad (1)$$

$$\log(\lambda_{it}) = \log(E_i) + \beta_0 + \beta_i^{'} x_i + u_i + v_i + \delta_t + \psi_{it},$$

where Table 3 presents the definitions of terms in this equation.

The regression coefficients and spatially unstructured random effect $\beta_{1 \ldots p} \sim N(0, \tau_\beta), u_i \sim N(0, \tau_u)$ are assumed to be zero-centered normally distributed

Table 3 Terms in the model

Variables	Explanation
Y_{it}	Disabled users counts
λ_{it}	Number of disabled passengers per areal unit/time slice
E_i	Expected number of people arriving at each areal unit, defined as $E_i = \frac{\sum_{i=1}^{n} Y_i}{\sum_{i=1}^{n} P_i} P_i$, where P_i is the total number of passengers in each areal unit
$\beta_{1 \ldots p}$	Regression coefficients
β_0	Intercept term
u_i	Spatially unstructured random effect
x_i	Covariates
v_i	Spatially structured random effect
δ_t	Temporal effect
ψ_{it}	Spatiotemporal interaction term

with precisions $\tau_u \sim Ga(a,b)$ with $a=b=0.001$ and $\tau_\beta=10^{-6}$. An unbounded uniform prior distribution is placed on the intercept term $\beta_0 \sim U(+\infty, -\infty)$.

An intrinsic Gaussian conditional autoregressive (ICAR) prior (Besag et al. 1995) is placed for the spatial effects v_i, while two alternative priors are placed for the temporal effects: an unstructured $\delta_t \sim N(0,\tau_\delta)$, for $t=1\ldots T$, and a random walk (RW) prior to reflect the notion that the temporal effect is correlated with its adjacent time units, with precision in both cases $\tau_\delta \sim U(0,1000)$.

The preceding parameter values reflect the choice of placing noninformative prior densities on the precision terms, so that inferences on the model's parameters are not affected by the choice of these values. In practice, different specifications of prior densities influence the posterior distribution to a varying degree, depending, among other factors, on sample size, model specification, etc. (Gelman et al. 2004). Checking the sensitivity of posterior inferences over alternative priors on random effects variances is considered to be a good way of assessing the robustness of hierarchical models (Gelman 2006). Changing the a, b values of the gamma distribution for τ_u from 0.001 to 0.01 resulted in small (<5) deviance information criterion (DIC) value differences, a fact that is considered to be an indication of model robustness under reparametrization (Spiegelhalter et al. 2002). A more complete sensitivity analysis by checking model performance under different prior specifications for the rest of the model's parameters will be tackled in future research.

For the purposes of capturing the combined spatiotemporal effect, Knorr-Held (1999) introduces a framework about the way these components could potentially interact. The nature of these interactions depends on the structure of the spatial and temporal weight matrices. For this study, the interaction effects were assumed to have no structure in space and time so that any random deviations from global space-time trends can be revealed, potentially providing evidence of the presence of a more complicated spatiotemporal structure:

$$p(\psi|\kappa_\psi) \propto exp\left(-\frac{\kappa_\psi}{2}\sum_{i=1}^{N}\sum_{t=1}^{T}\psi_{it}^2\right) \quad (2)$$

In Eq. (2), κ_ψ are the elements of matrix K_ψ defined as the Kronecker product of the matrices of the effects that are assumed to interact (Clayton 1996). In this case:

$$K_\psi = \mathrm{K}_u \otimes K_\delta = I \otimes I = I \quad (3)$$

Therefore, ψ_{it} acts as an unobserved covariate in space and time without any structure. Other authors (López-Quílez and Muñoz 2009) describe this structure as a global space-time heterogeneity effect.

The performance of the resulting models, in terms of their ability to explain the effects of the covariates, was compared relative to each other using the DIC value (Spiegelhalter et al. 2002). By generating samples from the predictive distribution and by comparing the generated samples with the observed values, the predictive power of the resulting models can be assessed. This was done using the predictive posterior loss (PPL) criterion with a mean squared predictive error loss function.

Four parallel Markov chains were run for 7,000 iterations each, discarding the first 2,000 as nonrepresentative of the posterior distribution. To allow for better mixing, 1 out of 20 samples was used to compute the posterior distribution. This results in a simulated sample size of 1,000 values that are used to approximate the marginal posterior distribution of each unknown parameter. The number of iterations was a compromise between total inference time and computer memory constraints. Although longer chains result in more precise estimates, in practice, the sampling can stop once the parameters converge to a stationary distribution. Convergence was achieved for the bulk of the parameters according to Geweke's Z-score criterion (Geweke 1991) and Gelman and Rubin's $\widehat{R}$ criterion (Gelman and Rubin 1992). The WinBUGS (Lunn et al. 2000) software, in combination with the R programming language (R2WinBUGS library [Sturtz et al. 2005]), was used to code the models and carry out the Markov chain Monte Carlo (MCMC) simulations.

4 Results

In all models, with the exception of PTAL, the effect of covariates was found to be statistically nonsignificant given that the zero value was within the 95 % credible intervals.

The spatially structured random effect portrayed in Fig. 4a reveals a statistically significant variability between the areal units clustered in the north of the borough,

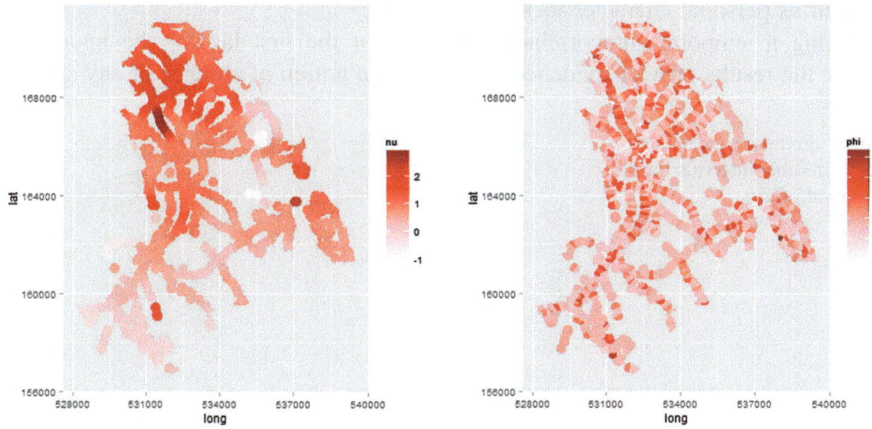

a Spatially structured random effect **b** Spatially unstructured random effect

Fig. 4 Posterior means of structured and unstructured spatial effects

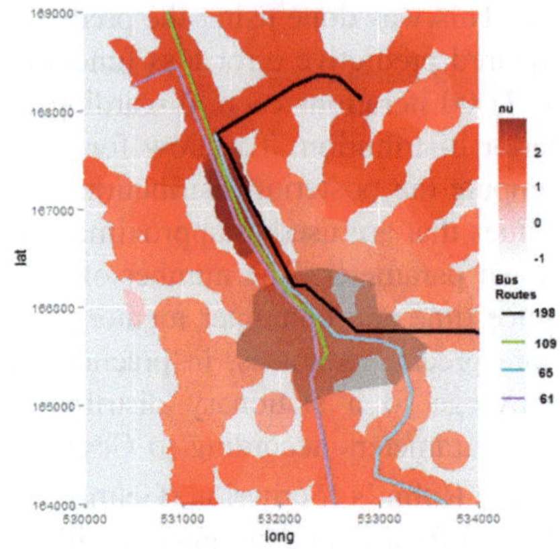

a Bus routes and Croydon city center (illustrated as a gray shaded region)

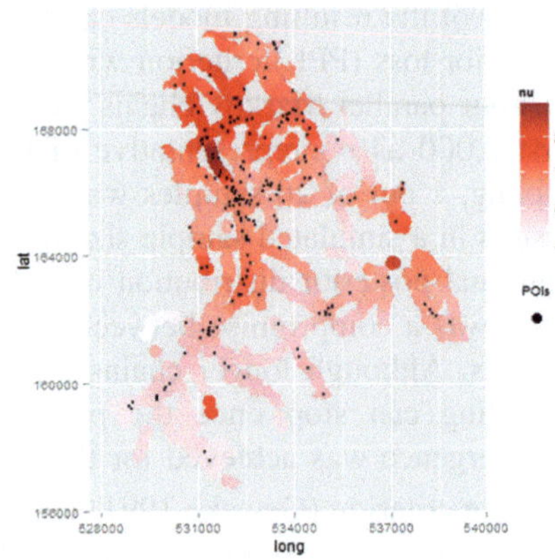

b Location of POIs

Fig. 5 Bus routes, Croydon city center, and POI locations

particularly where four bus routes overlap (Fig. 5a). Interestingly, the spatial concentration of disabled Oyster card users does not seem to be intense in Croydon's city center. This outcome could explain the reduced effect of the accessibility covariates because many of the POIs are located in the city center (Fig. 5b).

For the temporal aspect of the analysis, the lack of any time-related covariates led to the manifestation of a strong temporal pattern away from commuting to and from work rush hours (Fig. 6). This suggests need for further research about linking the observations with sociological covariates such as unemployment and poverty, as well as personal characteristics such as age.

Using a temporal autocorrelation function of the first lag for the interaction terms, the results provide evidence to support the notion of absence of any specific

Fig. 6 Posterior means and 95 % credible intervals for the temporal effects

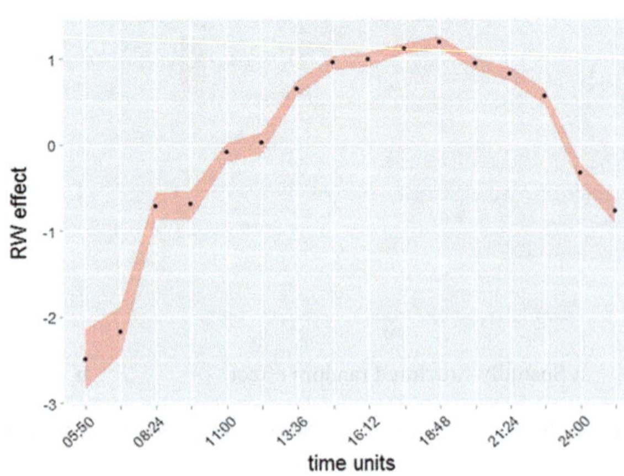

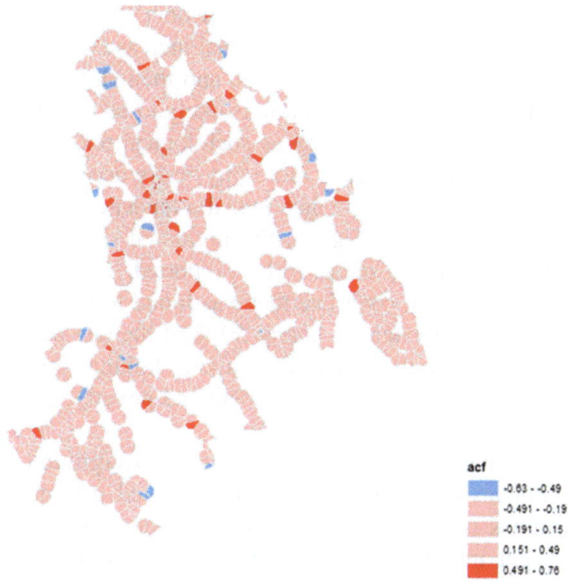

Fig. 7 Temporal autocorrelation for ψ_{it} interactions at the first lag. The 95 % confidence interval for the temporal autocorrelations was found to be ±0.49

Table 4 Model comparison criteria for all Oyster card models

Model	DIC	PPL (MSPE)
Unstructured δ	17,845	1.2076
RW δ	17,851	1.2081
Interactions	16,384	0.81

structure between the spatial and temporal components of the models for the bulk of the areal units. Although the temporal autocorrelations of the interaction effects are high (most probably due to the small number of temporal slices), for the bulk of the areal units they are statistically nonsignificant. Figure 7 indicates that little evidence exists supporting strong temporal (positive autocorrelation) or spatial interdependence (spatial clustering). Hence, the interaction term acts as a white-noise "pool" capturing any residual variation.

Finally, in terms of model comparison criteria, the model with the interaction term seems to outperform the other candidates (Table 4):

5 Conclusions

Different models were constructed in an attempt to explore disabled people's mobility patterns during an emergency, building simple to more complex specifications. In terms of model performance, the model having the spatiotemporal interaction term seems to outperform the rest of the models, both in its ability to explain the effect of the unknown parameters and in terms of predictive power. In all models, analysis revealed a distinct spatiotemporal pattern of disabled Oyster card counts. Specifically, a gradual increase of concentration occurs during the day, peaking in the afternoon hours, and then slowly declining in later hours.

The cause of this variation is likely to be found in the microscale variation of the built environment as well as sociological reasons. Are the bus stops/train stations with higher count rates more physically accessible, or do they simply serve more routes? Given that the data reveal a tendency to avoid the town center where most train stations are (Fig. 5), the notion of disabled people preferring to use buses rather than trains seems to be supported. This finding is also confirmed by examining the PTAL covariate that was found to have a decreasing contribution of $e^{0.71}$, which is equivalent to nearly 50 % of the rate of Oyster counts. This suggests that further research is required about linking the observations with sociological covariates such as unemployment and poverty, and personal characteristics such as age.

Furthermore, because many of the overground rail services were experiencing problems during the storm, the observed pattern being located away from those stations is sensible. In any case, the information can provide emergency planners insight into the dynamics of disabled people's concentration patterns during an emergency.

By utilizing the predictive posterior distribution, predictions can be made with fluctuated data conditioned on the original observations. This allows emergency managers to simulate the effects on the distribution of disabled people in case, for instance, of a station closure during an emergency given the current model's estimated parameters.

Finally, looking at the spatiotemporal interaction terms, the results provide evidence to support the notion of absence of any specific structure between the spatial and temporal components of the model for the bulk of the areal units. Hence, the interaction term acts as a white-noise "pool" capturing any residual variation between the spatial and temporal domain of the analysis. However, for 54 areal units, a statistically significant positive or negative temporal autocorrelation exists in the interaction values. Because the number of these areas hardly exceeds 5 % of the total 1,098 bus stops/rail stations, this may have been generated by chance alone. This should not be a surprise because the time period studied was too short to allow for any spatial or temporal deviations from the mean values to be captured.

6 Limitations and Future Work

This study attempted to use noninformative prior distributions to express uncertainty for the model's parameters. However, specification of truly vague prior distributions that have no effect on the posterior is difficult because the posterior depends on other factors such as sample size, model specification, choice of prior distribution family, proper versus improper priors, etc. In this study, model performance was invariant after reparametrizing the precision of the unstructured spatial random effect. The same analysis has to be performed on the rest of the parameters, so that the overall robustness of the model can be assessed.

To exploit the full power of Bayesian analyses, employing subjective prior distributions based on expert knowledge of both emergency managers and people with a disability could be insightful. For example, the amount of prior belief about covariates could be assessed by a qualitative study of the nature and type of daily activities of people with a disability. Likewise, emergency managers can assess the amount of spatial and temporal smoothing they require according to their needs and resources by imposing different beliefs on the spatial and temporal effects. Then, the amount of correspondence between the prior beliefs and the data can be assessed so that new assumptions can be made.

In this study, travel times and origin-destination aspects were not taken into account because the Oyster card for traveling by bus does not provide alighting information. Recently, TfL has begun to develop an algorithm that can infer alighting by coupling subsequent validations of Oyster cards together with live bus arrival information. Processing these data could reveal more interesting patterns to provide further insight into the spatiotemporal dynamics of people with a disability.

Although the total population using an Oyster card was indirectly included in the analysis via the exposure term, an analysis with a separate dataset for users who are not disabled, comparing differences, would be of interest. No doubt, the emerging patterns would provide emergency managers with information for a more holistic approach to disaster planning.

Furthermore, conducting an analysis for a noncrisis period would provide valuable information about concentration patterns of people with a disability, as well as a means of quantifying the expected number of people in transit relative to a crisis period.

Finally, the spatial neighborhood definition is a topic requiring further research. Implementing spatial continuity criteria other than the one implemented here to assess the sensitivity of inferences would be interesting. An example could be a distance weight function that would rank the neighbors with respect to a predefined threshold.

References

Abbott D, Porter S (2013) Environmental hazard and disabled people: from vulnerable to expert to interconnected. Disabil Soc 28(6):839–852

AIR Worldwide (2013) Event summary: European Windstorm Christian (St. Jude), update 2. http://alert.air-worldwide.com/EventSummary.aspx?e=723&tp=72&c=1. Accessed 1 June 2014

Alexander D (2011) Disability and disaster. In: Wisner B, Gaillard JC, Kelman I (eds) Handbook of hazards and disaster risk reduction. Routledge, London, pp 384–394

Banerjee S, Carlin BP, Gelfand AE (2004) Hierarchical modeling and analysis for spatial data. Chapman & Hall/CRC, London

Besag J, Green P, Higdon D, Mengersen K (1995) Bayesian computation and stochastic systems. Stat Sci 10(1):3–41

Bivand RS, Pebesma EJ, Gomez-Rubio V (2008) Applied spatial data analysis with R, vol 747248717. Springer, New York

Bivand RS, Portnov BA (2004) Exploring spatial data analysis techniques using R: the case of observations with no neighbors. In: Anselin L, Florax R, Sergio J (eds) Advances in spatial econometrics: methodology, tools and applications. Springer, Berlin, pp 121–142

Church RL, Marston JR (2003) Measuring accessibility for people with a disability. Geogr Anal 35 (1):83–96

Clayton DG (1996) Generalised linear mixed models. In: Gilks WR, Richardson S, Spiegelhalter DJ (eds) Markov Chain Monte Carlo in practice. Chapman & Hall, London, pp 275–301

Gelfand AE, Diggle P, Guttorp P, Fuentes M (eds) (2010) Handbook of spatial statistics. CRC Press, London

Gelman A (2006) Prior distributions for variance parameters in hierarchical models. Bayes Anal 1:1–19

Gelman A, Carlin JB, Stern HS, Rubin DB (2004) Bayesian data analysis, 2nd edn. Chapman & Hall/CRC, London

Gelman A, Rubin DB (1992) Inference from iterative simulation using multiple sequences. Stat Sci 7(4):457–472

Geweke J (1991) Evaluating the accuracy of sampling-based approaches to the calculation of posterior moments. Vol 196. Federal Reserve Bank of Minneapolis, Research Department, Minneapolis

HMGovernment (2008) Identifying people who are vulnerable in a crisis: guidance for emergency planners and responders. http://www.gov.uk/government/uploads/system/uploads. Accessed 1 June 2014

Jolly D, Priestley M, Matthews B (2006) Secondary analysis of existing data on disabled people's use and experiences of public transport in Great Britain. Disability Rights Commission. http://www.enil.eu/wp-content/uploads/2012/07/Secondary-analysis-of-existing-data-on-disabled-people%E2%80%99s-use-experiences-of-public-transport-in-Great-Britain_2006.pdf. Accessed 5 Mar 2014

Kailes JI (2002) Emergency evacuation preparedness; taking responsibility for your safety: a guide for people with disabilities and other activity limitations. Available via UNISDR. http://www.preventionweb.net/educational/view/8344. Accessed 22 Sept 2014

Kailes JI, Enders A (2007) Moving beyond "special needs": a function-based framework for emergency management and planning. J Disabil Policy Stud 17(4):230–237

Knorr-Held L (1999) Bayesian modelling of inseparable space-time variation in disease risk. Stat Med 19(17–18):2555–2567

Lathia N, Quercia D, Crowcroft J (2012) The hidden image of the city: sensing community well-being from urban mobility. Pervasive computing. Springer, Berlin, pp 91–98

London Councils (2014) Am I eligible to apply for a Freedom Pass? http://www.londoncouncils.gov.uk/services/freedompass/eligibility/default.htm. Accessed 15 Sept 2014

London's Poverty Profile (2014) Adult limiting illness or disability by "borough." http://www.londonspovertyprofile.org.uk/indicators/topics/health/adult-ill-health-by-borough/. Accessed 20 Nov 2014

López-Quílez A, Muñoz F (2009) Review of spatio-temporal models for disease mapping. http://www.uv.es/~famarmu/doc/Euroheis2-report.pdf. Accessed 13 Sept 2014

Lunn DJ, Thomas A, Best N, Spiegelhalter D (2000) Winbugs—a Bayesian modelling framework: concepts, structure, and extensibility. Stat Comput 10(4):325–337

May AD, Leake GR, Berrett B (1991) Provision for disabled people in pedestrian areas. Highways Transp 38(1):12–18

McGuire LC, Ford ES, Okoro CA (2007) Natural disasters and older US adults with disabilities: implications for evacuation. Disasters 31(1):49–56

Metz WC, Hewett PL, Muzzarelli J, Tanzman E (2002) Identifying special-needs households that need assistance for emergency planning. Int J Mass Emerg Disasters 20(2):255–281

Morrow BH (1999) Identifying and mapping community vulnerability. Disasters 23(1):1–18
Norwood F (2011) Promising practices for evacuating people with disabilities. http://www.ohsu.edu/xd/research/centers-institutes/institute-on-development-and-disability/public-health-programs/upload/Promising-Practices-final-1-21-2011.pdf. Accessed 10 Aug 2014
Openstreetmap.org (2014) Openstreetmap. http://www.openstreetmap.org/. Accessed 5 July 2014
Rooney C, White GW (2007) Consumer perspective narrative analysis of a disaster preparedness and emergency response survey from persons with mobility impairments. J Disabil Policy Stud 17(4):206–215
Smith F, Jolley E, Schmidt E (2012) Disability and disasters: the importance of an inclusive approach to vulnerability and social capital. Available via CRHnet. http://www.crhnet.ca/sites/default/files/library/Smith.Jolley.Schmidt.2012.Disability%20and%20disasters.pdf. Accessed 15 Sept 2014
Spiegelhalter DJ, Best NG, Carlin BP, Van Der Linde A (2002) Bayesian measures of model complexity and fit. J Royal Stat Society, Ser B: Stat Methodol 64(4):583–639
Sturtz S, Ligges U, Gelman A (2005) R2winbugs: a package for running winbugs from r. J Stat Soft 12(3):1–16. http://www.jstatsoft.org
Transport for London (2010) Measuring public transport accessibility levels: PTALS summary. http://data.london.gov.uk/documents/PTAL-methodology.pdf. Accessed 16 Sept 2014
Transport for London (2012) Understanding the travel needs of London's diverse communities: disabled people. https://www.tfl.gov.uk/cdn/static/cms/documents/understanding-the-travel-needs-of-london-diverse-communities.pdf. Accessed 1 Aug 2014
Twigg J, Kett M, Bottomley H, Tan LT, Nasreddin H (2011) Disability and public shelter in emergencies. Environ Hazards 10(3–4):248–261
Willink R, Lira I (2005) A united interpretation of different uncertainty intervals. Measurement 38 (1):61–66
Wisner B (2002) Disability and disaster: victimhood and agency in earthquake risk reduction. In: Rodrigue C, Rovai E (eds) Earthquakes. Routledge, London

Terra Populus: Challenges and Opportunities with Heterogeneous Big Spatial Data

David Haynes, Suprio Ray and Steven Manson

Abstract Big geospatial data have unique challenges not associated with the greater big data community, namely, that raster and vector analytical approaches have evolved along two separate paths. Terra Populus is a next-generation spatial database repository that focuses on the integration of heterogeneous big data. When accessing Terra Populus through a web interface, users are able to transform microdata, vector, and raster datasets into user-requested formats for analysis. By providing this framework, Terra Populus lowers the barriers for researchers examining human-environment interactions.

Keywords Terra populus • High-performance spatial analysis • Microdata • Raster • Vector • PostgreSQL • PostGIS • Apache spark • Parquet

1 Introduction

Spatial data volume has exploded in recent times, driven by technological advances in data collection and the emergence of novel spatiotemporal applications in diverse knowledge domains. These trends have ushered in a new era of geocomputation possibilities. In response, a number of specialized spatial data analysis systems have emerged, such as CyberGIS (Wang 2010), SpatialHadoop (Eldawy and Mokbel 2013, 2015), GeoMesa (GeoMesa 2014), and GeoTrellis (GeoTrellis 2014). However, each of these approaches caters to one of the two spatial data formats: vector or raster. Future high-performance spatial data analysis systems must offer an

D. Haynes (✉) · S. Manson
Minnesota Population Center, University of Minnesota, Minneapolis, USA
e-mail: dahaynes@umn.edu

S. Manson
e-mail: manson@umn.edu

S. Ray
Computer Science Department, University of New Brunswick, New Brunswick, USA
e-mail: suprio@unb.ca

D.A. Griffith et al. (eds.), *Advances in Geocomputation*,
Advances in Geographic Information Science,
DOI 10.1007/978-3-319-22786-3_11

integrated approach to manage and analyze data stored in these formats. Terra Populus (Minnesota Population Center 2015b), a National Science Foundation project, is the next stage of big spatial data—heterogeneous big spatial data and the strength of this project is its ability to integrate heterogeneous data types.

2 Terra Populus

The Terra Populus project demonstrates the emerging abilities of a next-generation spatial data repository. It identifies, acquires, and develops various data sources, ranging from historic handwritten census forms to current satellite observations of the earth. The aim of the project is to preserve these data and make them Internet accessible so that they are readily available for scholars, students, policy makers, and members of the public.

Terra Populus (TerraPop) will become the largest curated source for global human population- and sensor-derived environmental data. It draws on allied datasets from the Minnesota Population Center, namely, the National Historical Geographic Information System (NHGIS) and Integrated Public Use Microdata Series projects (IPUMS) (Minnesota Population Center 2015a). NHGIS is the largest population database about the United States, with 265 billion data points (Minnesota Population Center 2011). IPUMS is the largest population survey database in the world, with records for more than half a billion individuals.

Currently the Terra Populus repository contains microdata and raster and vector datasets spanning the globe. Microdata are stored and accessed using Apache Spark's Parquet database (Parquet 2014). Meanwhile, PostgreSQL's PostGIS extension is used to store and query both vector and raster data types (PostGIS 2005).

Microdata are hierarchical fixed-width text files, where each line represents an individual person or set of household characteristics. Individual characteristics describe age, education, employment by industry, and so on. Household characteristics describe items such as urban or rural residence, electricity, water source, and floor type.

The Terra Populus vector data collection is the first comprehensive, historically accurate map of spatiotemporal multilevel administrative geographic units that can be readily linked to census data. Globally accurate boundary data are sparse at the second and third administrative levels (e.g., counties and block groups), and relatively incomplete when examining the temporal component. A number of projects are attempting to remedy this situation, including the Global Administrative Unit Layers (GAUL), the United Nations Second Administrative Level Boundaries (SALB), and the Global Administrative Areas (GADM) database produced by Hijmans et al. (2011). Terra Populus has inventoried and analyzed these sources and now provides multilevel administrative historic geographic boundaries (Kugler et al. 2015). Researchers using Terra Populus are able to link boundary data to appropriate population datasets.

The raster data collection is rapidly growing, and each new dataset enriches the archive. Our raster datasets are obtained from government research agencies and academic research groups. Initial processing of the raster datasets requires them to be downloaded and mosaicked together for a complete global spatial extent. Therefore, when users request a study area, they obtain a seamless raster without needing to manage tiles. TerraPop provides access to over 300 rasters containing data about agriculture, land cover, and climate. In the future, we will integrate and release datasets about earth science, gridded population, and biodiversity.

3 Terra Populus User Interface

TerraPop has a web interface that allows users to interact with and create extracts from the data collection. Through the web interface, data.terrapop.org, users are able to build custom data extracts through one of three work flows. Currently the TerraPop supports microdata, vector, and raster extract work flows. Figure 1 portrays the data transformations that are capable within the TerraPop framework.

Users may browse variables and datasets within a work flow, selecting ones in which they are interested for inclusion in their extract. Users in a particular work flow, such as microdata, are not limited only to variables and datasets with the microdata format. Instead, the system guides users through any choices required to transform or contextualize a raster dataset onto a microdata record. When users submit their extract request, the application constructs an extract definition, which contains datasets, variables, and operations that the users specified, and passes the definition to the spatial database for processing. The system's logic determines the data that need be extracted, selected, or calculated, and then assembles them into files. The files are compressed into a zip archive, and an e-mail is generated providing the users with links to download the data.

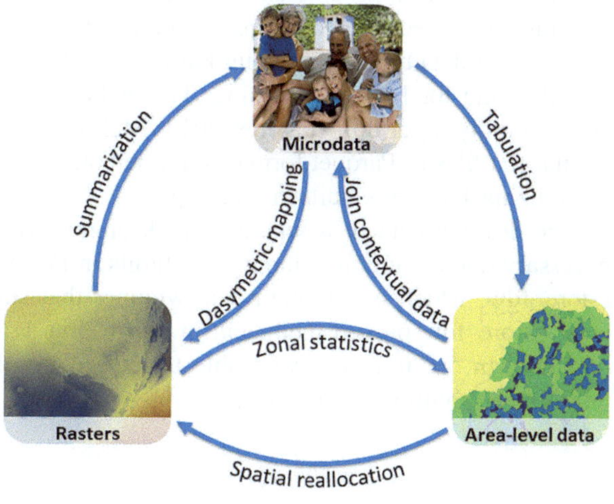

Fig. 1 Terra Populus data transformation processes

4 Terra Populus's High-Performance Architecture

> Enabling research, learning, and policy analysis by providing integrated spatiotemporal data describing people and their environment—Terra Populus mission statement

TerraPop fulfills its mission statement by providing an easily accessible web platform for integrated scientific data. When examining the literature, however, we were surprised by the lack of fully featured existing architectures that could be readily implemented. Briefly, we discuss the platform architectures we have developed that provide a high-performance computation environment for big heterogeneous data types.

4.1 Microdata Integration

The microdata in the TerraPop collection are obtained from national censuses and similar surveys of individuals and households. The variety of attributes associated with each microdata record make it a rich resource for analysis. Individual records can be tabulated to generate numerous subpopulations (e.g., number of males with an annual salary greater than $75,000 who do not have a college degree). Terra-Pop's web interface allows a user to define a tabulation query with having to construct a structured query language (SQL) query. Instead a user will choose one or more datasets on which to construct a query. Next, variables of interest are selected, and aggregations can be implemented. Currently, Terra Populus uses geographic aggregations for generating region-specific variables. Additional constraints or filters (e.g., Sex = F) can be applied to chosen variables for custom tabulation.

The TerraPop's microdata tabulation engine employs Apache Spark's Parquet columnar storage database because it allows storage, retrieval, selection, and filter operations to be done efficiently. Parquet allows for columnar storage in which columns are typecasted to Parquet's inbuilt data types, allowing for greater compression per data type. Employing Parquet provides a high compression ratio while still allowing for fast data fetching. This is because of Parquet's record shredding and assembly algorithm (Parquet 2014; Melnik et al. 2010). Although converting regular flat files to Parquet format is a time-consuming process, efficiency is gained by utilizing Parquet's built-in data types.

The development of a separate Spark engine for tabulating the microdata was necessary due to current column store limits in PostgreSQL. PostgreSQL leverages cstore_fdw, which is a foreign data wrapper that supports column storage. However, cstore_fdw proved inefficient in our testing as it has an upper limit of 1,600 columns. In our tests we were able to exceed that limit of columns through the unioning of multiple of datasets.

4.2 High-Performance Computation of Vector and Raster Data

The critical methodological challenge of big heterogeneous spatial data for human-environment systems is dealing with their size along with their varying spatiotemporal complexity. Of particular interest is the development of better methods for storage, retrieval, and analysis of large volumes of spatial and spatiotemporal data. Our experience indicates that existing approaches are simply overwhelmed by large-scale spatial datasets. One of the key challenges in spatial computing is integrating the fragmented array of conceptual and software approaches. Spatial data are most commonly stored in two formats, vector and raster, and each has evolved along a parallel but largely separate path over the past 30 years.

Our commitment to free and open source software led to us use PostgreSQL and its extension PostGIS, which is a popular database platform for spatial computation. However, the PostgreSQL platform does not support any native parallel query processing, and therefore our current system needs adaptation for high-performance computing. Significant work is needed to exploit multiple cores on both work station and supercomputing clusters for high-performance computing.

When PostgreSQL was first developed it only operated on a single core; even when placed on servers with multiple cores, it only utilizes a single core per query. Multiple projects have made an effort to scale PostgreSQL onto a cluster of machines. GridSQL was one of the first notable projects to make such an effort, followed by Stado (2011) and Postgres-XC. Current projects under development are CitusDB and Postgres-XL. Terra Populus has extensively tested these projects and determined that they do not support parallel spatial processing.

In our previous work, Niharika[1] (Ray et al. 2013), we demonstrated a sharding-based technique to parallelize spatial queries in a cluster of machines where a PostgreSQL database instance is run on each node. SpatialStado integrates the main concepts behind Niharika into Stado's original framework. By doing this, SpatialStado exploits a cluster of nodes, each of which hosts a PostgreSQL/PostGIS database instance. In our previous work, we presented a preliminary evaluation of SpatialStado for a dataset consisting of line and polygon objects from the TIGER California datasets (Haynes et al. 2015).

The queries listed in Table 1 are for spatial joins from the Jackpine spatial database benchmark (Ray et al. 2011). The queries are expressed in SQL with some of the spatial predicates adopted by the Open Geospatial Consortium (OGC). The results show that SpatialStado achieves near-linear speedup with two nodes, which bodes well for moving the system to multiple nodes.

The current implementation of SpatialStado uses a variant of round-robin declustering. The declustering algorithm produced 1,024 spatial partitions after processing the dataset in Table 1. The physical storage and management of the

[1]Niharika is a distributed spatial query processing framework that allows for partitioning and load balancing of spatial datasets.

Table 1 Comparison of query times: SpatialStado versus PostgreSQL

Query (acronym)	PostgreSQL (seconds)	SpatialStado (seconds)	Speedup
Polygon overlaps polygon (Aw_ov_Aw)	77.3	53.5	1.37
Polyline touches polygon (ED_to_Al)	452.9	246.0	1.84
Polyline crosses polyline (Ed_cr_Ed)	1,693.2	1,022.0	1.65

partitions in SpatialStado is done by taking advantage of PostgreSQL's sharding feature (PostgreSQL 2015). In our work, we extended the SpatialStado's "create table" SQL statement to specify spatial declustering parameters, such as the number of partitions to be created, the declustering method, and a label for the declustering scheme. To execute a spatial join, the labels of two tables being joined must match. This mechanism allows the same spatial dataset to be partitioned using different declustering schemes.

SpatialStado's performance demonstrates improvements in scaling the analysis of vector datasets. Currently, we are working on implementing the support for the raster data type within SpatialStado and implementing a sharding technique for raster datasets. Another challenge we are investigating is the development of a query optimizer for SpatialStado that generates "ideal query plans" while executing spatial join queries (Haynes et al. 2015). Our work here demonstrates that modest changes to existing open source architecture can be used to provide a platform for analyzing heterogeneous big spatial data.

5 Conclusion

The integration of big heterogeneous spatial data is the next step in the big spatial data revolution. However, a critical challenge in working with big heterogeneous spatial data is developing better methods for storage, retrieval, and analysis of these data. In particular, no comprehensive high-performance computation solutions exist that natively deal with both raster and vector data formats.

Beyond the challenges of parallelization lie more complicated problems in which one location affects another (such as land-water interactions in ecosystem services modeling), requiring intracluster communication. To that end, we plan to support transparent, background parallel computing and to extend our work to develop algorithms such as (1) dynamic spatial declustering with both vector and raster data; (2) parallel spatial join query processing with both vector and raster datasets; and (3) parallel geospatial operations on vector/raster/tabular datasets, including dasymetric mapping, pycnophylactic interpolation, geostatistics, spatial analytics, and network analytics. Finally, we argue that the development of scalable spatial functions and high-performance spatial architectures enables end users to delegate data-processing issues while allowing a greater range of derived products that may lead to a better understanding of the problems at hand.

The scalable processing of heterogeneous big spatial data will be essential to take the next leap in geocomputation. The Terra Populus project aims to address pertinent research challenges and to build the next-generation spatial database repository that facilitates these research needs.

References

Eldawy A, Mokbel MF (2013) A demonstration of spatialhadoop: an efficient mapreduce framework for spatial data. In: Proceedings of the VLDB endowment, from the 39th international conference on very large data bases, Riva del Garda, Italy, vol 5, no 12, pp 1230–1233. Curran Associates, Inc., Red Hook, NY

Eldawy A, Mokbel MF (2015) SpatialHadoop: a MapReduce framework for spatial data. In: 2015 IEEE 31st international conference on data engineering, Seoul, South Korea, April 13–17. http://ieeexplore.ieee.org/xpl/mostRecentIssue.jsp?punumber=7109453

GeoMesa (2014) GeoMesa documentation. http://www.geomesa.org/documentation

GeoTrellis (2014) GeoTrellis documentation. http://geotrellis.io/documentation

Haynes D, Ray S, Manson SM, Soni A (2015) High performance analysis of big spatial data. In: 2015 IEEE International Conference on Big Data, Santa Clara, CA, October 29–November 1, 1953–1957

Hijmans R, Nell G, Arnel R, Maunahan A, Wieczorek J, Kapoor J (2011) Global administrative areas. GADM v2 global shapefile. http://biogeo.ucdavis.edu/data/gadm2/gadm_v2_shp.zip

Kugler TA, Van Riper DC, Manson SM, Haynes D, Donato J, Stinebaugh K (2015) Terra Populus: workflows for integrating and harmonizing geospatial population and environmental data. J Map Geogr Librar 11:180–206

Melnik S, Gubarev A, Long JJ, Romer G, Shivakumar S, Tolton M, Vassilakis T (2010) Dremel: interactive analysis of web-scale datasets. In: Proceedings of the VLDB endowment, pp 330–339

Minnesota Population Center (2011) National historical geographic information system: version 2.0. https://nhgis.org/

Minnesota Population Center (2015a) Integrated public use microdata series, International: version 6.4 (machine-readable database). https://international.ipums.org/international/

Minnesota Population Center (2015b) Terra Populus. http://www.terrapop.org

Parquet (2014) Parquet documentation. https://parquet.apache.org/documentation/latest/

PostGIS (2005) PostGIS. http://postgis.net/

PostgreSQL (2015) PostgreSQL partitioning. http://www.postgresql.org/docs/9.4/static/ddl-partitioning.html

Ray S, Simion B, Brown AD (2011) Jackpine: a benchmark to evaluate spatial database performance. In: 2011 IEEE 27th international conference on data engineering, April 11–16, Hannover, Germany, pp 1139–1150

Ray S, Simion B, Brown AD, Johnson R (2013) A parallel spatial data analysis infrastructure for the cloud. In: Proceedings of the 21st ACM SIGSPATIAL international conference on advances in geographic information systems, Orlando, FL, November 5–8, pp 284–293

Stado (2011) Stado: Wiki. https://wiki.postgresql.org/wiki/Stado

Wang S (2010) A CyberGIS framework for the synthesis of cyberinfrastructure, GIS, and spatial analysis. Ann Assoc Am Geogr 100(3):535–557

Part II
Spatial Analysis: Methods and Applications

Spatial analysis has been, and continues to be, a core research area in geocomputation. Whereas the research aim of a number of conference papers was to make a new methodological contribution to the literature, other conference papers sought to contribute novel applications. Hong et al. develop a new location coverage model for a commercial drone delivery system for small-sized products. Their model identifies optimal locations of recharging stations to construct a feasible delivery network. Two other papers investigate interactions of information. Yuan examines distance decay effects in a gravity model context for mass media data and geo-tagged social media data in China. That is, she investigates how a distance decay effect plays a role in information interaction. Gao et al. explore how physical space for, say, a city can be divided based on the information interaction and human movement using individual-level mobile phone call records in Senegal. Ahmadi et al. investigate space and temporal clusters of ozone pollution in the Dallas-Fort Worth metropolitan area using a hierarchical clustering method and then formulate a regression-based spatial forecasting model for ozone level. Sivasailam and Cummings examine whether different species of plants that provide critical resources for indigenous people in southern Guyana are associated with the locations of their villages and how those plants are spatially distributed vis-à-vis the village centers.

Two additional papers use a dynamic time-warping technique for climate search and classification at the global level (Netzel and Stepinski) and for identification of similarities in multiple attributes of geographical units such as countries (Piburn et al.). Parmentier et al. propose a new framework to furnish insights into how to investigate geographical phenomena in a space-time context. Specifically, they analyze when and how space process modeling can produce a better outcome than time process modeling. Montgomery et al. implement a soft computing logic, the logic scoring of preference method, for evaluating landscape suitability for agricultural production, extending GIS-based multicriteria evaluation methods. Finally, Nara et al. investigate the performance of context recognition of moving object data in space. They analyze imaging and tracking data in neurosurgical operations with optical flow algorithms and trajectory clustering techniques; their results show such information furnishes an improvement of context awareness in terms of surgical operations.

A Deviation Flow Refueling Location Model for Continuous Space: A Commercial Drone Delivery System for Urban Areas

Insu Hong, Michael Kuby and Alan Murray

Abstract Drones, which refer to a range of small-sized unmanned aerial vehicles propelled by multiple rotors, recently have been utilized for various purposes, such as military, surveillance, photography, and entertainment. Delivery service for small products is one of their potential applications, and optimal path planning is essential for operational efficiency of such a delivery service. Because a drone's movement is not limited to existing transportation networks, path planning needs to be conducted in continuous space while taking into account obstacles for flight. However, due to the limited flight range of battery-powered drones, multiple recharging stations are required in large urban areas to complete delivery without running out of power. In this chapter, we present a new coverage model that can optimize the location of recharging stations for delivery drones as well as ensure construction of a feasible delivery network that connects the stations and covered demand based on continuous space shortest paths. A heuristic solution technique is utilized for the optimization of station locations. Application results show the effectiveness of our model for construction of a drone delivery network that covers a large urban area.

Keywords Drone • Coverage location model • Euclidean shortest path

---

I. Hong (✉)
Department of Geology and Geography, West Virginia University, 98 Beechurst Ave., Morgantown, WV 26506, USA
e-mail: insu.hong@mail.wvu.edu

M. Kuby
School of Geographical Sciences and Urban Planning, Arizona State University, Coor Hall 5515, 975 S. Myrtle Ave., Tempe, AZ 85287, USA
e-mail: mikekuby@asu.edu

A. Murray
Department of Geography, University of California at Santa Barbara, Santa Barbara, CA 93106, USA
e-mail: amurray@ucsb.edu

D.A. Griffith et al. (eds.), *Advances in Geocomputation*,
Advances in Geographic Information Science,
DOI 10.1007/978-3-319-22786-3_12

1 Introduction

Drones, or small-sized, battery-powered, unmanned aerial vehicles propelled by multiple rotors, have been in the news in recent years. They increasingly are utilized for purposes ranging from military to surveillance, photography, and entertainment, and civilian usages of drones are increasing rapidly in the public and private sectors (Finn and Wright 2012; Clarke 2014).

Among potential civilian applications, drone delivery service for small packages is drawing interest in the public and private sectors. Several private companies and public agencies around the world have proposed or tested drones for delivering packages for various purposes (Clarke 2014). Drones have potential uses for fast and low-cost deliveries for short distances and for benefiting areas of poor transportation infrastructure, such as small islands, especially for emergency situations.

For a drone package delivery system, flight path planning for delivery and return requires a global route planning approach, which generates a graph that represents collision-free paths for a drone (Barraquand et al. 1992; Quinlan and Khatib 1993). Because a drone moves in airspace, its path is not necessarily limited to a fixed transportation network. However, such barriers as obstacles and flight-restricted zones may affect a drone's flight path. Therefore, considering the location of obstacles for continuous space movement is essential for drone route derivation methods.

To provide delivery service over a large area, a method to extend a battery-powered drone's limited flight range needs to be considered, together with spatial optimization approaches for such service facilities to maximize service coverage while maintaining a feasible delivery network. Battery recharging stations offer a potential solution. A new location model that optimizes the spatial configuration of recharging stations while considering feasibility of a delivery network is necessary.

In this chapter, we propose a new location model for a commercial drone delivery system in an urban environment. A recently developed obstacle-avoiding path derivation technique is utilized for route construction and distance measurement. A coverage optimization model is presented for locating recharging stations and building a delivery network with a spatial heuristic solution technique. Application results are presented to demonstrate the capability and efficiency of the new location model and solution technique.

2 Route Derivation: A Convex Path Algorithm

Route planning for drones needs to take into account several considerations. The movement of a drone is not necessarily confined to a fixed transportation network. However, barriers that impede a drone's flight must be addressed, such as physical features (e.g., high-rise buildings and mountains) and flight-restricted zones established for safety and security reasons. Therefore, a path derivation method that is able to avoid obstacles is required for drone route planning.

Another consideration that needs to be addressed is the dimension of a flight path. Although a drone may follow a three-dimensional (3D) path, such a path can be simplified to two dimensions. If a drone maintains its altitude except for takeoff and landing, which can be assumed for maximization of battery efficiency, its route becomes a two-dimensional (2D) path. Therefore, a path derivation method for 2D space can be applied to route planning for a drone.

This obstacle-avoiding shortest path in continuous space has been referred to as the Euclidean shortest path (ESP), and a number of derivation methods have been developed (Lozano-Pérez and Wesley 1979; Asano et al. 1986; Hershberger and Suri 1993; Mitchell 1999). Recently, Hong and Murray (2013a, b) developed a new algorithm, referred to as the convex path algorithm, for efficient derivation of the ESP. The convex path algorithm exploits spatial knowledge and geographic information systems (GIS) functionality to construct a minimum-sized graph that guarantees inclusion of the ESP. The notions of a convex hull and a spatial filtering technique are utilized for efficient identification of obstacles that impact on the ESP for given origin and destination points, as well as graph construction. Figure 1 presents an example of an obstacle-avoiding route from an origin to a destination.

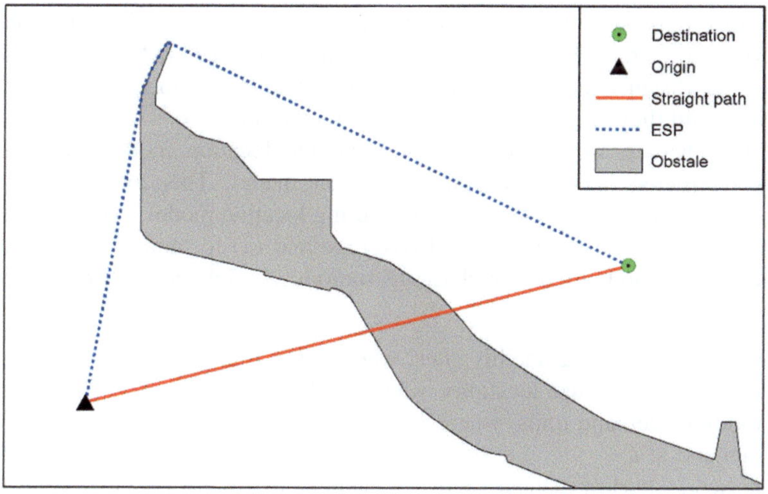

Fig. 1 An example of ESP route planning

3 Distance-Restricted Maximal Coverage Location Model

To provide delivery service over a large area that exceeds a drone's maximum flight range, battery recharging stations for the drone offer a potential solution. In this chapter, we present a location model for such recharging stations to maximize delivery service coverage with a given number of stations and warehouses, while maintaining a feasible and efficient delivery network.

In this model, a delivery network is constructed representing recharging stations as nodes. The ESP between two nodes is included as an edge if the length of the ESP is shorter than a given drone's maximum flight distance carrying a full payload. The delivery coverage of the network is derived by calculating the ESP distance from each station to each demand point. If the ESP distance to a demand point from any recharging station is within a drone's delivery threshold that ensures safe return to the station after delivery, the demand is considered to be covered. Since a drone will fly through the network of recharging stations to reach a demand, each station is considered to be a service-providing facility. A key consideration here is finding an optimal configuration of the stations that can maximize feasible service coverage while minimizing delivery flight distance for any covered demand location.

To construct a location model for efficient drone delivery network, we assumed the following: (1) a drone departs from a warehouse with a fully charged battery and returns to the same warehouse after delivery; (2) a drone makes single-package deliveries; (3) with payload, a drone's flight range is reduced to half of its remaining fuel; (4) the distance metric is the ESP distance; and (5) the location of warehouses is known, and they are included as recharging stations.

In this chapter, we propose a new coverage location model, referred to as distance-restricted maximal coverage location model. This model is based on Church and ReVelle's (1974) maximal covering location model. Our model has two objectives: (1) to maximize demand coverage and (2) to minimize average flight distance from warehouses to recharging stations via delivery networks. Consider the following notion:

j, k indices of potential facility sites, where $j, k = 1, 2, \ldots, m$
l index of warehouse locations, where $l = 1, 2, \ldots, r$
i index of demand units, where $i = 1, 2, \ldots n$
h_i demand at i
s_{ij} network shortest distance between i and j
p number of stations
r number of warehouses
f_{max} flight range with maximum payload
f_0 Flight range with empty payload
N_i {a set of sites that can cover demand i}
M_j {a set of sites that are within f_{max} from site j}
X_j $\begin{cases} 1, & \text{if a facility is located at potential site j} \\ 0, & \text{if not} \end{cases}$

Z_i $\begin{cases} 1, & \text{if a demand i is covered by at least one facility} \\ 0, & \text{if not} \end{cases}$

C_{lj} $\begin{cases} 1, & \text{if a site j is connected to warehouse l} \\ 0, & \text{if not} \end{cases}$

Objective function

$$maximize \sum h_i Z_i \frac{r(p-r)}{\sum_{l=1}^{r} \sum_{j=1}^{m} X_j s_{lj}} \tag{1}$$

Subject to

$$\sum_{j \in N_i} X_j \geq Z_i \quad \forall i \tag{2}$$

$$\sum_{k \in M_j} X_k - X_j \geq 0 \quad \forall j \tag{3}$$

$$C_{lj} = X_j \quad \forall l, j \tag{4}$$

$$\sum_{j=1}^{m} X_j + \sum_{l=1}^{r} X_l = p \tag{5}$$

$$X_l = 1 \quad \forall l \tag{6}$$

$$X_j, Z_i, C_{lj} = \{0, 1\} \tag{7}$$

Objective function (1) is to maximize covered demand while minimizing average network distance from each warehouse to each selected facility site. Constraint (2) defines coverage. Constraint (3) ensures minimum connectivity constraints to be satisfied, to prevent isolated stations that are separated from warehouses. Constraint (4) is source connectivity constraint, which ensures the connection of each demand to a warehouse via the delivery network. Constraints (5) and (6) are for considering given warehouses.

4 A Heuristic Solution Technique: Simulated Annealing with a Greedy Algorithm

To obtain a solution for the distance-restricted maximal coverage location model, a heuristic solution technique that utilizes spatial knowledge was developed. Simulated annealing (Kirkpatrick 1984) was used to obtain high-quality heuristic solutions within a reasonable amount of computing time. A greedy algorithm is utilized to generate feasible solutions. An interchange algorithm (Teitz and Bart 1968) improves the quality of solutions by the greedy algorithm.

Novel in this approach is the utilization of spatial knowledge for efficient evaluation of candidate sites while preserving feasibility. In the greedy process, only candidate sites that can be reached from a current solution set are evaluated to facilitate the process. Once a solution set and corresponding delivery network are generated, the interchange algorithm improves the solution quality while maintaining the feasibility of the network. The interchange process also uses spatially restricted candidates but more strategically. If a station is critical for preserving the feasibility of the network, the interchange algorithm evaluates candidates around it that are able to maintain the connectivity of stations.

Simulated annealing (Kirkpatrick 1984) is applied for the distance-restricted coverage model to prevent the solution process from stopping at a local optimum. To improve solution quality, we enhanced the simulated annealing with a solution memory. It "remembers" the best solution identified but accepts inferior solutions based on temperature condition. However, if a resulting solution after termination is inferior to the memorized one, the stored best solution is selected as the final result.

This spatial simulated annealing derives a solution as follows: (1) an initial solution is generated at random but distance restrictions and warehouse connectivity are considered; (2) a given number of stations are randomly removed from the preceding solution; (3) a new solution is generated using the greedy algorithm; (4) the new solution is improved using the interchange algorithm; (5) acceptance of the new solution is determined based on simulated annealing criteria; and (6) steps 2–5 are repeated until the termination condition is satisfied.

5 Application Results

To assess efficiency and solution quality of the distance-restricted maximal coverage location model for a drone delivery system, a test application in a large urban area environment is analyzed. A part of the Phoenix, Arizona, metropolitan area is utilized for this test application. For demand representation, centroids of census blocks are used, and a total of 32,940 demand points are included in the application. For candidate sites for recharging stations, 500 points are randomly selected from the demand locations, including three warehouses. For obstacles, federal control lands (mostly parks) and a five-mile radius buffer around airports are included. The flight range of a drone is assumed to be ten miles with no payload and five miles with a full payload. Therefore, demand coverage of each station is 3.3 miles, which ensures the safe return of a drone to the station after finishing its delivery. The heuristic solution technique was implemented in Python 2.7 using the open source spatial library. The analysis was carried out on Intel i7 CPU with 8 GB memory.

Figure 2 shows a solution with 25 stations, including three warehouses. Demand is represented by the intensity of red in the background. This solution covers 91.5 % of the total population in this area and takes 1.353 s to compute.

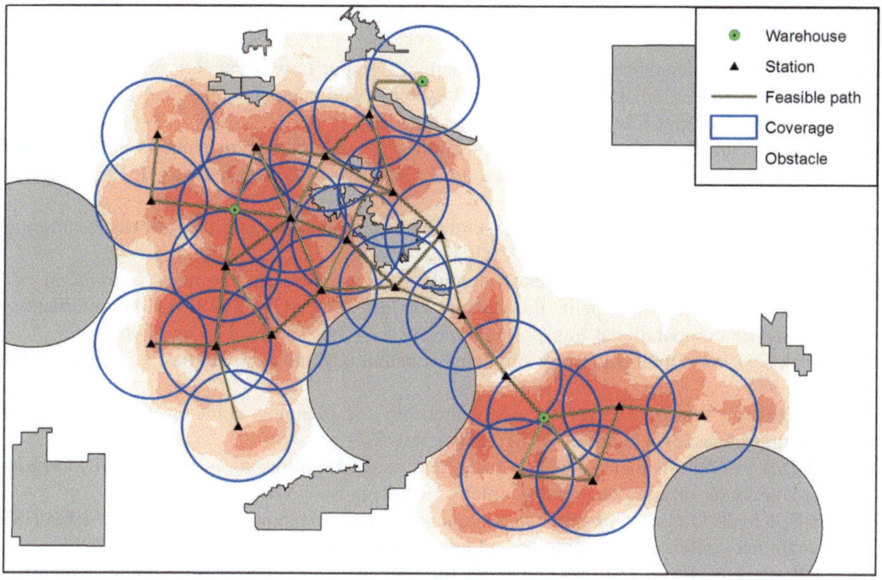

Fig. 2 A solution for 25 recharging stations in the Phoenix area

6 Conclusions

The distance-restricted maximal coverage location model shows significant potential for optimizing drone delivery service in large urban areas. Because this is ongoing research, additional investigations are required to construct a more efficient drone delivery network. For model parameters, more obstacles and flight-restricted zones will be included to reflect the current development of Federal Aviation Administration regulations and other safety and security issues pertaining to drone operations in populated areas. Also, flight range and corresponding delivery threshold values may be adjusted to consider fast-developing drone technology.

The spatial heuristic solution technique will be improved to produce better-quality solutions, as well as to process efficiently larger datasets. Moreover, for assessing solution quality produced by the heuristic technique, a commercial solver will be used to derive the global optimum solution, comparing it to other heuristic applications with different model parameters.

References

Asano T, Guibas L, Hershberger J, Imai H (1986) Visibility of disjoint polygons. Algorithmica 1 (1):49–63

Barraquand J, Langlois B, Latombe J-C (1992) Numerical potential field techniques for robot path planning. IEEE Trans Syst Man Cybern 22(2):224–241

Church R, ReVelle C (1974) The maximal covering location problem. Papers Reg Sci 32(1):101–118
Clarke R (2014) Understanding the drone epidemic. Comput Law Secur Rev 30(3):230–246
Finn RL, Wright D (2012) Unmanned aircraft systems: surveillance, ethics and privacy in civil applications. Comput Law Secur Rev 28(2):184–194
Hershberger J, Suri S (1993) Efficient computation of Euclidean shortest paths in the plane. 34th annual symposium on foundations of computer science, 1993, proceedings. IEEE, Palo Alto, CA, pp 508–517
Hong I, Murray AT (2013a) Efficient measurement of continuous space shortest distance around barriers. Int J Geogr Inf Sci 27(12):2302–2318
Hong I, Murray AT (2013b) Efficient wayfinding in complex environments: derivation of a continuous space shortest path. Proceedings of the sixth ACM SIGSPATIAL international workshop on computational transportation science. ACM, Tampa, FL, pp 61–63
Kirkpatrick S (1984) Optimization by simulated annealing: quantitative studies. J Stat Phys 34(5–6):975–986
Lozano-Pérez T, Wesley MA (1979) An algorithm for planning collision-free paths among polyhedral obstacles. Commun ACM 22(10):560–570
Mitchell JSB (1999) Geometric shortest paths and network optimization. In: Sack JR, Urrutia J (eds) Handbook of computational geometry. Elsevier, New York, pp 633–702
Quinlan S, Khatib O (1993) Elastic bands: connecting path planning and control. In: 1993 IEEE international conference on robotics and automation, proceedings, IEEE, pp 802–807
Teitz MB, Bart P (1968) Heuristic methods for estimating the generalized vertex median of a weighted graph. Oper Res 16(5):955–961

Exploring the Spatial Decay Effect in Mass Media and Location-Based Social Media: A Case Study of China

Yihong Yuan

Abstract The rapid development of big data analytics provides tremendous possibilities to investigate large-scale patterns in both the spatial and temporal dimensions. In this research, we utilize a unique open dataset, the Global Database on Events, Location, and Tone (GDELT), and a geotagged social media dataset (Weibo) to analyze connections between Chinese provinces. Specifically, this study constructs a gravity model to compare the distance decay effect between the GDELT data (i.e., mass media data) and the Weibo data (i.e., location-based social media [LBSM] data). The results demonstrate that mass media data possess a weaker distance decay effect than LBSM data for Chinese provinces. This study generates valuable input to interpret regional relations in a fast-growing, developing country—China. It also provides methodological references to explore urban relations in other countries and regions in the big data era.

Keywords Mass media · Global database on events, location, and tone (GDELT) · Location-based social media (LBSM) · Gravity model · Distance decay

1 Introduction

The rapid development of techniques and theories in the big data era has introduced new challenges and opportunities to analyze a large amount of social media data available online (Gao et al. 2012; Eagle et al. 2009; Liben-Nowell et al. 2005; Wu et al. 2014). The widespread use of smart phones, which are equipped with sensors allowing users to locate themselves instantly, inserts another crucial aspect into this development: location. Researchers have defined location-based social media (LBSM) as social network sites (SNS) that include location information (Roick and Heuser 2013; Elwood et al. 2012). These data are user-generated and geolocated,

Y. Yuan (✉)
Department of Geography, Texas State University, San Marcos, TX 78666, USA
e-mail: yuan@txstate.edu

D.A. Griffith et al. (eds.), *Advances in Geocomputation*,
Advances in Geographic Information Science,
DOI 10.1007/978-3-319-22786-3_13

and contain varying types of contextual information (e.g., text, videos, and images.) Therefore, they can be utilized as potential resources to characterize spatial interactions and social perceptions of place (Malleson and Birkin 2014; Hasan et al. 2013; Gao et al. 2012).

Many tremendous changes have occurred in the traditional media industry over the past few decades. In particular, the rapid development of new media, such as video games and online news columns, also has allowed a new paradigm to emerge in human behavior modeling and pattern recognition. For researchers in this field, these newly available data sources offer ever-increasing opportunities to conduct data mining and knowledge discovery tasks (Li and Liu 2003; Masand et al. 1992; Mazzitello et al. 2007). However, compared to social media, traditional mass media is characterized by the significance and aggregated nature of associated events (Liebert and Schwartzberg 1977; Klapper 1968). As such, mass media data often are suitable for investigating the aggregated pattern of an urban system. However, very few empirical studies have compared corresponding spatial patterns extracted from mass media data and LBSM data.

The distance decay effect has been a hot topic in many research fields, such as migration and transportation (e.g., the decay of traffic flows between locations) (Rodrigue et al. 2013). Researchers have employed different models to investigate how distance decay affects the magnitude of interaction between geographic entities. Among all alternative models, researchers commonly use the gravity model (Sen and Smith 1995) due to its effectiveness in predicting the degree of interaction, the simplicity of its equation, and its ability to deal with flows in both directions (Hardy et al. 2012). In this research, we apply an open source dataset, the Global Database on Events, Location, and Tone (GDELT) to analyze connections between Chinese provinces in terms of mass media. The fields of communication, history, and political science, among others, have widely explored GDELT's continuous compilation of print, broadcast, and web-news media events (Leetaru and Schrodt 2013; Yonamine 2013), but the spatial element of this dataset has not been investigated sufficiently. This chapter compares the magnitude of the distance decay effect in the GDELT data with a dataset from a Chinese social media website (Weibo.com) based on the gravity model. We focus on demonstrating the effectiveness of utilizing both mass media and LBSM data to reveal geographic patterns. This can be considered a data preprocessing strategy for pattern recognition and outlier identification in multiple areas, such as urban planning, sociology, and political geography. Our results also provide valuable input for policy makers to interpret the dynamic nature of interregion relations in different datasets.

2 Datasets

This research primarily utilizes two datasets: (1) the main dataset GDELT, consisting of more than a quarter-billion event records from 1979 to the present capturing what has happened/is happening worldwide in multiple columns, such as the source, actors, time, and approximated location of recorded events; (2) a complementary dataset from the Chinese social networking site Weibo,[1] to compare the distance decay effects in mass media and LBSM. The remainder of Sect. 2 illustrates the two datasets in detail.

2.1 The Main Dataset: GDELT

This research utilizes a CAMEO-coded[2] open dataset, GDELT (Schrodt 2012). For consistency, we use the data from January 1, 2014, to May 20, 2014, in this analysis. Previous studies utilized this dataset as a valuable resource for modeling societal-scale behavior and beliefs across all countries of the world (Leetaru and Schrodt 2013; Shook et al. 2012; Yuan and Liu 2015). For example, a study conducted by Jiang and Mai (2014) analyzes the strength of links between countries based on the GDELT dataset. They also explored bilateral and multilateral events in certain countries from the same dataset. Other research by Yonamine (2013) predicts the level of conflict in Afghanistan by incorporating multiple sociopolitical factors, such as drug prices, unemployment levels, and ethnic diversity. Yet other researchers have looked into the connections and differences between GDELT and other event-based datasets, such as the Integrated Crisis Early Warning System (ICEWS), which is an early-warning system designed to help policy makers predict a variety of international events and crises. They conducted a side-by-side comparison of the data quality, quantity, design scheme, and many other features of these two datasets, and concluded that although the efficiency of each dataset mainly depends on the research questions to be answered, GDELT has many more events per country per unit of time than ICEWS.

In GDELT, for instance, for a news report entitled "An Artist in Shanghai Sold His Painted Box Room to the Sifang Art Museum in Nanjing," the associated geographic locations of Actor 1, Actor 2, and the actual action (i.e., sold) is demonstrated in Table 1. Here we consider records only when two actors are explicitly identified and geotagged in China.

Note that GDELT measures the detailed level of the spatial information by a field named Geo_Type. This field specifies the geographic resolution of each location and holds one of the following values: 1 = COUNTRY (country level), 2 = USSTATE (a United States [US] state), 3 = USCITY (a US city or landmark), 4 = WORLDCITY (a city or landmark outside the US), 5 = WORLDSTATE (an Administrative Division 1[3] outside the US—roughly equivalent to a US state) (Leetaru and Schrodt 2013; Yuan and Liu 2015). Because this research is conducted

[1]www.weibo.com.

[2]Conflict and Mediation Event Observations (CAMEO) is a framework for coding event data.

[3]From the GDELT codebook (http://data.gdeltproject.org/documentation/GDELT-Global_Knowledge_Graph_Codebook.pdf).

Table 1 A sample record from GDELT*

Event date	Actor 1_Geo	Actor 2_Geo	Action_Geo
2014-01-28	Shanghai, China	Nanjing, Jiangsu, China	Shanghai, China

*Due to page limits, only fields related to this research are displayed

Fig. 1 Chinese provinces (the Paracel Islands are omitted for simplicity)

at the province level, we should consider only records with Geo_Type = 4 or Geo_Type = 5 (Fig. 1, map created in Mercator projection).

2.2 *Complementary Datasets*

Besides the main GDELT dataset, we utilize a complementary dataset to compare the distance decay effects in mass media and LBSM. This dataset covers 3 million users of the Chinese social networking site, Weibo, a microblogging website functionally similar to Twitter. The records were obtained from the official Weibo application program interface (API) between May 1, 2014, and May 20, 2014. Each record captures such attributes as the geographic coordinates (e.g., volunteered geographic information from the built-in positioning module of smart phones), date, time, and user identification (ID).[4]

3 Methodology and Preliminary Results

This section presents the model-construction procedure and our preliminary results. As discussed in Sect. 1, the major objective of this research is to compare the magnitude of the distance decay effects in mass media and LBSM by investigating how distance affects interregional connection. Although researchers have applied various techniques to analyze spatial interaction in many research fields, such as transportation (e.g., traffic flows between locations) and migration (e.g., relocation flows between countries) (Rodrigue et al. 2013; Lewer and Van den Berg 2008), the gravity model furnishes one of the most commonly used descriptions of this phenomena because of its effectiveness in predicting the degree of interaction and its algebraic simplicity (Rodrigue et al. 2013; Hardy et al. 2012; Sen and Smith 1995). This study also adopts the gravity model. The analysis involves the following two steps.

3.1 Data Preprocessing

First, we calculate the frequency of co-occurrence between each pair of Chinese provinces in GDELT. The frequency is noted as I_{ij}, which denotes the frequency of provinces i and j appearing as the two actors' locations in the same news record. We also processed the Weibo data to identify the Chinese province associated with each geotagged post.

3.2 Model Construction

As discussed in Sect. 1, this research utilizes the gravity model to examine the distance decay effect (Eq. 1):

$$I_{ij} = K\frac{P_i^{\beta_1} P_j^{\beta_2}}{D_{ij}^{\beta_3}}, \tag{1}$$

[4]Here user IDs are long integers generated by Weibo.com and are not directly connected to any personally identifiable information (PII), unless the users volunteer to make such information publicly accessible.

where P_i and P_j are the conceptual sizes (relative importance) of provinces i and j, D_{ij} represents the great circle distance separating the geographic centroids of i and j, and I_{ij} denotes the interaction/connection between i and j. β_1 and β_2 indicate how the conceptual sizes of two countries contribute to the interaction term I_{ij} (Austin 1963). β_3 (distance friction coefficient) investigates the role of distance. Here we construct two gravity models to investigate the role of the friction of distance in the two datasets (GDELT and Weibo). The specific parameters are:

GDELT: I_{ij} The frequency of co-occurrence of provinces i and j in news records
P_i The total occurrence of province i in news records
P_j The total occurrence of province j in news records

Weibo: I_{ij} The number of unique users who have checked in at their locations in both provinces i and j
P_i The number of unique users who have checked in at their locations in province i
P_j The number of unique users who have checked in at their locations in province j

Note that the connection between two provinces in the Weibo data can be defined from various perspectives; for example, the co-appearance of two province names in the same Weibo post. However, the primary functionality of Weibo.com is to share moments of one's personal life, and hence users rarely publish posts that explicitly discuss two province names. Also, this research focuses on comparing spatial patterns from mass media and geotagged social media; thus, we define connections based on the individual footprint of Weibo users. In other words, the Weibo dataset measures physical movements while the GDELT dataset measures information interactions. However, one of the objectives of this research is to investigate the representativeness of Weibo data in modeling physical mobility, because other studies demonstrated the presence of strong spatial biases when utilizing LBSM data to model spatial behavior. Flickr is an example, where photo-uploading activities do not exhibit a significant distance decay effect because people tend to upload photos when they travel to faraway destinations.

Based on the preceding definitions, we calculated the best fit for coefficients β_1, β_2, β_3 based on Poisson regression (Fig. 2). Table 2 indicates the fitted values and the goodness of fit (R^2) for both datasets. Because R^2 is scale-free, the constant K does not affect our models.

Results for the two datasets demonstrate distinct patterns for the distance decay effect. For the GDELT dataset, the distance decay effect (based on great circle distances between provinces) is weak ($\beta_3 = 0.0516$, pseudo-$R^2 = 0.758$), whereas the Weibo dataset shows a much stronger distance decay effect ($\beta_3 = 1.054$, pseudo-$R^2 = 0.882$). Compared to the β_3 values obtained for several related studies

Table 2 Fitted β values and pseudo-R^2 of Poisson regression models

	β_1	β_2	β_3	Pseudo-R^2
GDELT	0.826	0.826	0.0516	0.758
Weibo	0.799	0.799	1.054	0.882

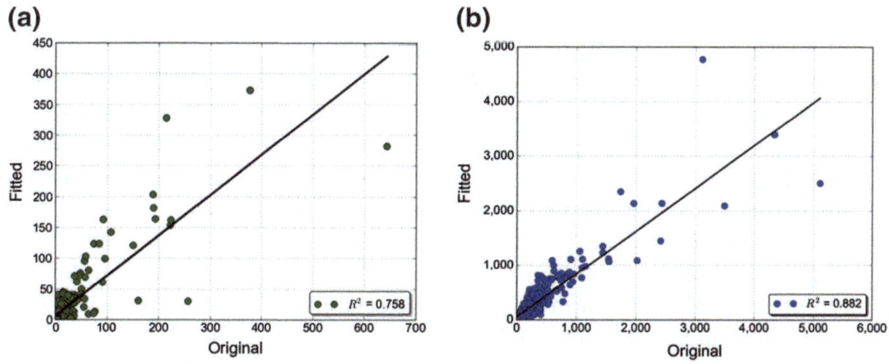

Fig. 2 **a** Observed and fitted I_{ij} (GDELT); **b** observed and fitted I_{ij} (Weibo)

—0.2 for Chinese province name co-occurrences on web pages (Liu et al. 2014), 1.59 for banknote trajectories (Brockmann and Theis 2008), and 1.75 for individual mobility patterns by mobile phone data (Gonzalez et al. 2008)—our study further confirms that mass media data reveal a weaker distance decay effect than LBSM data for Chinese provinces. Additionally, β_1 and β_2 values also indicate interesting patterns regarding the role of conceptual sizes (P_i, P_j) in determining the magnitude of this interaction. P_i and P_j play a more important role in GDELT data. Note that, unlike the gravity of trade where the interaction term is directional, the interaction term I_{ij} here is not bilateral in either dataset. For example, the frequency of co-occurrence of provinces *i* and *j* in news records is equivalent to the frequency of co-occurrence of provinces *j* and *i* in news records; thus, the roles of provinces *i* and *j* are exchangeable, resulting in identical β_1 and β_2 estimates.

Finally, the different aspects of uncertainty involved in this study are noteworthy, including but not limited to

- *Natural variability in human activities*: Although human mobility seems to be highly predictable (Yuan et al. 2012; Song et al. 2010; Gonzalez et al. 2008), randomness is an inevitable part of human motion.
- *Potential impact of spatial autocorrelation*: Many researchers have made a distinction between spatial association (autocorrelation) and spatial interaction in the geography field. Interaction primarily refers to movement of tangible entities; therefore, it is less related to correlation. However, several previous studies also argued that spatial interaction models are a special case of a general model of spatial autocorrelation (Fischer et al. 2010; Getis 1991). Hence, the impact of spatial autocorrelation on the gravity models constructed in this research needs to be examined in future research (Chun and Griffith 2011).
- *Inaccuracy/imprecision due to the limitation of available data*: Positional inaccuracy, sampling resolution, and imprecision contribute to the uncertainty of a data source. For instance, the precision of geotagged Weibo posts is strongly related to the strength of an available global positioning system (GPS) signal.

Also, the sampling resolution is unevenly distributed, as many users may not post their locations to Weibo on a regular basis.

- *Imperfection of models and algorithms*: As Box and Draper (1987, p. 424) state: "Essentially, all models are wrong, but some are useful." The results of this study are also highly impacted by the chosen models and algorithms. For the GDELT dataset, actors are georeferenced automatically based on various text mining and machine learning algorithms, naturally introducing potential inaccuracy into the location data. In this research, the gravity model is adopted to interpret interregion connections in China; however, the application of different models inevitably has an impact on the uncertainty of results. For example, an interesting future direction could be to simulate and estimate the conceptual sizes (P_i and P_j) from an inverse gravity model and compare the results to the current model where P_i and P_j are predefined.

4 Conclusion

The study summarized in this chapter employed GDELT and Weibo data to assess the connection between Chinese provinces. We examined the distance decay effects in two types of datasets (mass media and LBSM) for interregional patterns. The fitted β_3 values demonstrate that mass media data (GDELT) indicate a weaker distance decay effect than geotagged social media data (Weibo) for Chinese provinces. Unlike the Flickr dataset discussed in Yuan and Liu (2015), which indicates a very weak distance decay effect (as users are more likely to post photos when they travel faraway), the geotagged Weibo data still demonstrate a strong distance decay effect. This finding suggests an interesting direction for future research, namely, to examine the distance decay effect in various social network sites based on their functionalities. We also demonstrated the effectiveness of applying GDELT and big data analytics to investigate informative patterns in interdisciplinary studies. One potential explanation for the low β_3 value in the GDELT data is China's status as a developing country with a strong central government. Its capital, Beijing, has a significant impact on all other provinces, regardless of the distance separating them. However, this research does not aim to provide an in-depth interpretation of these findings from a political perspective. Rather, it proposes a method to discover patterns that can provide insights in different research fields.

Future research directions include extending this method to other countries and regions to test its robustness. GDELT provides a rich data source to analyze international relations at various spatial scales, such as investigating the connections between different countries. Future studies also can look into the correlation between connection strength and various demographic variables, such as population and economic status.

References

Austin LC (1963) Review of shaping the world-economy, Tinbergen. J Int Aff 17(2):221
Box GEP, Draper NR (1987) Empirical model-building and response surfaces. Wiley series in probability and mathematical statistics. Applied probability and statistics. Wiley, New York
Brockmann D, Theis F (2008) Money circulation, trackable items, and the emergence of universal human mobility patterns. IEEE Pervas Comput 7(4):28–35
Chun Y, Griffith DA (2011) Modeling network autocorrelation in space–time migration flow data: an eigenvector spatial filtering approach. Ann Assoc Am Geogr 101(3):523–536. doi:10.1080/00045608.2011.561070
Eagle N, Pentland A, Lazer D (2009) Inferring friendship network structure by using mobile phone data. Proc Natl Acad Sci USA 106(36):15274–15278. doi:10.1073/pnas.0900282106
Elwood S, Goodchild MF, Sui DZ (2012) Researching volunteered geographic information: spatial data, geographic research, and new social practice. Ann Assoc Am Geogr 102(3):571–590. doi:10.1080/00045608.2011.595657
Fischer M, Reismann M, Scherngell T (2010) Spatial interaction and spatial autocorrelation. In: Anselin L, Rey SJ (eds) Perspectives on spatial data analysis: advances in spatial science. Springer, Berlin, pp 61–79. doi:10.1007/978-3-642-01976-0_5
Gao H, Tang J, Liu H (2012) Exploring social-historical ties on location-based social networks. Paper presented at the 6th international AAAI conference on weblogs and social media, Dublin, Ireland, June 4–7, 2012
Getis A (1991) Spatial interaction and spatial autocorrelation: a cross-product approach. Environ Plan A 23:1269–1277
Gonzalez MC, Hidalgo CA, Barabasi AL (2008) Understanding individual human mobility patterns. Nature 453(7196):779–782. doi:10.1038/Nature06958
Hardy D, Frew J, Goodchild MF (2012) Volunteered geographic information production as a spatial process. Int J Geogr Inf Sci 26(7):1191–1212. doi:10.1080/13658816.2011.629618
Hasan S, Zhan X, Ukkusuri SV (2013) Understanding urban human activity and mobility patterns using large-scale location-based data from online social media. In: UrbComp 13, Chicago, 2013, Chicago, August 11, 2013
Jiang L, Mai F (2014) Discovering bilateral and multilateral causal events in GDELT. Paper presented at the international conference on social computing, behavioral-cultural modeling, and prediction, Washington, DC, April 2–4, 2014
Klapper JT (1968) Effects of mass-media-depicted violence—a review of research findings. Am J Orthopsychiat 38(2):310
Leetaru K, Schrodt P (2013) GDELT: global data on events, language, and tone, 1979–2012. Paper presented at the international studies association annual conference, San Diego, CA
Lewer JJ, Van den Berg H (2008) A gravity model of immigration. Econ Lett 99(1):164–167. doi:10.1016/j.econlet.2007.06.019
Li X, Liu B (2003) Learning to classify texts using positive and unlabeled data. Paper presented at the 18th international joint conference on artificial intelligence, Acapulco, Mexico, August 3–9, 2003
Liben-Nowell D, Novak J, Kumar R, Raghavan P, Tomkins A (2005) Geographic routing in social networks. Proc Natl Acad Sci USA 102(33):11623–11628. doi:10.1073/pnas.0503018102
Liebert RM, Schwartzberg NS (1977) Effects of mass-media. Annu Rev Psychol 28:141–173
Liu Y, Wang FH, Kang CG, Gao Y, Lu YM (2014) Analyzing relatedness by toponym co-occurrences on web pages. Trans GIS 18(1):89–107. doi:10.1111/Tgis.12023
Malleson N, Birkin M (2014) New insights into individual activity spaces using crowd-sourced big data. Paper presented at the ASE bigdata/socialcom/cybersecurity conference, Stanford, CA, May 27–31, 2014
Masand B, Linoff G, Waltz D (1992) Classifying news stories using memory based reasoning. Paper presented at the 15th annual international ACM SIGIR conference on research and development in information retrieval, Copenhagen, Denmark, June 21–24, 1992

Mazzitello KI, Candia J, Dossetti V (2007) Effects of mass media and cultural drift in a model for social influence. Int J Mod Phys C 18(9):1475–1482

Rodrigue J-P, Comtois C, Slack B (2013) The geography of transport systems, 3rd edn. Routledge, Abingdon, Oxon

Roick O, Heuser S (2013) Location based social networks—definition, current state of the art and research agenda. Trans GIS 17:763–784

Sen AK, Smith TE (1995) Gravity models of spatial interaction behavior. Advances in spatial and network economics. Springer, Berlin

Schrodt P (2012) Conflict and Mediation Event Observations event and actor codebook V.1.1b3. [http://eventdata.psu.edu/cameo.dir/CAMEO.Manual.1.1b3.pdf]

Shook E, Leetaru K, Cao G, Padmanabhan A, Wang S (2012) Happy or not: generating topic-based emotional heatmaps for culturomics using CyberGIS. Paper presented at the 8th IEEE international conference on eScience, Chicago, October 8–12, 2012

Song CM, Qu ZH, Blumm N, Barabasi AL (2010) Limits of predictability in human mobility. Science 327(5968):1018–1021. doi:10.1126/science.1177170

Wu L, Zhi Y, Sui ZW, Liu Y (2014) Intra-urban human mobility and activity transition: evidence from social media check-in data. PLoS One 9(5):e97010. doi:10.1371/journal.pone.0097010

Yonamine JE (2013) Predicting future levels of violence in Afghanistan district using GDELT. UT Dallas, Technical Report. http://data.gdeltproject.org/documentation/Predicting-Future-Levels-of-Violence-in-Afghanistan-Districts-using-GDELT.pdf

Yuan Y, Liu Y (2015) Exploring inter-country connections in mass media: a case study of China. In: International conference on location-based social media, Athens, GA, 2015

Yuan Y, Raubal M, Liu Y (2012) Correlating mobile phone usage and travel behavior—a case study of Harbin, China. Comput Environ Urban Syst 36(2):118–130

Uncovering the Digital Divide and the Physical Divide in Senegal Using Mobile Phone Data

Song Gao, Bo Yan, Li Gong, Blake Regalia, Yiting Ju and Yingjie Hu

Abstract In this research, we first aim at developing data analytics that can derive insights about how people from different regions communicate and connect via mobile phone calls and physical movements. We uncover the digital divide (geographical segregation of phone communication patterns) and the physical divide (geographical limits of human mobility) in Senegal. The research also demonstrates that the chosen spatial unit and temporal resolution can affect the community detection results of spatial interaction graphs when analyzing human mobility patterns and exploring urban dynamics in the mobile age. We find that the daily detection has generated a more stable partition structure than an hourly one, while monthly changes also exist over time. The presented framework can help identify patterns of spatial interaction in both cyberspace and physical space with phone call detailed records in some regions where census data acquisition is difficult, especially in African countries.

Keywords Digital divide • Physical divide • Community detection • Mobile phone data • Spatial interaction

1 Introduction

The mobile phone call detail records (CDRs) distributed within the framework of the "Data for Development" challenge in Senegal were run through several processes that intended to anonymize all source users' information while still providing sufficient and meaningful data to researchers (de Montjoye et al. 2014). For example, the hourly site-to-site traffic data for cell phone sites are beneficial for analyzing dynamic digital communication patterns at the spatial resolution of a cell phone tower's coverage or at other aggregated regional scales. In addition,

S. Gao (✉) · B. Yan · L. Gong · B. Regalia · Y. Ju · Y. Hu
STKO Lab, Department of Geography, University of California, Santa Barbara
CA 93106, USA
e-mail: sgao@geog.ucsb.edu

D.A. Griffith et al. (eds.), *Advances in Geocomputation*,
Advances in Geographic Information Science,
DOI 10.1007/978-3-319-22786-3_14

individual-based records provide opportunities to study human mobility patterns at both the individual level and the geographically aggregated level, which has been a hot topic in the literature (Gonzalez et al. 2008; Song et al. 2010; Kang et al. 2012; de Montjoye et al. 2013). For this research, we first aim at developing data analytics that can derive insights about how people from different regions communicate and connect via phone calls and physical movements. While many studies exist applying community detection techniques based on graph theory to identify the spatial connectivity and characteristics of regions, social segregation, or functional zones of a city using mobile phone data (Ratti et al. 2010; Gao et al. 2013a; Amini et al. 2014; Chi et al. 2014), few researchers have addressed the spatiotemporal resolution issue (Cheng and Adepeju 2014). The chosen spatial unit (e.g., cell-based, region-based) or temporal resolution (e.g., hour, day, week, month) might affect the results of analyzing human mobility and urban dynamics in the mobile age (Gao 2015).

To this end, we discuss the impact of changing the spatial analysis unit and temporal resolution when detecting community patterns of spatial interaction in both cyberspace and physical space extracted from one-year CDRs in Senegal.

2 Methods

Two types of weighted graphs can be built based on the given CDRs. Let *G_CallFlow (V, E)* denote a weighted undirected graph of phone call flows among different spatial units (*S*) where cellular or administrative sites (e.g., regions, departments, arrondissements[1]) are transformed into graph nodes (*V*) while communication flows among places are represented as weighted edges (*E*). Let W_{ijt} represent the total phone calls between a spatial unit i and another spatial unit j during the time interval t (by hour, day, or month). As an example of one selected node accompanied by its links in a graph, Fig. 1 shows the monthly phone call flows that connect the capital city Dakar to other arrondissements in Senegal.

Similarly, let *G_MobilityFlow (V, E)* be a weighted undirected graph of human movement flows in physical space, and let M_{ijt} represent the total volume of movement flow between a spatial unit i and another spatial unit j during time interval t, including the movement flows both from i to j and from j to i. Note that although we can build the weighted directed graph of spatial interaction by adding the direction of flows, it is not required for community detection operations.

In the study of complex networks, a community is defined as a subset of the entire graph, where nodes within the same community are densely connected and grouped together. The identification of such divisions in a graph is called community detection. Newman and Girvan (2004) propose a modularity metric to

[1]Arrondissement is usually a level of administrative division under department in Francophone countries.

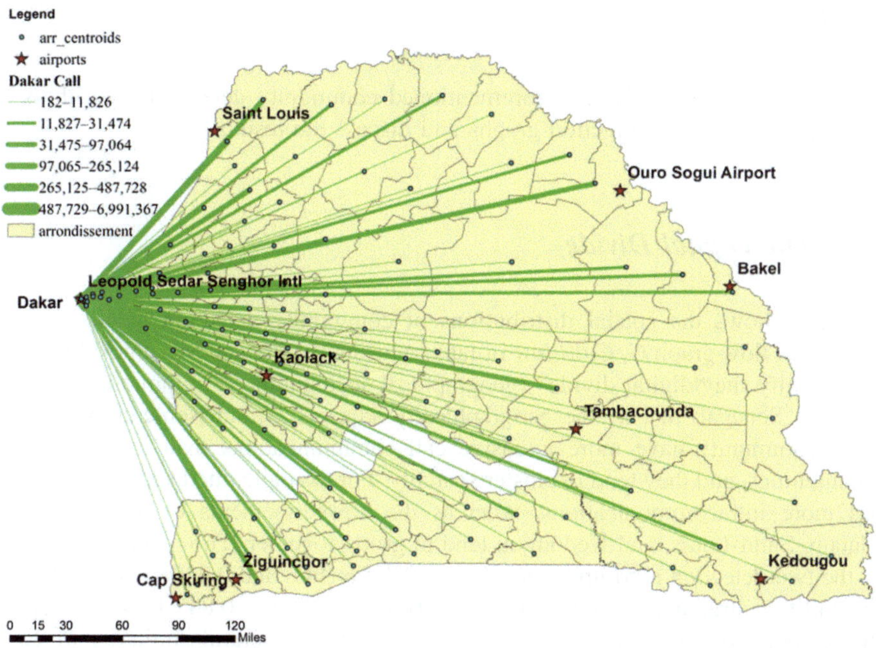

Fig. 1 Visualizing the phone call interactions between the capital Dakar and other arrondissements in Senegal

evaluate the quality of a particular division into communities within a graph. Modularity compares a proposed partition to a null model in which connections between nodes are random. The larger the modularity value is, the more robust (stable) the detected community structure is. We apply two popular techniques for community detection in our work: (1) a fast-greedy (FG) modularity maximization algorithm (Clauset et al. 2004) that merges pairs of communities iteratively and always chooses the pair that yields the maximum increase in the overall modularity; and (2) a multilevel (ML) algorithm (Blondel et al. 2008) in which nodes are moved between communities such that each node makes a local choice that maximizes its own contribution to the modularity score and can unfold a complete hierarchical community structure in multiple steps.

For each type of weighted graph (*G_CallFlow* or *G_MobilityFlow*), we process the data for different spatial and temporal resolutions, and then identify the communities for each graph by maximizing the modularity value. To compare the similarity of different scenarios of the community detection results, we calculate the normalized mutual information (NMI) index proposed by Danon et al. (2005) to measure the similarity between different partitions. The NMI value is in the range between zero and one. The higher the NMI value is, the more similar the graph partitions are.

3 Results

In this section, we apply the aforementioned community detection algorithms to two types of spatial interaction graphs and discuss the results.

3.1 The Digital Divide

Figure 2 shows the spatial distributions of community detection results for the phone call flow graph *G_CallFlow* in January using the FG and ML algorithms. We can identify the "digital divide" (geographical segregation of communication patterns) in Senegal; that is, the geographically adjacent arrondissements within the same community have more intensive call communications than those of inter-communities, and they tend to cluster spatially. For example, the Dakar region itself has more intracommunications, whereas the arrondissements of Tambacounda, Matam, Saint-Louis, and Kedougou tend to group together. The modularity values of the two detection algorithms are similar 0.4396 (FG) and 0.4408 (ML), whereas the partition structures have a high similarity value (NMI = 0.84). Figure 3 depicts the temporal changes of modularity values and the structural similarity of community detection results of phone call flows among arrondissements in different months. The month-to-month similarity matrix shows that more similar community structures are detected from October to December (pink and white color grids at the top right corner of Fig. 3b, c).

Considering the temporal resolution effect, we also apply these two detection algorithms to the hourly and daily aggregated phone call flow graphs, and compare the modularity values as well as the partition structures. Figure 4 demonstrates the hourly and daily changes of modularity values. The modularity value reaches a maximum during the hour 07–08, which represents the most stable community structure, whereas the value gets lower at night, which indicates a relatively

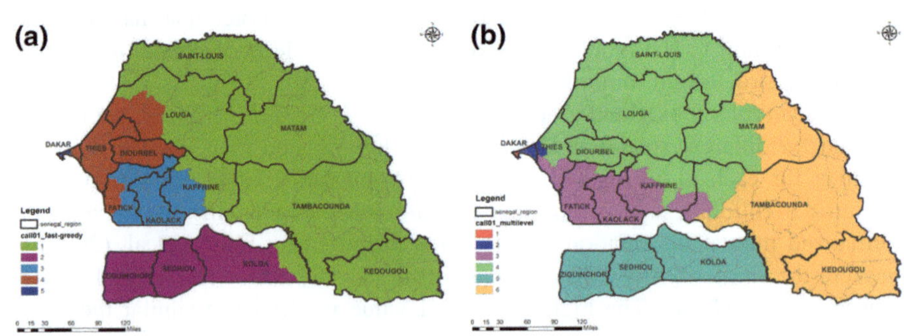

Fig. 2 Community detection of phone call flows at the arrondissement level in January using the two algorithms: **a** FG and **b** ML

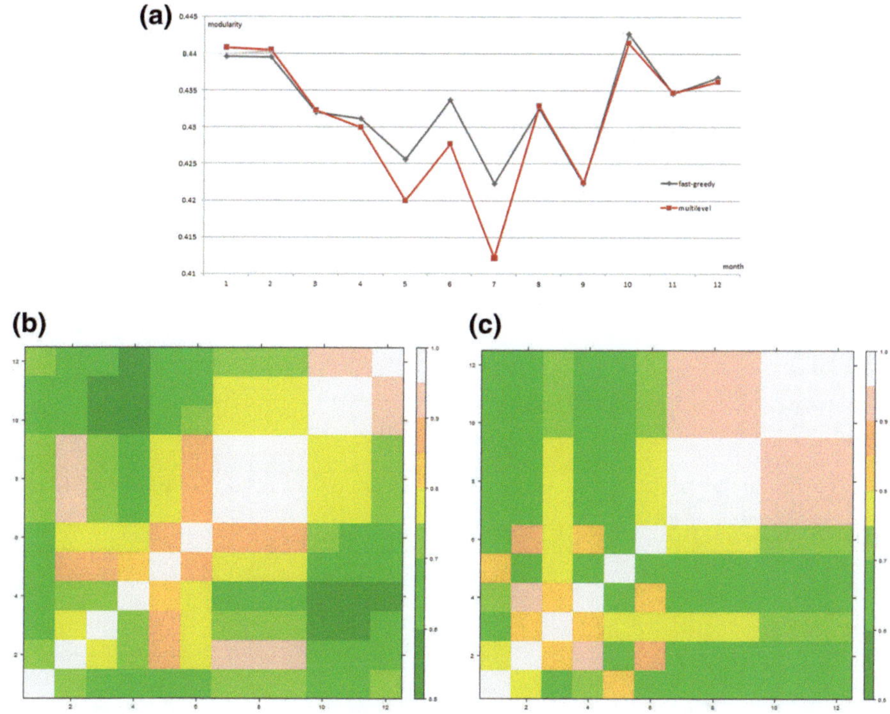

Fig. 3 **a** The modularity values for the community detection results of *G_CallFlow* in different months; **b** a month-to-month similarity matrix for the FG partition results; and **c** a month-to-month similarity matrix for the ML partition results

unstable community structure (Fig. 4a). The daily detection has generated a more stable partition structure over time, although we can still identify the unstable community structure in the first days of January, which might result from irregular mobile phone call patterns on New Year's Day (Fig. 4b). No significant difference exists between the temporal changes of community structure using the FG and ML detection algorithms.

3.2 *The Physical Divide*

Another type of spatial interaction network is generated by phone users' physical movement; more detailed discussion can be found in Gao et al. (2013a). Figure 5a, b show the spatial segregation of monthly human mobility flow graph *G_MobilityFlow* at the cellular site scale by applying two community detection algorithms (FG and ML). The term "physical divide" in this context is used to represent such human mobility patterns within limited geographical space. Not surprisingly, the

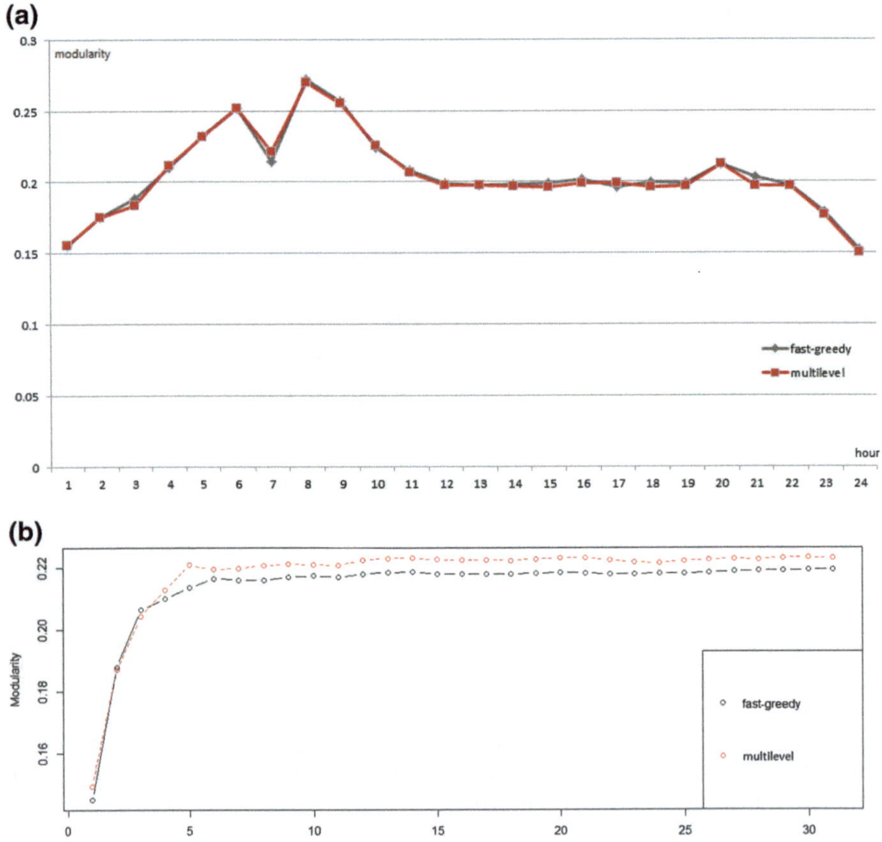

Fig. 4 The temporal variability of modularity in January by **a** hour and **b** day

spatially adjacent cellular sites are more likely to be grouped together based on mobility flows, although several abnormal grouping patterns occur across space. For example, the northeastern blue community along the country boundary tends to have more cross-site mobility flows because of highway connections along the border. In addition, we found that the modularity values ($M_{FG} = 0.7260$, $M_{ML} = 0.7248$) based on a site-to-site mobility graph are larger than those for the partition results of an arrondissement-to-arrondissement mobility graph ($M'_{FG} = 0.4396$, $M'_{ML} = 0.4408$) as shown in Fig. 5c, d. From the geographical context perspective, such physical divide patterns might be associated with terrain barriers (Fig. 5e), streets network centrality (Fig. 5f) (Gao et al. 2013b), or other natural environment and socioeconomic factors. The temporal changes and structural similarity of graph partition results for monthly mobility graphs have also been studied in this work (see Fig. 6a, b). In general, intraseason similarity tends to be higher than interseason similarity.

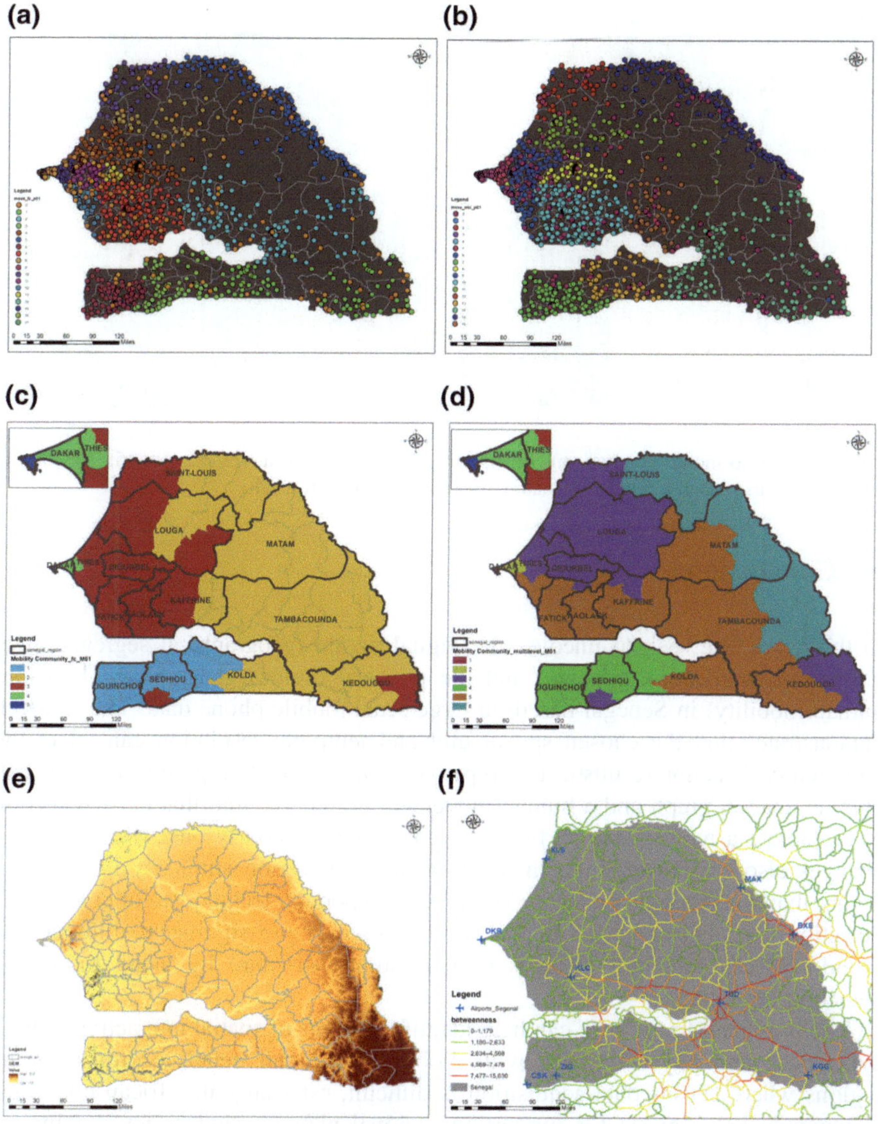

Fig. 5 Community detection of monthly mobility flows **a** at the site-to-site scale using FG; **b** at the site-to-site scale using ML; **c** at the arrondissement-to-arrondissement scale using FG; **d** at the arrondissement-to- using ML; **e** a terrain elevation map in Senegal; and **f** a map of highway network centrality

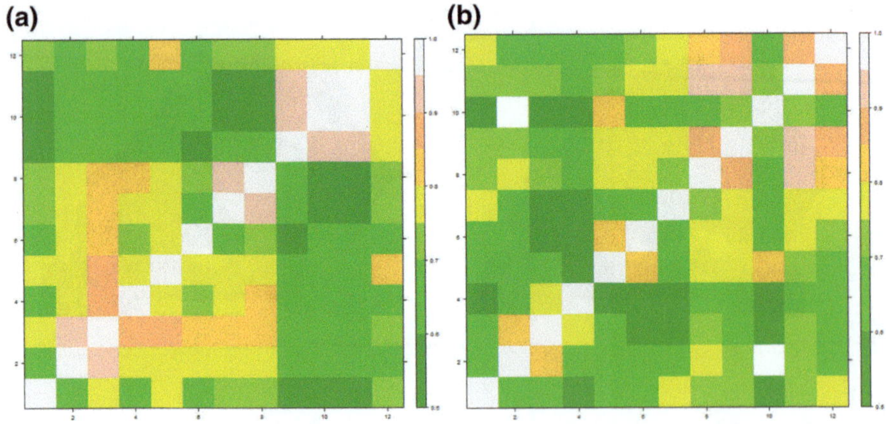

Fig. 6 Month-to-month comparisons on mobility flow graphs: **a** a similarity matrix for the FG partition results; and **b** a similarity matrix for the ML partition results

4 Conclusions

In this work, we seek to uncover the digital divide (geographical segregation of phone communication patterns) and the physical divide (geographical limits of human mobility) in Senegal based on large-scale mobile phone data. The research demonstrates that the chosen spatial unit and temporal resolution can affect the community detection results of two types of spatial interaction graphs: a phone call communication graph and a human movement graph. We find that daily detection can generate a more stable partition structure than an hourly one, while monthly changes also exist over time. In addition, the intraseason similarity is generally higher than interseason similarity. We apply two popular techniques for community detection (i.e., FG and ML) in this work. However, no significant difference is found between temporal changes of community structure by using these two detection algorithms.

The presented framework can help identify patterns of spatial interaction in both cyberspace and physical space with mobile phone call detail records in some regions where census data acquisition is difficult, especially in African countries. A potential value exists for supporting regional planning and policy making by mining large-scale geospatial datasets, although there has been some debate on whether to use mobile phone data or other big data analytics because of geo-privacy concerns. Related research in this direction (e.g., geospatial data anonymization) might attract more attention from both academic researchers and industry engineers.

References

Amini A, Kung K, Kang C, Sobolevsky S, Ratti C (2014) The impact of social segregation on human mobility in developing and industrialized regions. EPJ Data Sci 3(1):6

Blondel VD, Guillaume JL, Lambiotte R, Lefebvre E (2008) Fast unfolding of communities in large networks. J Stat Mech: Theory Exp 2008(10). doi:10.1088/1742-5468/2008/10/P10008

Cheng T, Adepeju M (2014) Detecting emerging space-time crime patterns by prospective STSS. In: Proceedings of the 12th international conference on geocomputation. http://www.geocomputation.org/2013/papers/77.pdf

Chi G, Thill JC, Tong D, Shi L, Liu Y (2014) Uncovering regional characteristics from mobile phone data: a network science approach. Pap Reg Sci. doi:10.1111/pirs.12149

Clauset A, Newman ME, Moore C (2004) Finding community structure in very large networks. Phys Rev E 70(6). doi:10.1103/PhysRevE.70.066111

Danon L, Diaz-Guilera A, Duch J, Arenas A (2005) Comparing community structure identification. J Stat Mech Theory Exp 2005(09). doi:10.1088/1742-5468/2005/09/P09008

de Montjoye YA, Hidalgo CA, Verleysen M, Blondel VD (2013) Unique in the crowd: the privacy bounds of human mobility. Sci Rep 3. doi:10.1038/srep01376

de Montjoye YA, Smoreda Z, Trinquart R, Ziemlicki C and Blondel VD (2014) D4D-Senegal: the second mobile phone data for development challenge. http://arxiv.org/abs/1407.4885v2

Gao S (2015) Spatio-temporal analytics for exploring human mobility patterns and urban dynamics in the mobile age. Spat Cogn. Comput 15(2):86–114. doi:10.1080/13875868.2014.984300

Gao S, Liu Y, Wang Y, Ma X (2013a) Discovering spatial interaction communities from mobile phone data. Trans GIS 17(3):463–481. doi:10.1111/tgis.12042

Gao S, Wang Y, Gao Y, Liu Y (2013b) Understanding urban traffic-flow characteristics: a rethinking of betweenness centrality. Environ Plan 40(1):135–153. doi:10.1068/b38141

Gonzalez MC, Hidalgo CA, Barabasi AL (2008) Understanding individual human mobility patterns. Nature 453(7196):779–782

Kang C, Ma X, Tong D, Liu Y (2012) Intra-urban human mobility patterns: an urban morphology perspective. Physica A 391(4):1702–1717

Newman ME, Girvan M (2004) Finding and evaluating community structure in networks. Phys Rev E 69(2). doi:10.1103/PhysRevE.69.026113

Ratti C, Sobolevsky S, Calabrese F, Andris C, Reades J, Martino M, Claxton R, Strogatz SH (2010) Redrawing the map of Great Britain from a network of human interactions. PloS one 5 (12). doi:10.1371/journal.pone.0014248

Song C, Qu Z, Blumm N, Barabási AL (2010) Limits of predictability in human mobility. Science 327(5968):1018–1021

References

Application of Spatio-Temporal Clustering For Predicting Ground-Level Ozone Pollution

Mahdi Ahmadi, Yan Huang and Kuruvilla John

Abstract Ground-level ozone is an air pollutant, and as such negatively impacts human health and the environment. The complexity of the physical process of ozone formation makes ambient ozone concentration difficult to predict accurately. In this chapter, clustering techniques and multiple regression analyses are used to construct a simply interpretable forecasting model. Time series of ozone and meteorological variables in the Dallas–Fort Worth area for 12 years at 14 monitoring stations were acquired and processed. First, K-means cluster analysis was performed on ozone time series to specify data-driven ozone seasons at each station. Next, spatial hierarchical clustering was performed to find ozone zones in the area during each ozone season recognized in the previous step. Finally, a multiple linear regression was executed with meteorological variables and ozone in each zone. For ozone forecasting, temperature, solar radiation, wind speed, and previous ozone values were used because ozone is temporally autocorrelated. Monitoring stations in each temporal and spatial cluster show consistent behavior, which makes ozone forecasting possible even when a station is off. Results show high accuracy of ozone forecasting coupled with ease of interpreting the link between meteorology and ozone behavior. Also, clustering results are useful to understand the temporal and spatial patterns of the ozone dynamics in the area.

Keywords Ozone forecasting · Temporal clustering · Spatial clustering · Linear regression

M. Ahmadi (✉) · Y. Huang · K. John
University of North Texas, 1155 Union Circle #311098, Denton, TX 76203-5017, USA
e-mail: MahdiAhmadi@unt.edu

Y. Huang
e-mail: Yan.Huang@unt.edu

K. John
e-mail: Kuruvilla.John@unt.edu

D.A. Griffith et al. (eds.), *Advances in Geocomputation*,
Advances in Geographic Information Science,
DOI 10.1007/978-3-319-22786-3_15

1 Introduction

Ozone is a highly reactive gas with proven negative impacts on humans (WHO 2003). Ground-level ozone pollution is formed by a chain of photochemical reactions in the presence of nitrogen oxides (NO_X) and reactive volatile organic compounds (VOCs). Tropospheric ozone formation is a complex process that displays strong seasonal and diurnal patterns, with higher concentration during the summer and in the afternoon (Seinfeld and Pandis 2012). Ozone prediction is difficult because of the complexity of its formation process and spatiotemporal variations in both meteorological factors and precursors.

Several statistical techniques have been developed to account for the effect of meteorological factors and to predict ozone (Lou Thompson et al. 2001; Schlink et al. 2003). The simplest model is multiple linear regression that assumes an additive linear relationship to link ozone concentration to meteorological factors (Kuntasal and Chang 1987; Feister and Balzer 1991; Abdul-Wahab et al. 1996; Katsoulis 1996; Dueñas et al. 2002). Although linear models are easy to interpret, they have poor accuracy due to the nonlinear effects of predictor variables. Also, methods such as principal component analysis, artificial neural networks (Sousa et al. 2007; Al-Alawi et al. 2008), and clustering and classification techniques (Bruno et al. 2004; Lengyel et al. 2004) are proposed to reduce the dimensionality of the problem and to make better predictions (Sahu et al. 2007; Kovač-Andrić et al. 2009; Sahu and Bakar 2012; Austin et al. 2014). Ozone forecasting can be performed more effectively if its temporal and spatial patterns are known. In this chapter, clustering techniques and multiple linear regression analysis are used to explain the patterns of ozone in the Dallas–Fort Worth (DFW) area and to develop a linear model for ozone prediction.

2 Method

The objective of this work is to perform clustering of ozone time series and spatial clustering of ozone monitors to create better input for linear regression analysis. Our data-mining tasks (Fig. 1) involve the following:

Step I: Acquire a dataset.
Step II: Preprocess the data.
Step III: Perform a K-means cluster analysis on time series of 8 h average daily maximum ozone concentration; the goal here is to establish ozone seasons.
Step IV: Evaluate the overlap between ozone seasons at different monitoring stations to determine the best split for the entire monitoring network (i.e., regional ozone seasons).

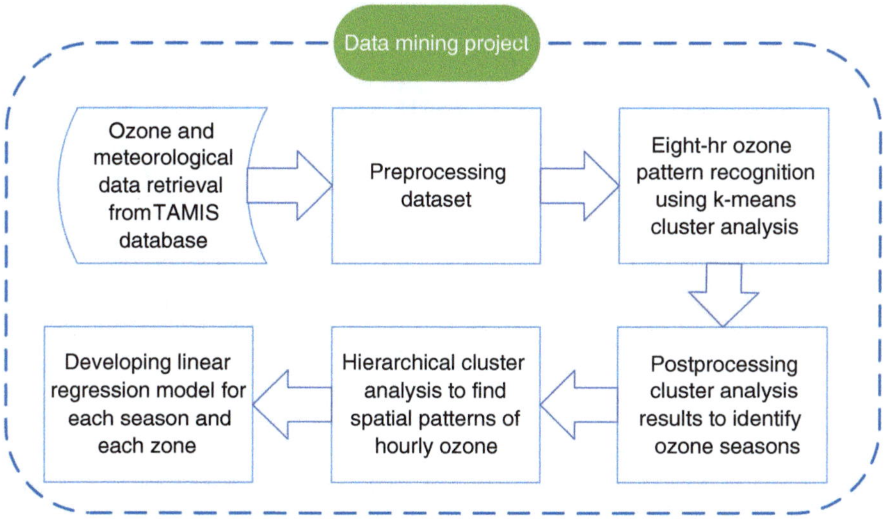

Fig. 1 A flow chart of the data-mining tasks

Step V: Perform hierarchical cluster analysis on 1 h ozone time series from all monitoring stations for each ozone season; the goal is to determine the best spatial clustering (ozone zones) for each ozone season.

Step VI: Develop a multiple linear regression model for each ozone zone in each season.

3 Dataset

The DFW Metroplex is the study area and includes both high-density urban and low-density rural regions. The DFW area, located in north-central Texas, is the largest metropolitan area in the South and the fourth largest by population in the United States (more than 7 million as of July 1, 2012). The volume of transportation and other industrial and residential activities in the presence of particular meteorological conditions has created air quality issues. Various DFW counties have failed to comply with the National Ambient Air Quality Standards (NAAQS) in relation to ozone since 1990.

Measurement data collected by Texas Commission on Environmental Quality (TCEQ) continuous air monitoring stations (CAMS) were used to build the dataset. Data were retrieved from the Texas Air Monitoring Information System (TAMIS), which is publicly available. Figure 2 presents the map of CAMS in the DFW area. The dataset includes 1 h measurement time series of ozone (O_3), ambient temperature (T), solar radiation (SR), and wind speed (W) for 12 years (2002–2013).

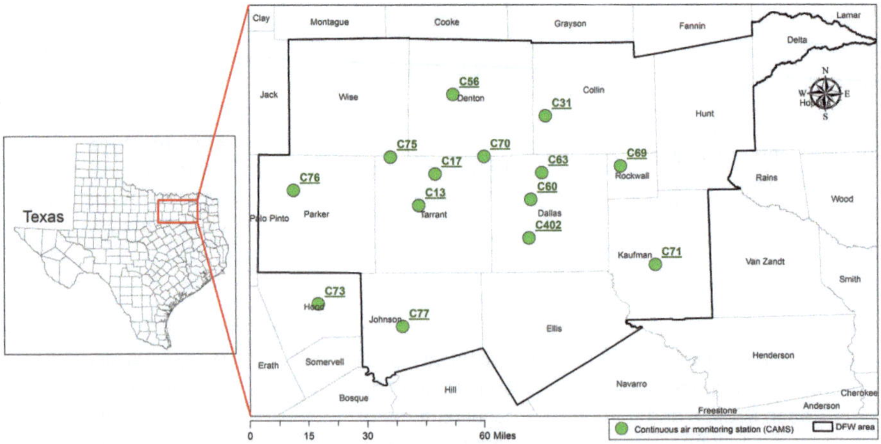

Fig. 2 A map of the study area with locations of continuous air monitoring stations (CAMS)

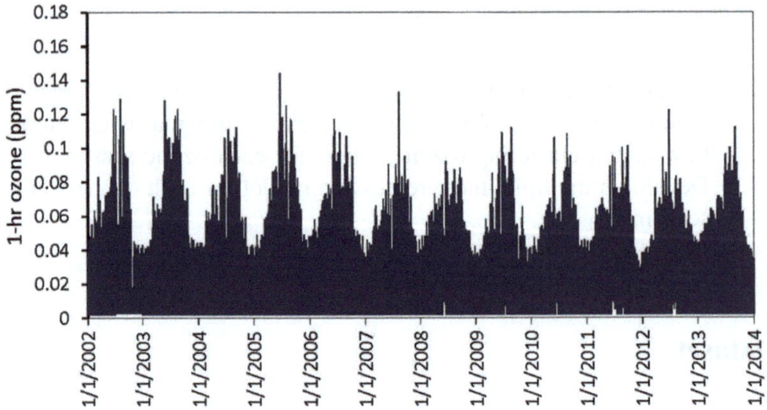

Fig. 3 A time series of 1 h ozone level at C13 CAMS

The dataset has approximately 5,886,720 entries. Figures 3, 4, 5, 6 and 7 portray time series of variables at C13 CAMS.

In the preprocessing step, any gap in a time series to equal or be less than 4 h was estimated by linear interpolation. Any day with more than four consecutive missing values was removed from the dataset. For example, the gap at the end of 2012 observed at C13 CAMS (see Figs. 3, 4, 5, 6) was removed from the analysis. A seasonal analysis is more sensible if a cluster analysis uses 8 h moving average ozone because it is a standard time unit measure for evaluating ozone levels for regulatory and air pollution control purposes (US EPA 2008). Therefore, in the second step of preprocessing, 8 h average time series were generated using a moving average technique. The original 1 h time series were kept for spatial cluster and linear regression analysis.

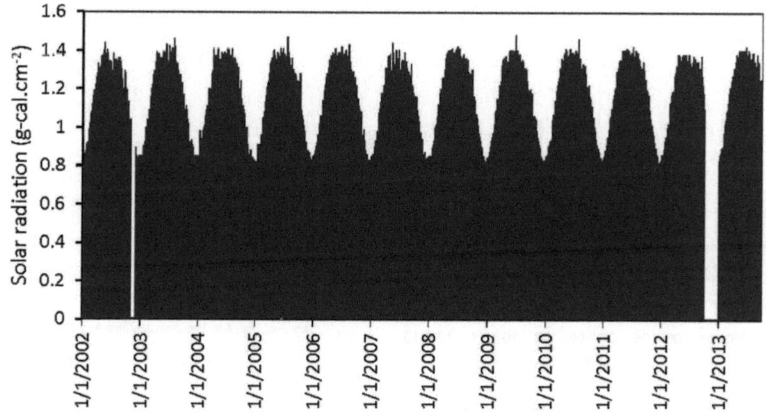

Fig. 4 A time series of 1 h solar radiation measured at C13 CAMS

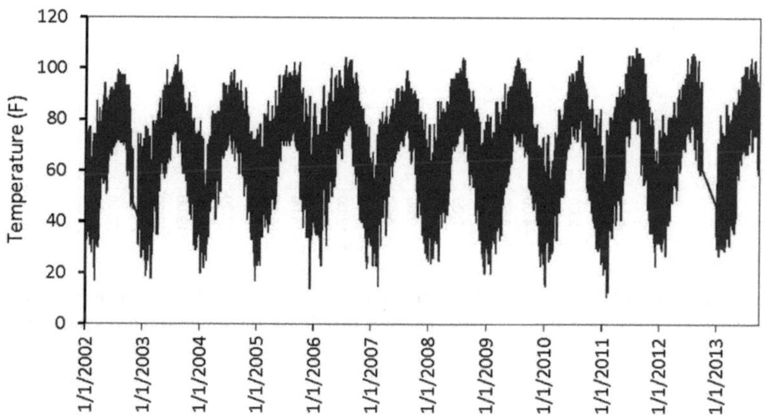

Fig. 5 A time series of 1 h temperature measured at C13 CAMS

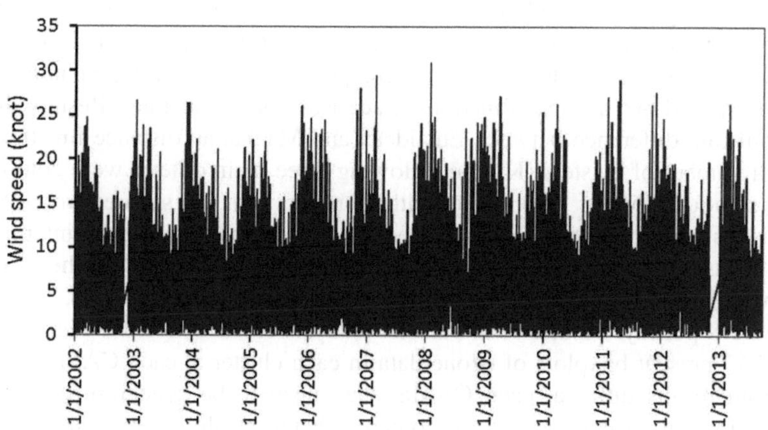

Fig. 6 A time series of 1 h wind measured at C13 CAMS

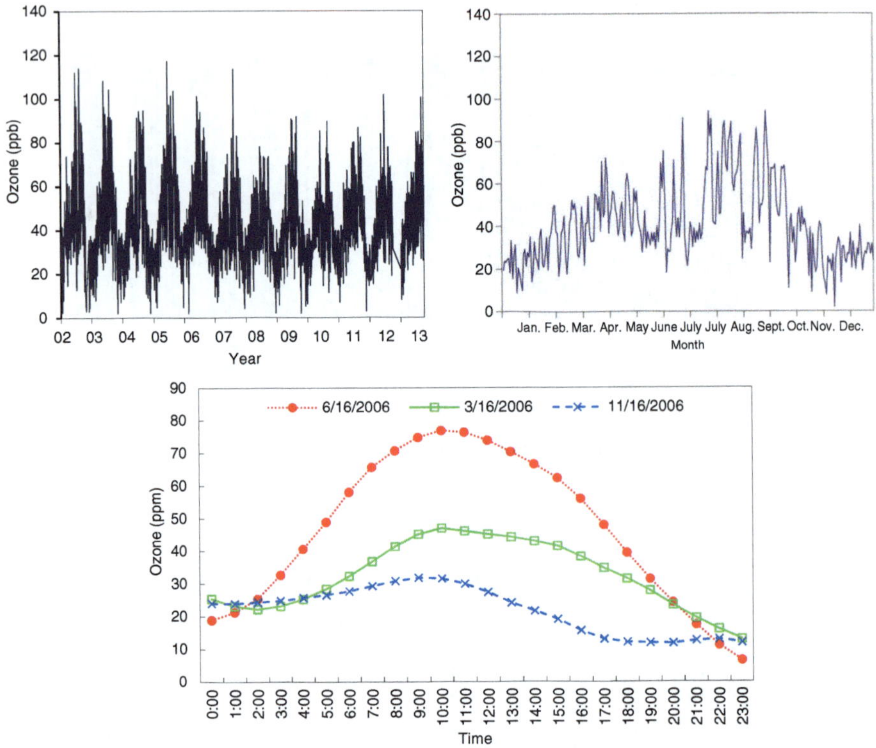

Fig. 7 A time series of 8 h average daily maximum ozone level (*top left*); a time series for one random year (*top right*); an 8 h average ozone level profile for three random days measured at C13 CAMS

4 Data Mining

After executing Steps I and II (data acquisition and preprocessing), for Step III, simple K-means cluster analysis was performed on time series of daily maximum 8 h average ozone. Cluster analyses were performed with different numbers of clusters (K) and two distance functions (Euclidean and Manhattan). Results showed no significant difference between Euclidean and Manhattan distance functions. To select a number of clusters (K), the following three main criteria were considered: (1) a solution with a reasonable within-cluster sum of square error (SSW); (2) clusters with minimum within-cluster variability; and (3) a highly interpretable solution based on knowledge of ozone pollution in the area. Based on these criteria, the optimal solution was achieved with four temporal clusters ($k = 4$).

Figure 8 portrays results of the cluster analysis for C13 CAMS. Figures 9, 10, 11 and 12 present boxplots of ozone data in each cluster at each CAMS. Because clustering results differ at each CAMS, they need to be synchronized to create global temporal clusters across the domain. Therefore, the average share of each

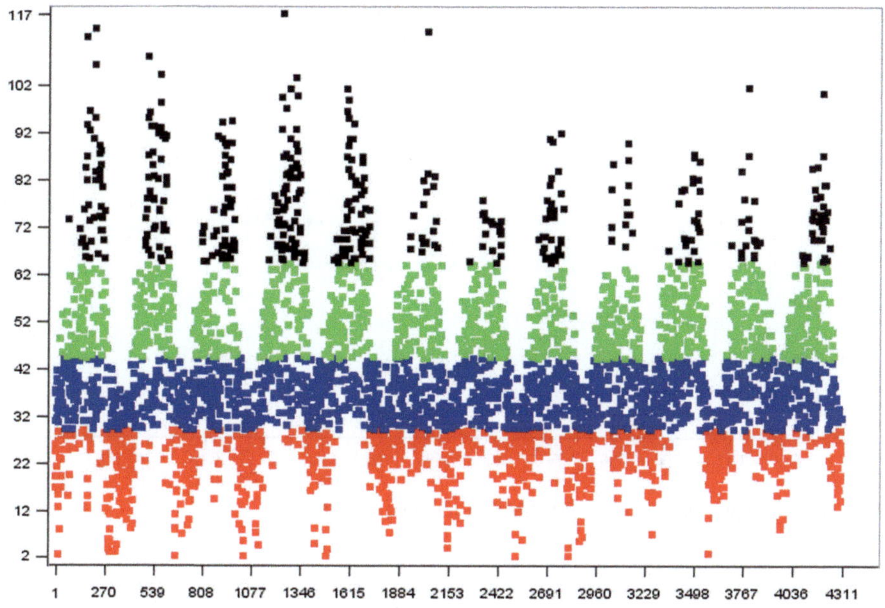

Fig. 8 Seasonal clusters of ozone at C13 CAMS using the K-means method

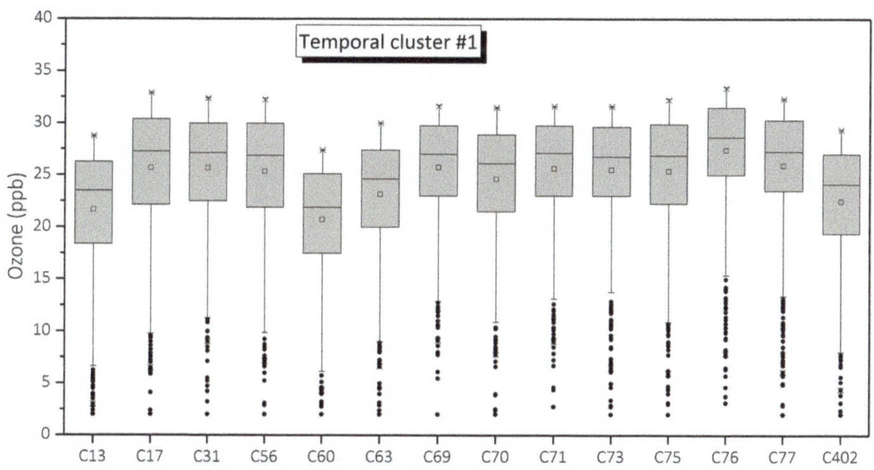

Fig. 9 A boxplot presentation of CAMS in temporal Cluster #1

cluster in each month at all CAMS was used in Step IV to transform ozone clusters into temporal clusters. Figure 13 shows the share of four ozone clusters over the year. Because Cluster 2 was not dominant in any month, it was removed from the

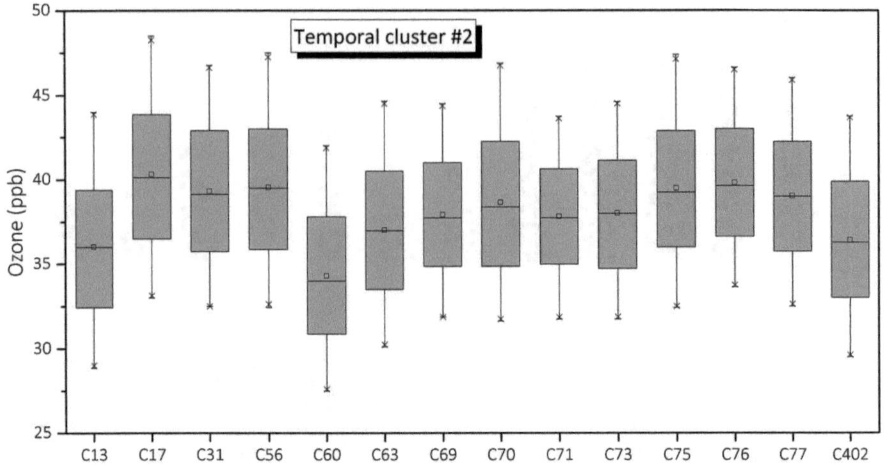

Fig. 10 A boxplot presentation of CAMS in temporal Cluster #2

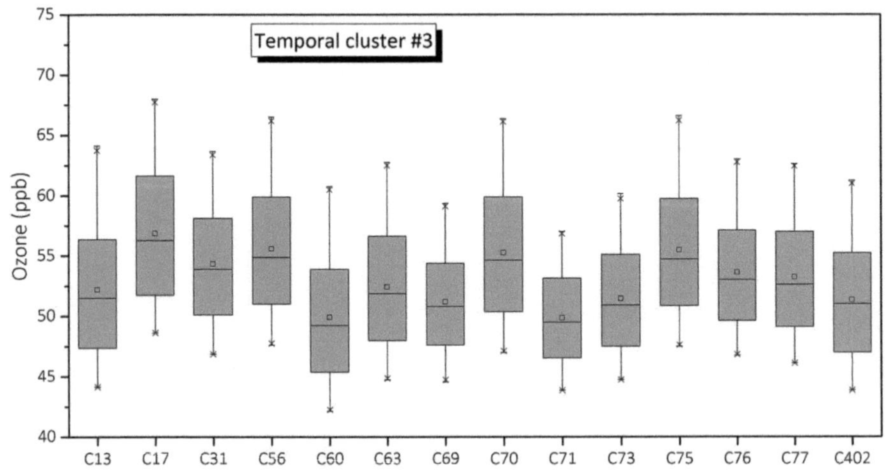

Fig. 11 A boxplot presentation of CAMS in temporal Cluster #3

temporal clustering results. The final result is three ozone seasons: low, moderate, and high (shown in Table 1).

In Step V, CAMS in each season were clustered based on similarities so that the spatial pattern of ozone behavior can be recognized. Agglomerative hierarchical cluster analysis was performed following Ward's (1963) method, where the criterion for making a new cluster is a within-cluster squared error increase when two clusters are merged. Figure 14 shows the hierarchical trees (i.e., dendrograms) and

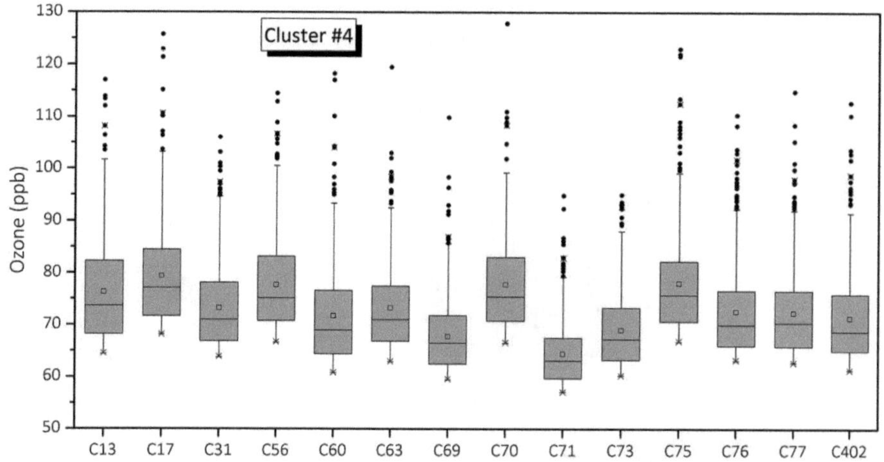

Fig. 12 A boxplot presentation of CAMS in temporal Cluster #4

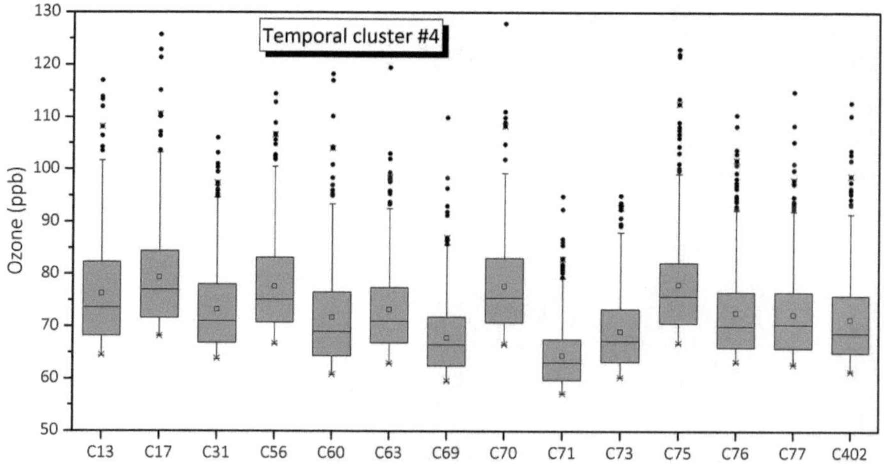

Fig. 13 Average share of four ozone clusters over the study period

Table 1 Ozone seasons resulting from ozone time series clustering

Temporal cluster	Season	Months			
#1	Low	Jan.	Feb.	Nov.	Dec.
#2	–	–	–	–	–
#3	Moderate	Mar.	Apr.	May	Oct.
#4	High	June	July	Aug.	Sept.

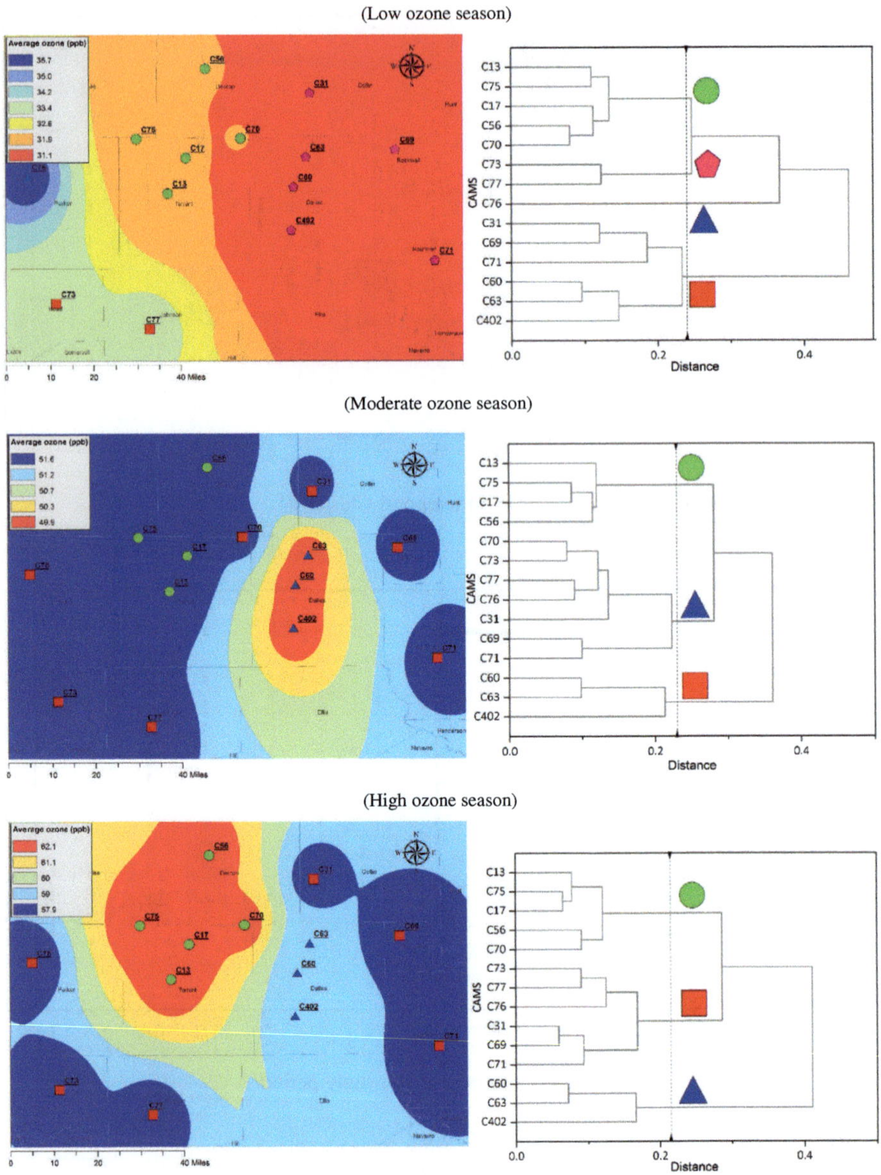

Fig. 14 Hierarchical cluster trees and average ozone concentration for low, moderate, and high (*top, middle, bottom*) seasons

clusters on maps of the area. The average value of ozone in each zone (i.e., cluster) was used with an inverse distance weighting function (Fortin and Dale 2005) in ArcGIS© software to produce the maps.

5 Ozone Forecasting

The last step is to develop a multiple linear regression model. Ozone time series is temporally autocorrelated (Rao et al. 1997); thus, an autoregressive model is also used. Because of the collinearity of meteorological variables and their multiplicative effect in ozone generation, natural logarithms of ozone values were used in the regression model (Rao and Zurbenko 1994). The general form of the multiple linear function for ozone forecasting at time (t) is given by

$$\mathrm{Log}[O_3]_t = \alpha T_{t-i} + \beta \mathrm{SR}_{t-j} + \chi W_{t-k} + \omega \mathrm{Log}[O_3]_{t-l} + \lambda(1) + \epsilon_t,$$

where α, β, χ, ω,. and λ respectively are the linear regression coefficients for T (temperature)t time $t-i$, SR (solar radiation) at time $t-j$, W (wind speed) at time $t-k$, and the logarithm of ozone at time $t-l$. The residual of the regression (i.e., observed value at *t*–predicted value at *t*) is represented by ϵ_t. In this step, the goal is to determine coefficients and time lags so that R^2 (the coefficient of determination) and RMSE (root mean square error) of the fitted equation are optimal. To determine the best time lag for each predictor variable, *i*, *j*,. and *k* were varied from –48 (indicating two-day forward lag) to +48 (indicating two-day backward lag), and R^2 and p-values were monitored. The highest R^2 and lowest p-values were observed when the lag for the meteorological variables was zero ($i = j = k = 0$). However, the best correlations were achieved when previous ozone values from 1 h ago were used ($l = 1$). Although higher-order autoregressive models might account for temporal autocorrelation of ozone time series better, the first-order model has more explanatory power and is simpler.

Time series of average 1 h ozone in three spatial clusters in the high season (shown in Fig. 13, bottom) were forecasted by three independent multiple linear regression models. Table 2 reports the parameter estimates of these linear regression models. Figure 15 portrays the scatterplots of predicted versus observed 1 h ozone concentration for three spatial clusters during the high ozone season.

The linear model developed for each cluster was applied to predict ozone; Fig. 16 portrays the results. Its high accuracy prediction, even without using time series of ozone precursors, suggests that this method can be used successfully for ozone forecasting. In general, there is a pattern of overprediction in ozone peaks that could be due to the stronger effect of previous values of ozone with regard to wind effect in the linear regression equations. Overprediction occurs because the effect of short-range ozone transportation (away from or toward a station) cannot be captured completely when the average value of wind speed is used. However, the overshoot is within a reasonable range and favors a better safety margin.

The proposed linear regression model does not account for spatial autocorrelation. Ozone concentration time series measured at fixed locations show various degrees of spatial autocorrelation (Diem and Comrie 2002; Diem 2003), with stronger correlation in the long-term and seasonal trends than in the synaptic variations (Rao et al. 1995; Rao et al. 1997; Ahmadi and John 2015). However, the

Table 2 Summary of multiple linear regression parameter estimates for spatial clusters in the high ozone season

Parameter	Spatial cluster A (●)			Spatial cluster B (■)			Spatial cluster C (▲)		
	Estimate	Standard error	p-value	Estimate	Standard error	p-value	Estimate	Standard error	p-value
α	0.002797	0.000311	0.000	0.000206	0.000146	0.219	0.002699	0.000328	0.000
β	0.359812	0.005260	0.000	0.241734	0.002520	0.000	9.652480	0.150081	0.000
χ	0.021030	0.000811	0.000	−0.000247	0.000375	0.524	0.027754	0.001041	0.000
ω	0.793794	0.002790	0.000	0.871374	0.002039	0.000	0.687160	0.003461	0.000
λ	0.176548	0.022300	0.000	0.356038	0.010856	0.000	0.311890	0.025280	0.000
R^2	0.827	–	–	0.890	–	–	0.827	–	–
RMSE	7.361	–	–	5.142	–	–	7.872	–	–

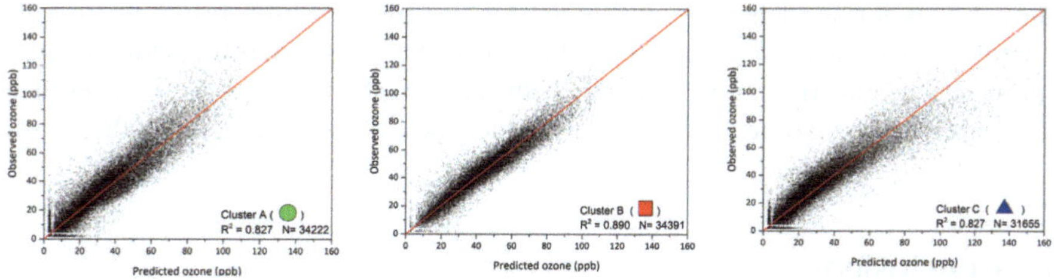

Fig. 15 Scatterplots of predicted versus observed ozone concentration for three spatial clusters in the high ozone season

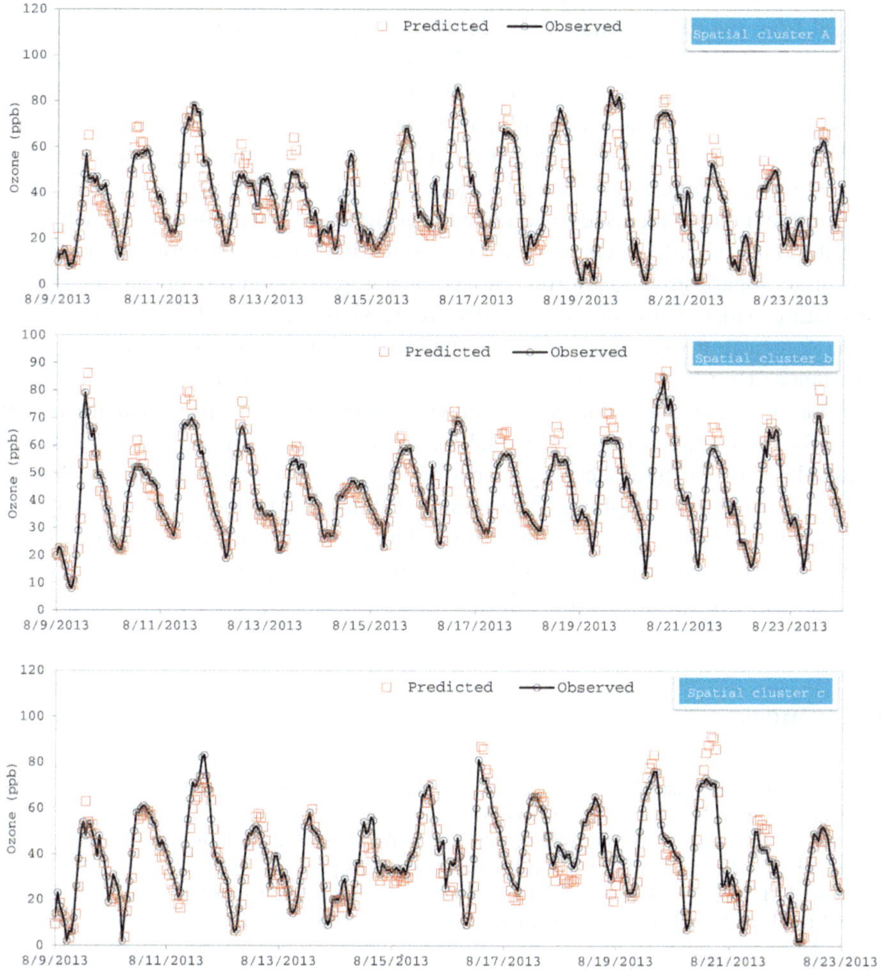

Fig. 16 Comparisons of observed and predicted 1 h ozone concentration in spatial clusters in the high ozone season

linear regression models were developed for one ozone season resulting from the temporal cluster analysis. Therefore, the seasonality of ozone time series is minimized because spatial autocorrelation between CAMS in each zone is a result of correlation among ozone synaptic terms, which are less significant than other terms.

6 Conclusion

In this research, multivariate data-mining techniques were used to increase ozone forecasting accuracy and to ease the interpretation of the model. Instead of categorizing time series to conventional seasons of the year, ozone seasons were derived from the measurement data. Temporal pattern recognition helps reduce the variability of ozone in each cluster. Moreover, hierarchical cluster analysis was performed on 14 monitoring stations in the area during each recognized ozone season to identify spatial patterns. Three ozone zones were recognized inside the DFW area during the high ozone season. The method presented here allows ozone forecasting for a zone even when all but one of the air monitoring stations are down. Measurement data were used to validate the accuracy of linear models in each zone.

Although a simple linear regression model for ozone time series is weak and inaccurate, logarithmic transformation and an autoregression model can improve its forecasting and explanatory power. The regression model in this research is for ozone forecasting only, and it does not test a hypothesis or perform analysis of covariance. However, the proposed temporal and spatial cluster analysis can be used for more advanced statistical analyses. Moreover, the proposed method is useful for conventional time series analysis, such as detrending or spectral analysis.

Results show a high accuracy of ozone forecasting using only meteorological variables. Therefore, applying data-mining techniques in the proposed way can increase the accuracy of estimating ozone. The proposed method is useful when a reliable and fast prediction of ozone concentration is required for both air quality management and information services.

References

Abdul-Wahab S, Bouhamra W, Ettouney H, Sowerby B, Crittenden BD (1996) Predicting ozone levels. Environ Sci Pollut Res 3:195–204

Ahmadi M, John K (2015) Statistical evaluation of the impact of shale gas activities on ozone pollution in North Texas. Sci Total Environ 536:457–467

Al-Alawi SM, Abdul-Wahab SA, Bakheit CS (2008) Combining principal component regression and artificial neural networks for more accurate predictions of ground-level ozone. Environ Model Softw 23:396–403

Austin E, Zanobetti A, Coull B, Schwartz J, Gold DR, Koutrakis P (2014) Ozone trends and their relationship to characteristic weather patterns. J Expo Sci Environ Epidemiol 25:532–542

Bruno F, Cocchi D, Trivisano C (2004) Forecasting daily high ozone concentrations by classification trees. Environmetrics 15:141–153

Diem JE (2003) A critical examination of ozone mapping from a spatial-scale perspective. Environ Pollut 125:369–383

Diem JE, Comrie AC (2002) Predictive mapping of air pollution involving sparse spatial observations. Environ Pollut 119:99–117

Dueñas C, Fernández M, Cañete S, Carretero J, Liger E (2002) Assessment of ozone variations and meteorological effects in an urban area in the Mediterranean Coast. Sci Total Environ 299: 97–113

Feister U, Balzer K (1991) Surface ozone and meteorological predictors on a subregional scale. Atmos Environ Part A: Gen Top 25:1781–1790

Fortin MJ, Dale MRT (2005) Spatial analysis: a guide for ecologists. Cambridge University Press, Cambrdige, UK

Katsoulis BD (1996) The relationship between synoptic, mesoscale and microscale meteorological parameters during poor air quality events in Athens, Greece. Sci Total Environ 181:13–24

Kovač-Andrić E, Brana J, Gvozdić V (2009) Impact of meteorological factors on ozone concentrations modelled by time series analysis and multivariate statistical methods. Ecol Inform 4:117–122

Kuntasal G, Chang TY (1987) Trends and relationships of O_3, NO_x and HC in the south coast air basin of California. JAPCA 37:1158–1163

Lengyel A, Héberger K, Paksy L, Bánhidi O, Rajkó R (2004) Prediction of ozone concentration in ambient air using multivariate methods. Chemosphere 57:889–896

Lou Thompson M, Reynolds J, Cox LH, Guttorp P, Sampson PD (2001) A review of statistical methods for the meteorological adjustment of tropospheric ozone. Atmos Environ 35:617–630

Rao S, Zurbenko I, Neagu R, Porter P, Ku J, Henry R (1997) Space and time scales in ambient ozone data. Bull Am Meteorol Soc 78:2153–2166

Rao ST, Zalewsky E, Zurbenko IG (1995) Determining temporal and spatial variations in ozone air quality. J Air Waste Manag Assoc 45:57–61

Rao ST, Zurbenko IG (1994) Detecting and tracking changes in ozone air quality. Air & Waste 44:1089–1092

Sahu SK, Bakar KS (2012) Hierarchical Bayesian autoregressive models for large space–time data with applications to ozone concentration modelling. Appl Stochast Models Bus Ind 28: 395–415

Sahu SK, Gelfand AE, Holland DM (2007) High-resolution space–time ozone modeling for assessing trends. J Am Stat Assoc 102:1221–1234

Schlink U, Dorling S, Pelikan E, Nunnari G, Cawley G, Junninen H, Greig A, Foxall R, Eben K, Chatterton T (2003) A rigorous inter-comparison of ground-level ozone predictions. Atmos Environ 37:3237–3253

Seinfeld JH, Pandis SN (2012) Atmospheric chemistry and physics: from air pollution to climate change. Wiley, Hoboken, NJ

Sousa S, Martins F, Alvim-Ferraz M, Pereira MC (2007) Multiple linear regression and artificial neural networks based on principal components to predict ozone concentrations. Environ Model Softw 22:97–103

US EPA (2008) National ambient air quality standards for ozone; final rule. 40 CFR Parts 50 and 58. Government Printing Office, Washington, DC

Ward JHJ (1963) Hierarchical grouping to optimize an objective function. J Am Stat Assoc 58:236–244

WHO (2003) Health aspects of air pollution with particulate matter, ozone and nitrogen dioxide: report on a WHO working group. World Health Organization, Regional Office for Europe, Bonn, Germany

Does the Location of Amerindian Communities Provide Signals About the Spatial Distribution of Tree and Palm Species?

Aravind Sivasailam and Anthony R. Cummings

Abstract This chapter examines the proximity of plants that provide critical ecosystem services to indigenous peoples of the Rupununi, southern Guyana, relative to their village centers. We explore the hypothesis that plants of greater importance to indigenous peoples' livelihood practices are distributed closer to village centers, focusing on multiple-use plants, species that provide either two or more ecosystem services of nontimber forest products, food for wildlife, and commercial timber. We consider multiple-use plants to be more important than single-use plants, and we expect them to be distributed closer to village centers. Using the von Thünen model as a theoretical basis, we measure the average Euclidian distance of multiple-use plants to village centers, and compare their proximity across 12 villages and two controls. Results suggest that plants associated with some form of traditional use for indigenous peoples are distributed closest to village centers. This finding supports the idea that plants with higher economic value are located closest to village centers. Our analysis also suggests the potential for resource use conflict with plants in a multiple-use class, and suggests that logging close to villages has implications for traditional practices. Implications from removal of such plants for logs include accessing medicines and hunting services. The potential for resource use conflict underscores the importance of employing spatial analysis tools in studying plant distribution and resource use allocation schemes. Our results support the idea that indigenous peoples and their livelihood practices favor forest structure and plant distributions of greater economic value closer to their villages.

Keywords Indigenous peoples · Multiple-use plants · Guyana · Von thünen model · Forest structure

---

A. Sivasailam · A.R. Cummings (✉)
Geospatial Information Sciences, School of Economic Political and Policy Sciences, University of Texas at Dallas, 800 W. Campbell Road, Richardson, TX 75080, USA
e-mail: Anthony.Cummings@utdallas.edu

A. Sivasailam
e-mail: aravindsivasailam@gmail.com

D.A. Griffith et al. (eds.), *Advances in Geocomputation*,
Advances in Geographic Information Science,
DOI 10.1007/978-3-319-22786-3_16

1 Introduction

The influence of neotropical indigenous peoples on their environment, in particular how their actions shape the nature of the forests within which they live, has long attracted the attention of scholars (e.g., Denevan 1992; Posey 1982). Yet much of the data about how indigenous peoples influence forest structure in favor of plants for food, medicines, and shelter has come from oral history (e.g., Forte 1996; Goeman 2008) and archaeology (Kristensen and Davis 2015). The data derived from oral history have been used to justify the protection of indigenous territories from commercial exploitation and in mechanisms geared at payment for ecosystem services. The advent of payment for ecosystem services initiatives, such as Reduced Emissions from Deforestation and Degradation (REDD +), and the continuous call for evidence-based ecosystem services (ES) mapping mean that oral history alone is not sufficient for determining areas of indigenous territory relevant for such programs. Using geographic information systems (GIS)-based tools to study plant distribution patterns provides critical and new insights into detecting indigenous people's presence within a landscape and areas that may be relevant for payment for ES initiatives such as REDD +.

With the foregoing in mind, this study examines whether the distribution of plants, trees, and palms, which are critical to Amerindian traditional practices, such as providing medicines, thatch for roof, building materials, and weaponry materials, on titled lands provides signals into Amerindian historical occupation of such areas.

To understand how plants with the highest economic and traditional value are distributed relative to village centers, we drew on GIS tools and von Thünen's model (Hall 1966; Fig. 1). By understanding the distribution of plants of highest traditional value, we gain insights into areas that should be protected from commercial logging and possible inclusion in REDD + activities where communities may want to avoid resource use conflicts.

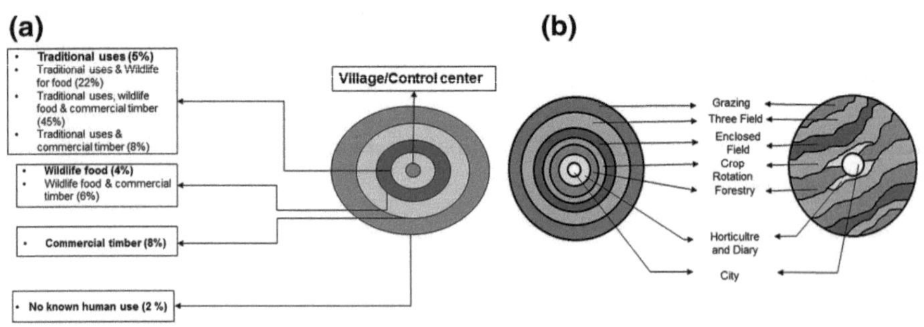

Fig. 1 **a** A hypothesized von Thünen model based on the classification of plants around Amerindian villages into their resource use classes. The proportion each class represents in our sample is provided in parenthesis. **b** A typical representation of von Thünen's model (*left*) with distribution of activities relative to a town, with modifications of the model for a river running through the town (*right*)

von Thünen's model was chosen as a theoretical basis because it suggests that, for a centrally located town, crops are grown in surrounding concentric circles that reflect their value as measured by the cost for transporting those products to the market. Therefore, crops that are more valuable, difficult to transport, and perishable are grown closer to a city center (Hall 1966). Thus, von Thünen's model implies that a commercial farmer picks a location for his/her activities based on transportation and land costs that allow for profits to be maximized (Venables and Limão 2002). Scholars have applied von Thünen's model to study the impacts of deforestation in tropical forests (e.g., Thomas 2007; Walker 2014) and of global trade and production patterns (Venables and Limão 2002). In our study, we examine whether this model can provide insights into how Amerindians plan their livelihood activities as reflected in the distribution of plants. Based on the ES associated with the plant species of the Rupununi, we developed a hypothetical von Thünen model (Fig. 1a) to describe the distribution of ES relative to village centers. The distribution of multiple-use plants, defined as species whose fruit, bark, leaves, stems, or other portions thereof are of interest to more than one group of forest users and dwellers (Cummings 2013), from the center of Amerindian communities was studied to address two overarching questions:

1. Can the proximity and presence of particular tree and palm species be attributed to the activities of indigenous peoples?
2. Does a relationship exist between plant species that provide key ES and distance to Amerindian villages; that is, as distance increases from the center of villages, does the presence of species of most critical economic importance to Amerindians decline?

2 Methodology

The analysis of the distribution of plants relative to Amerindian village centers was completed in the tropical forest and savannah biome of the Rupununi, southern Guyana. The Near Tool® in ArcGIS and a Python tool developed for this project were used to automate distance determinations to individual plants from village centers within a 12 km radius. The distance to plants in the various economic classes relative to village centers was compared across 14 Rupununi study sites.

2.1 *The Study Area*

This study was completed in the Rupununi, southern Guyana, located between 1°59' –4°43' N latitude and 58°29' –59°48' W longitude. The study area is located approximately 340 km from Guyana's capital city, Georgetown (Fig. 2) and is the homeland to the Cariban-speaking Makushi and Arawakan-speaking Wapishiana peoples (Colchester 1997). The Makushi are the dominant indigenous group of the North Rupununi, whereas the Wapishiana are dominant in the South Rupununi. The study area sits within a savannah/tropical forest biome (Read et al. 2010), and we sampled in a mixture of forest and forest-edge sites (Fig. 2). The study area is characterized by an annual rainy season between May and July that converts the lowland regions into a giant wetland. Both savannah and forested regions are flooded due to the seasonal rains, and indigenous peoples have included such areas in their descriptions of the Rupununi landscape (Cummings et al. 2015). The study area's elevation ranges from 30 m above mean sea level in the savannah areas, to 1,100 m above mean sea level in the mountainous regions. The area consists of both lowland and highland regions (Cummings et al. 2015), and the plants in our sample are distributed across both regions, with slightly higher densities in the lowlands. Similarly, the majority of study sites are located in the lowland regions (Fig. 2).

2.2 Collection of Spatial and Attribute Data About Multiple-Use Plants

The plants included in our sample are dispersed in four primary ways: zoochorically, autochorically, anemochorically, and hydrochorically. The majority of plants are dispersed zoochorically (Roosmalen 1985), reflecting the strong relationship that exists between animals and plants in the region. This strong relationship facilitates many indigenous livelihood practices, with swidden agriculture and traditional gathering practices allowing animals that are hunted for food to be accessed within forests closest to villages (Read et al. 2010).

To determine the distribution of plants relative to village centers, ninety-two 10 m wide belt transects, each 4 km long, were sampled across 14 study sites, 12 Amerindian villages, and 2 controls (Fig. 2; see Cummings 2013 for more details). At each study site, 8 randomly located transects were installed, and all trees greater than 25 cm diameter-at-breast-height and mature palms were sampled.

At the time of sampling, tree and palm species were classified into four economic use classes—*traditional use, wildlife food, commercial timber*, and *no known human uses*—based on common names and traditional knowledge of plant usage. Where two or more economic uses intersected in a single species, such a plant was defined as multiple-use, reflecting the ES associated with that species. Four multiple-use classes—*commercial timber and traditional use*, *wildlife food and traditional use, wildlife food and commercial timber,* and *wildlife food, commercial timber, and traditional use*—emerged based on this classification.

A total of 33,457 plants, comprising 165 species, were classified into the four multiple-use classes, three single-use classes, and no known uses. Multiple-use species dominated the sample (see Fig. 1), with the class *wildlife food*, *commercial timber*, and *traditional use* accounting for 45 % of plants and 41 of the 165 species.

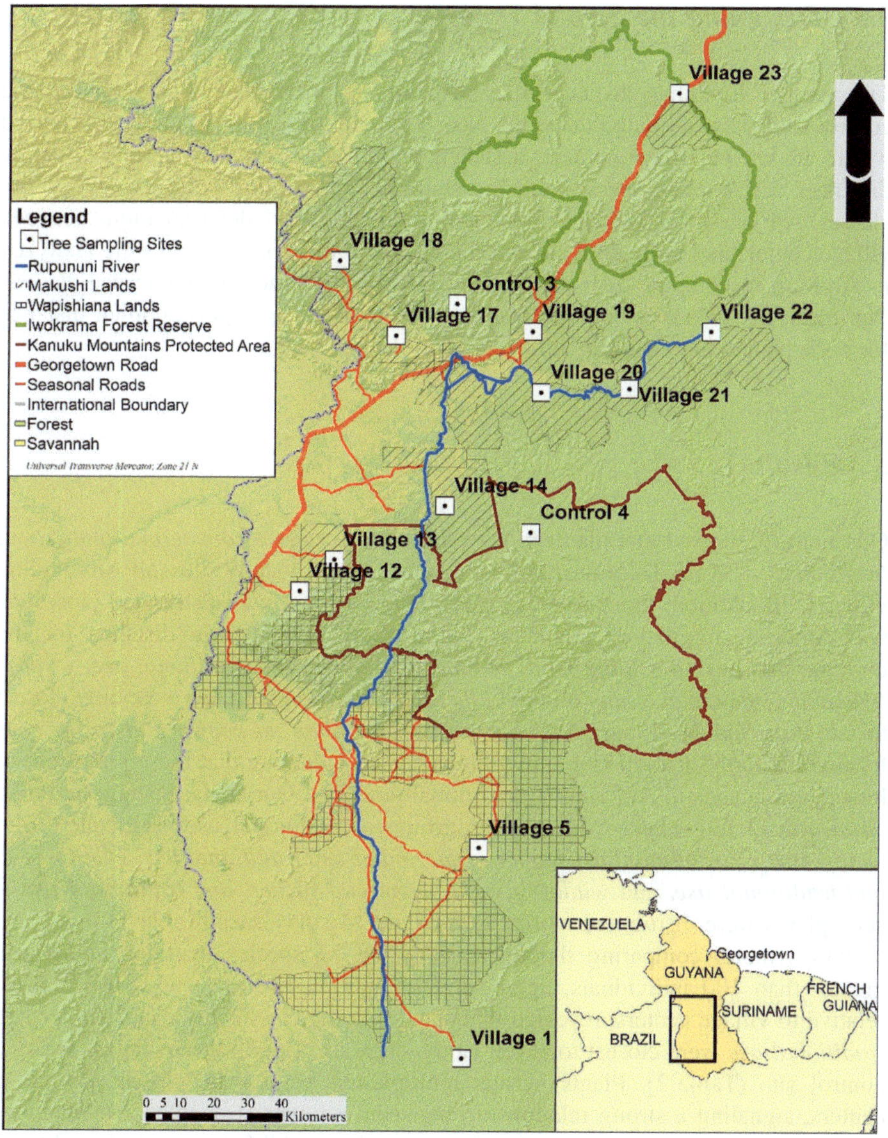

Fig. 2 The study area in Guyana

Given the strong relationship that exists between indigenous peoples and their forests, we expect plants associated with traditional uses for indigenous peoples to be distributed closer to village centers. Because commercial timber is a relatively new endeavor within the indigenous landscape, we expected plants with this economic usage to be distributed farther away from village centers.

2.3 Designing the Spatial Dataset

The average Euclidian distance of each plant to Amerindian village centers and control sites was computed using the Near Tool® in ArcGIS. As sampling occurred within 12 km of village centers, a search radius of 12 km was used to compare distances across plant species and the study area by multiple-use and single-use classes. A Python tool was developed to automate distance determinations between village centers and individual plants. The results of this computation were analyzed in ArcMap to compare distance by multiple-use and other economic use classes. The mean distance was computed and assigned as the average Euclidian distance for each plant and economic value classes.

3 Results

Our analysis showed that plants in the multiple-use classes *commercial timber* and *traditional use* were distributed closest to four of the 14 study sites, all Amerindian villages. In addition, the single-use class, *wildlife food*, was distributed closest to two Amerindian villages (Table 1). Interestingly, the average distance for the multiple-use classes *wildlife food, commercial timber* and *traditional use*, *wildlife food and commercial timber* and *wildlife food and traditional use* were only closest to one Amerindian village. As we expected, the class *no known human use* was located the farthest away from any village center, signaling that these plants are of lesser significance to Amerindian traditional practices. Our analysis suggested that, when considering plants with some attribute of traditional uses associated with them—that is, ***traditional use***, *commercial timber* and ***traditional use***, *wildlife food and* ***traditional use,*** and *wildlife food, commercial timber, and* ***traditional use***—such plants were distributed closest to 8 of the 14 study sites (Table 1).

Furthermore, comparing the distance to the 73 species that had abundance greater than 100 individuals, aggregated to multiple-use and single-use classes, relative to village centers, revealed that species in the class *commercial timber* and *traditional use* were closest to seven study sites, six Amerindian villages, and one control site (Table 2). Plants in this multiple-use class were closest to village centers, signaling a strong relationship between indigenous peoples and plants of traditional importance. Plants in the single-use class *wildlife food* were found to be closest for three study sites, two Amerindian villages, and one control site, indicating the importance of species that support wildlife hunted by Amerindians to their villages. This analysis, by contrast to the one that looked at the entire sample, placed plants in the class *traditional use* farthest away from village centers (Table 2). However, when considering plants with some attribute of traditional use associated with them—that is, *traditional use, commercial timber and traditional use, wildlife food and traditional use*, and *wildlife food, commercial timber and*

Table 1 Average Euclidian distances based on the aggregated economic class types

Economic type	Commercial timber	Traditional use	Wildlife food	No known human use	Commercial timber and traditional use	Wildlife food and traditional use	Wildlife food and commercial timber	Wildlife food, commercial timber, and traditional use
Village 13	7,476	10,169	9,070	**5,873**	7,143	7,485	7,321	7,760
Village 14	8,082	9,224	**7,935**	7,943	8,040	8,221	8,746	80,889
Village 17	8,274	**3,430**	7,408	11,843	5,887	8,745	11,423	10,272
Village 18	6,396	6,249	**5,349**	6,925	6,468	6,715	7,823	7,421
Village 23	**6,379**	6,912	6,896	7,666	7,298	6,924	6,859	7,100
Village 22	6,640	6,848	4,683	5,621	**4,258**	6,888	4,761	7,003
Village 21	8,059	8,834	8,415	8,270	**6,894**	7,637	7,259	7,981
Village 19	7,077	7,125	6,330	7,361	7,454	6,599	**5,767**	6,964
Village 20	8,972	8,823	10,283	8,913	**7,835**	8,529	9,448	8,038
Village 12	10,430	10,189	9,868	10,218	9,688	**9,543**	10,490	10,477
Control 4	6,540	6,796	6,069	6,717	5,868	6,273	**5,732**	6,792
Control 3	7,376	**6,637**	6,790	8,067	7,110	7,295	7,868	7,105
Village 5	7,888	7,983	8,312	8,750	**6,292**	7,199	8,315	8,880
Village 1	6,781	7,168	9,170	7,118	7,004	6,946	7,095	**6,686**
Avg. dist.	7,598	7,599	7,613	7,949	**6,946**	7,500	7,779	7,898

Note: Numbers in bold indicate closest classes

Table 2 Average Euclidian distances based on individual species that had abundance greater than 100 individuals

Economic type	Commercial timber	Traditional use	Wildlife food	Commercial timber and traditional use	Wildlife food and traditional use	Wildlife food and commercial timber	Wildlife food, commercial timber, and traditional use
Village 13	7,125	10,726	**2,774**	3,921	7,566	3,582	6,107
Village 14	8,574	9,065	6,712	**6,193**	7,522	8,508	7,391
Village 17	5,095	6,297	8,410	4,655	**3,735**	7,809	4,060
Village 18	6,141	5,915	**3,833**	5,243	5,104	4,506	5,550
Village 23	**4,711**	5,078	4,826	5,573	5,033	5,653	5,942
Village 22	5,951	7,003	6,650	5,387	6,973	**3,866**	5,945
Village 21	7,337	5,413	6,005	**4,789**	5,764	5,590	6,005
Village 19	7,269	6,619	4,863	**4,191**	5,299	5,560	5,761
Village 20	7,417	**4,557**	4,846	5,278	7,996	4,760	7,286
Village 12	8,784	8,399	7,317	**4,961**	9,242	8,998	8,214
Control 4	6,277	6,238	**3,376**	4,654	5,275	6,139	5,655
Control 3	6,976	6,954	5,015	**4,454**	5,713	7,244	5,745
Village 5	7,555	7,709	6,909	**5,933**	7,521	7,899	6,625
Village 1	7,253	7,478	6,088	7,199	6,838	6,719	**6,082**
Avg. dist.	6,890	6,961	4,137	**5,174**	6,399	6,202	6,169

Note: Numbers in bold indicate closest classes

traditional use—such plants were distributed closest to 8 of the 14 study sites (Table 2).

4 Discussion/Conclusions

Our results suggest an agreement with the hypothesized von Thünen model, with plants having greatest value to indigenous peoples located closer to their homes (Tables 1 and 2). Two findings contained in our results with respect to plant distribution of higher traditional value were particularly informative. First, we found that plants with some form of traditional uses associated with them were distributed closest to village centers (see Tables 1 and 2). Second, our results showed that plants providing food for wildlife were the next closest overall to village centers. These two findings suggested that the distribution of plants can be attributed to the activities of indigenous peoples, favoring species that support traditional uses, including hunting. These findings suggest that indigenous peoples seek to invest less effort in moving across their landscape to access plants that support their livelihood activities. The close proximity of plants that provide food for wildlife to village centers is supportive of the view that less effort is required for traditional practices. Cummings (2013) and Roosmalen (1985) proposed that for wildlife species that are hunted by indigenous peoples, most depend on such plants for food, and Read et al. 2010) suggested that most hunting occurs within 6 km of village centers. Based on these findings, indigenous peoples appear to have had some influence on how the forest structure around them is shaped.

In terms of resource use conflict and the allocation of areas for payment for ES initiatives, perhaps our most important finding is that plants in the class *traditional use and commercial timber,* when taken as a category by itself, were distributed closest to Amerindian villages (Tables 1 and 2). This finding introduces the potential for resource use conflict. Removing plants in this class for logs, for example, has implications for traditional practices. Given our results, extractive industries and REDD + -related initiatives should be located away (6–8 km) from village centers to avoid potential resource use conflict, including traditional hunting. These findings also emphasize the importance of using spatial analysis methods to study plant distribution, because our results can complement oral history and archaeology in helping communities to plan resource extraction activities.

Our results (Table 1) also suggest that distance from village centersis important, with plants of lesser significance to indigenous people's livelihood practices distributed farther away. However, while our results suggest strong agreement with von Thünen's model and imply that Amerindian communities favor species with specific traditional/economic value, the presence of plants is also impacted by edaphic, hydrologic, and other factors. These factors were not considered in this project and will be the subject of future work.

This work represents a first analysis of the impacts of distance on the distribution of plants in the Rupununi landscape. As observed in hunting practices in the

Rupununi (Read et al. 2010), our results suggest that the proximity of plants with higher economic value to village centers is influenced by the ease with which people can move through the landscape to access ES. Similar to von Thünen's consideration of transportation costs determining the nature of activities that occur closer to a town's center (Hall 1966), our analysis suggests that the ease of movement through a landscape impacts where plants of traditional value are located. Because factors such as elevation and plant density influence indigenous peoples' movement within a landscape, our finding that the most important plants are located closest to where people live was not surprising.

This initial probe considers only Euclidian distance, with the impact of factors such as soils and hydrology on plant distribution excluded. Amerindian village centers may have changed over time as communities moved closer to schools and other social services over the years; this also was not considered here and remains an area that future research will address. The importance of plant species abundance was not included in our analysis; rather, we focused on the collective economic value of plants in specific classes to define links between traditional uses and plant distributions. Considering individual plant species and their role in driving traditional practices and providing wildlife food, for example, may yield different outcomes. Such themes will be the subject of future research.

5 Future Work

von Thünen's model considers how transportation and production costs, such as land rent, determine the distribution of economic activities. Future work will consider the location of farm sites and whether village centers have moved over the years. In addition, the demand and supply of nontimber forest products based on the Amerindian economy will be considered in future work. Additionally, using population sizes of individual plant species relative to village centers and data about how Amerindians' transportation mechanisms, currently and in the past, were developed will be used to provide insights into how plants are distributed across the study area.

Acknowledgements Funding for this project was made possible by the US National Science Foundation BE/CNH Grant 0837531 led by Dr. Jose Fragoso. Any opinions, findings, and conclusions or recommendations expressed in this material are those of the author(s) and do not necessarily reflect the views of the National Science Foundation. Drs. Michael Tiefelsdorf, Jane Read, and Han Overman provided support at various stages in the development of this chapter for which we are grateful.

References

Colchester M (1997) Guyana: fragile frontier. World rainforest movement, forest peoples programme. Gloucestershire, UK

Cummings AR (2013) For logs, for traditional purposes and for food: identification of multiple-use plant species of northern Amazonia and an assessment of factors associated with their distribution. Doctoral dissertation, Department of Geography, Syracuse University. http://surface.syr.edu/etd/17

Cummings AR, Read JM, Fragoso JMV (2015) Utilizing Amerindian hunters' descriptions to guide the production of a vegetation map. Int J Appl Geospatial Res 6(1):118–142

Denevan WM (1992) The pristine myth: the landscape of the Americas in 1492. Ann Assoc Am Geogr 82:369–385

Forte J (1996) Makusipe Komanto Iseru: sustaining Makushi way of life. North Rupununi District Development Board, Guyana

Goeman M (2008) (Re)mapping indigenous presence on the land in native women's literature. Am Q 60(2):295–302

Hall P (ed) (1966) von Thünen's isolated state: an English edition of Der Isolierte Staat by Johan Heinrich von Thünen. Pergamon Press, Oxford

Kristensen TJ, Davis R (2015) The legacies of indigenous history in archaeological thought. J Archaeol Method Theory 22:512–542

Posey DA (1982) Keepers of the forest. New York Bot Gard Mag 6(1):18–24

Read JM, Fragoso JMV, Silvius KM, Luzar J, Overman H, Cummings A, Giery ST, Flamarion de Oliveira L (2010) Space, place, and hunting patterns among Indigenous peoples of the Guyanese Rupununi region. J Lat Am Geogr 9(3):213–243

Roosmalen M (1985) Fruits of the Guianan flora. Institute of Systematic Botany, Utrecht University Netherlands

Thomas TS (2007) A global analysis of tropical forests from a von Thünen perspective. Companion paper for the policy research report on forests, environment, and livelihoods. World Bank, Washington

Venables AJ, Limão N (2002) Geographical disadvantage: a Heckscher–Ohlin–von Thünen model of international specialization. J Int Econ 58(2):239–263

Walker R (2014) Sparing land for nature in the Brazilian Amazon: implications from location rent theory. Geogr Anal 46(1):18–36

World Climate Search and Classification Using a Dynamic Time Warping Similarity Function

Pawel Netzel and Tomasz F. Stepinski

Abstract We present a data-mining approach to climate classification and analysis. Local climates are represented as time series of climatic variables. A similarity between two local climates is calculated using the dynamic time warping (DTW) function that allows for scaling and shifting of the time axis to model the similarity more appropriately than a Euclidean function. A global grid of climatic data is clustered into 5 and 13 climatic classes, and the resultant world-wide map of climate types is compared to the empirical Köppen–Geiger classification. We also present a concept of climate search—an interactive, Internet-based application that allows retrieval and mapping of world-wide locations having climates similar to a user-selected location query.

Keywords Climate classification · Dynamic time warping · Climate search · Clustering · Similarity

1 Introduction

Global climate classification (CC) schemes discretize a multitude of local climates (LCs) across the earth's land surface in order to identify several typical climate types and to map their geographical extents. Thus, a CC scheme simplifies spatial variability of climates and makes analysis of their temporal change easier. CCs are used to study relationships between climate and other elements of environment, including the biota, hydrology, and agriculture. They also are used to provide visualization of global climate datasets and to illustrate climate change.

Classical CC schemes, such as, the popular Köppen–Geiger classification (KGC) (Köppen 1936; Kottek et al. 2006), rely on heuristic decision rules reflecting a body

P. Netzel · T.F. Stepinski (✉)
Space Informatics Lab, University of Cincinnati, Cincinnati 45221-0131, USA
e-mail: stepintz@uc.edu

P. Netzel
e-mail: netzelpl@ucmail.uc.edu

D.A. Griffith et al. (eds.), *Advances in Geocomputation*,
Advances in Geographic Information Science,
DOI 10.1007/978-3-319-22786-3_17

of environmental and geographical research to identify climate types and to delineate boundaries between them. The KGC has been widely used, and has become a de facto standard for global CC. However, from a modern perspective, the KGC appears to be rather arbitrary and not consistent with today's preference for data-oriented approaches to classification problems.

A data-oriented approach is feasible because of the availability of large, world-wide climate datasets. A climate dataset is a large collection of geographical locations for which climate variables are given. The coordinates of these locations constitute a geographic space; they are not used directly to find climate types. Climate variables constitute a climate space, and are used to find climate types through a clustering process. Clustering divides climate space into mutually exclusive and collectively exhaustive clusters in a way that maximizes intra-cluster homogeneity of LCs and their inter-cluster heterogeneity between cluster exemplars. Cluster exemplars are identified with climate types. Once clusters are found, their members are reassembled in geographic space, resulting in a map of climate types.

Previous studies about global classification of climates via clustering (Zscheischler et al. 2012; Metzger et al. 2012; Zhang and Yan 2014) relied on a generic, not a climate domain-specific, set of clustering techniques that includes: representing LCs by vectors of climate variables, and using Euclidean distance to calculate dissimilarity between LCs. However, a LC is an annually repeating pattern of weather conditions, and thus is more naturally represented by a time series rather than by a vector. A time series representation of a LC takes into consideration month-to-month sequencing information that vector representation does not provide. Given a time series representation of LCs, the dynamic time warping (DTW) distance (Berndt and Clifford 1994), rather than Euclidean distance, is the most logical choice for measuring dissimilarities between two LCs.

In this chapter, we present a new, data-oriented approach to a global CC, one that is based on a time series representation of LCs and on using the DTW as a distance function. We also introduce a new analytical tool—climate search—as a means of studying similarities of climates across the world. Climate search (CS) is an interactive, internet-based application that retrieves locations having climates similar to a user-identified query. It works using a paradigm of similarity search. This paradigm was previously used in a spatial domain for finding patterns of land cover similar to a query pattern (Stepinski et al. 2014), but here we apply it to a temporal domain for finding LCs (time series) similar to a query climate.

2 Data and Methods

In the following sections, we present input data and their relevant pre-processing steps, the method of calculating the similarity of LC and CC, and the method used to compare CC results.

2.1 Data Source

We use a gridded climatic dataset available from the WorldClim project (Hijmans et al. 2005). The following monthly variables are used: air temperature (T), maximum air temperature (T_{max}), minimum air temperature (T_{min}), and total precipitation (P). These are the only WorldClim variables available as monthly variables. All data are long-term averages calculated from measurements taken between 1950 and 2000. The 30 arc-second grid, having spatial extent of (180° W, 60° S) – (180° E, 90° N), is obtained by interpolating measurements from world-wide networks of climate stations. The geographic distribution of locations of these stations can be found in Hijmans et al. (2005). The original reference system of this dataset is WGS84 (EPSG:4326).

2.2 Data Preprocessing

WorldClim data are preprocessed differently, depending on application to either global CC or CS.

For CC, the data first were reprojected to the Mollweide projection (EPSG:54009). This allows clustering of cells having near equal areas (Usery and Seong 2001). We need equal area cells for evaluating similarity between different classifications (Cannon 2012). To reduce the size of the grid, we resampled the Mollweide grid to spatial resolution 75 km × 75 km, resulting in 213 × 482 grid cells, of which 23,979 represent land surface and the rest represent water (nodata).

For CS, the data first were reprojected to the spherical Mercator projection (EPSG:3857). By using this projection, the data and the results can be visualized in a web browser using Google or Bing maps as background for convenient reference. Reprojected data were resampled to a spatial resolution of 4 km × 4 km. This resolution is fine enough to distinguish between LCs, but, at the same time, coarse enough so that the time of search across the entire world is only about 40 seconds per query.

2.3 Variables and Their Normalization

A LC is defined as a 12-month-long time series constructed from climate data at a particular grid cell. The data first are corrected to remove a phase shift caused by sun position change during the year. With this correction, the clustering algorithm can find climate types that occur simultaneously in both northern and southern hemispheres. We use the following three climate variables:

- air temperature, T, a measure of average thermal conditions;
- precipitation, P, a measure of humidity of the climate and
- air temperature range, $dT = T_{max} - T_{min}$, a measure of thermal conditions, variability.

CC and CS can be conducted using either two variables (T, P) or three variables (T, P, dT). The two-variable classification (T, P) adheres to the KGC protocol of describing climate types using variables derivable only from the values of monthly averages of temperature and precipitation. By adding the third variable, the classification also takes into account an in-month variability of thermal conditions.

The ranges of the three selected variables are as follows:

- air temperature: (−50, 40) [°C],
- air temperature range: (0, 25) [°C] and
- precipitation: (0, 1550) [mm/month].

For the variables to contribute equally to the value of dissimilarity between two LCs, they need to be normalized to a common range of [0,1]. This is a standard way to normalize climate variables (see, e.g., Zhang and Yan 2014); we refer to such normalization as "global" and denote it by the letter g. Another problem is the distribution of the values of these variables. In particular, P has a distribution that is highly skewed toward large values. This means that an overwhelming number of normalized values of precipitation are very small, resulting in diminishing influence of P on the overall value of dissimilarity. To prevent this from happening, we introduce a "modified" normalization of P (denoted by the letter l):

$$P \leftarrow \begin{cases} \frac{P}{350}, & \text{if } P \leq 350 \\ 1, & \text{if } P > 350 \end{cases} \tag{1}$$

With this definition, the top 1 % of the highest values of P ($\geq$350 mm/month) are normalized to one, the remaining 99 % have a better behaving distribution of their values, and the influence of precipitation on the value of dissimilarity is restored.

A time series representing a LC is a multivariate time series. Each cell (location) is described by a time series of 12 two-dimensional (or three-dimensional) vectors V_i, $i, 1, \ldots, 12$:

$$LC = (V_1, V_2, ..., V_{12}),$$

where

$$V_i = \begin{cases} (T_i, P_i), & \text{in the case of two variables} \\ (T_i, P_i, dT_i) & \text{in the case of three variables} \end{cases} \tag{2}$$

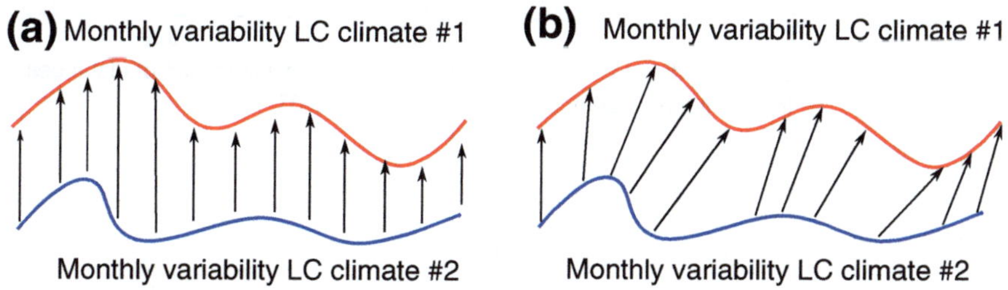

Fig. 1 Two ways of calculating distance between time series: **a** Euclidean distance and **b** DTW

2.4 Dissimilarity Measure

We use the DTW algorithm (Rabiner and Juand 1993) to calculate a dissimilarity (also referred to as a "distance") between any two LCs. DTW is widely used in calculating distances between time series.

The main idea of the DTW is to find an optimal alignment of two time series before calculating a distance between them. This alignment minimizes the sum of distances between the elements of the two series. The difference between Euclidean distance and the DTW distance (Fig. 1) is that DTW synchronizes two time series, thus finding two LCs similar even if there are some time shifts in their variability.

The matrix of distances between two series consists of Euclidean distances between each pair of their constituent vectors. For example, the $\{2, 3\}$ element of such a matrix is the Euclidean distance between a vector of climatic variables characterizing the month of February in the first series, and a vector of climatic variables characterizing the month of March in the second series. The general description of the DTW is as follows: for two time series v_1, v_2 with length N, the DTW algorithm finds alignments $s_1, s_2, ..., s_M$ and $t_1, t_2, ..., t_M$, such that

$$\begin{cases} s_1 = t_1 = 1, \\ s_M = t_M = N, \\ 0 \leq s_{k+1} - s_k \leq 1 & \text{for } k = 1, 2, ..., M-1, \\ 0 \leq t_{k+1} - t_k \leq 1 & \text{for } k = 1, 2, ..., M-1. \end{cases} \tag{3}$$

and

$$DTW(v_1, v_2) = \min_{s,t} \sum_{i=1}^{M} \|v_{1s_i} - v_{2t_i}\| \tag{4}$$

where $\|.\|$ is the Euclidean distance between the two vectors. Note that the length of the alignment sequence $M \geq N$ because the alignments do not have to be one-to-one. Thus, DTW finds a minimum value path crossing the matrix of distances from the

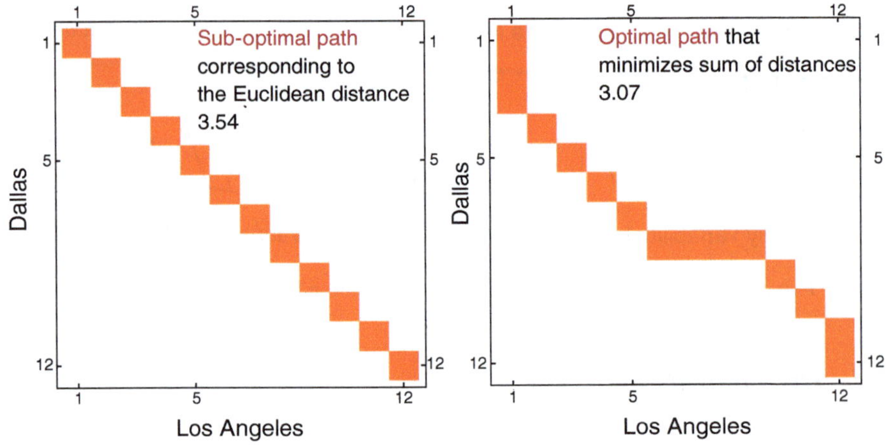

Fig. 2 Paths of summation of distances in a distance matrix: Euclidean distance (*left*) and the DTW distance (*right*)

upper left-hand corner to the lower right-hand corner of the matrix. This is illustrated in Fig. 2, with the left panel showing a path corresponding to the Euclidean distance (which can, but does not have to, be minimal), and the right panel showing the minimal path corresponding to the DTW distance. Finding an optimal time warping path is calculated efficiently using dynamic programming (Rabiner and Juand 1993).

Figure 3 shows an example of two locations for which a value of dissimilarity between their climates is very different, depending on whether it is calculated using Euclidean distance or the DTW. Two-dimensional time series are shown as an illustration. Each time series is depicted as a three-dimensional curve showing dependence of temperature and precipitation on time (month). The curves in both panels are identical, but the dotted lines, which show pairings of months used to calculate

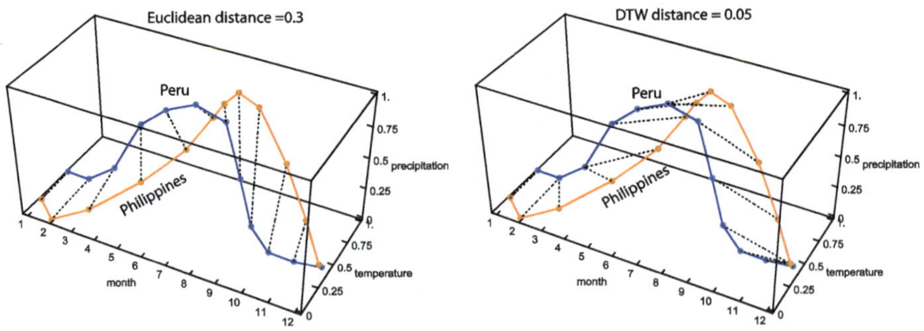

Fig. 3 Example of climates at two locations for which the value of dissimilarity calculated using the DTW measure is significantly different from the value of dissimilarity calculated using Euclidean distance

a dissimilarity value, are different. In the case of Euclidean distance, corresponding months are compared, for example, January to January, whereas in the case of the DTW, non corresponding months can be compared. As a result, the DTW dissimilarity is six times smaller than the Euclidean dissimilarity. According to Euclidean distance, the two climates are not very similar because a month-to-month comparison of temperature and precipitation shows significant differences. However, according to the DTW, the two climates are similar because their climatograms (blue and orange curves in Fig. 3 have similar topologies despite differences in a phase and a stretch. Therefore, human perception of climate is expected to be similar in the two locations.

2.5 Clustering Methods and CCs Comparisons

To explore how CC depends on the number of variables, normalization protocol, a dissimilarity function, clustering method, and a number of clusters, we calculate a large number of CCs corresponding to the combination of these parameters. There are two sets of variables ($[T, P]$ and $[T, P, dT]$), two protocols of normalizations (global and modified), two dissimilarity functions (Euclidean and DTW), two clustering methods (see the following paragraph), and two values for the number of clusters (see ensuing discussion). These result in 32 possible classifications plus the two KGC for two different numbers of clusters.

We used two different clustering methods: hierarchical clustering with Ward linkage (Ward 1963) and partition around medoid (PAM) (Kaufman and Rousseeuw 1987). Both methods use only the values of dissimilarities between LCs; thus, the first step in using them is to calculate a $23{,}979 \times 23{,}979$ matrix of dissimilarities between climates in all locations. We do not attempt to find an optimal number of clusters; rather, we assume either 5 or 13 clusters, which corresponds to the number of clusters at the first two levels of the KGC.

To quantify a degree to which two classifications partition the land surface into similar or dissimilar climatic zones, we follow Zscheischler et al. (2012) in calculating an information theoretic index called the V-measure (Rosenberg and Hirschberg 2007), which measures the degree of association between two different ways of delineating climatic zones. Briefly, the V-measure evaluates spatial association between two sets of climatic zones using two criteria: homogeneity and completeness. An association satisfies the homogeneity criterion if for all climatic zones in the first classification, each zone contains only locations that have a single climatic label in the second classification. An association satisfies the completeness criterion if for all climatic zones in the second classification, each zone contains only locations that have a single climatic label in the first classification. Only two identical classifications satisfy completely both criteria. In the case of different classifications, the V-measure calculates a degree to which these two criteria are satisfied. The V-measure is given by computing the harmonic mean of homogeneity and completeness measures. The range of V is [0,1], with 1 indicating a perfect correspondence between two classifications. We use a value (1-V) to quantify dissimilarity between two classifications (not to be confused with dissimilarity between two LCs).

3 Climate Classifications

We calculate a value of (1-V) between each pair of 34 classifications (32 classifications based on our clustering method and the two KGCs), resulting in a 34×34 matrix of distances between classifications. Figure 4 shows the heat map (Wilkinson and Friendly 2009)—a graphical representation of this matrix with rows and columns rearranged so the classifications most similar to each other are next to each other in the matrix. The black-to-white color gradient indicates values of (1-V) from small (similar classifications) to large (dissimilar classifications). Classifications are numbered from 1 to 34, the first two being the KGCs with 13 and 5 climate types,

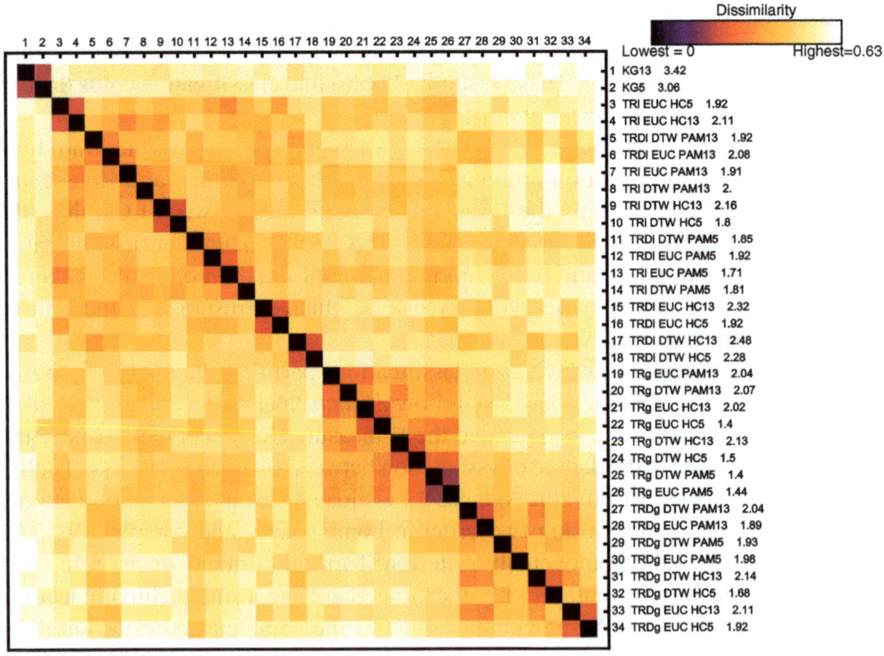

Fig. 4 A heat map illustrating a V-measure-based comparison between 34 different CCs. The color gradient indicates dissimilarities between pairs of classifications, from small to large. The numbers after classification names are the values of the Davies-Bouldin index (see Sect. 3 details)

respectively. The remaining classifications are labeled to indicate the choice of free parameters used; for example, TPdTl DTW PAM5 indicates a classification obtained using variables T, P, dT, modified normalization (l), the DTW dissimilarity function, the PAM clustering algorithm, and 13 clusters. Similar classifications are identified in the heat map as square reddish blocks located on the diagonal.

First, we note that for each protocol (the number of variables, normalization and dissimilarity function) hierarchical clusterings (including KGCs) into 5 and 13 classes are similar. Hierarchical clusterings subdivide more broadly defined climate types (those resulting from dividing a dataset into 5 clusters) into constituent, more narrowly-defined climate types (those resulting from dividing a dataset into 13 clusters). The result of this hierarchy is the high spatial correspondence between partitionings.

Second, we note that, for a given number of variables, classifications obtained using global (g) normalization are similar. This is especially true for classifications based on only two variables. The g normalization reduces the influence of P on the clustering of LCs, resulting in de facto delineation of thermal zones. No grouping is observed for classifications that use "modified" (l) normalization.

Objective determination of the "best" classification is not possible due to lack of validation data. We certainly cannot treat the KGC as the ground truth. Each classification delineates a climate dataset differently, based on the method used. The classification maps reflect definitions of what constitute a climate and how climate similarity is described. Various classifications can be consider the "best" depending on a given point of view or a specific application. From the point of view of homogeneity of climates within climate types, all clusterings are similarly homogeneous and more homogeneous than the KGCs. The degree of homogeneity of climates within a given clustering is measured using the Davies–Bouldin (DB) index (Davies and Bouldin 1979). The values of the DB index are given in Fig. 4 following the names of classifications. The smaller the value of the DB index, the better the homogeneity of climate types in a classification.

All 32 maps of climate types resulting from our classifications are not possible to show here. We have selected only four maps for side-by-side comparison. Figure 5 shows a comparison between the TPdTl DTW PAM5 and the KG5. The KGC shows the five well-established climate types: tropical A, arid B, temperate C, continental D, and polar E. The TPdTl DTW PAM5 shows five unnamed climate types obtained via clustering. A visual comparison of the two maps indicates that climatic types of these two classifications can be matched to each other, although spatial extents of matched types vary.

Figure 6 (top panel) shows a comparison between the TPdTl DTW PAM13 and the KG13. The KGC shows the 13 well-established climate types: tropical rainforest Af, tropical monsoon Am, tropical savanna Aw, arid desert BW, arid steppe BS, temperate dry summer Cs, temperate dry winter Cw, temperate without dry season Cf, continental dry summer Ds, continental dry winter Dw, continental without dry season Df, polar tundra ET, and polar frost EF. The TPdTl DTW PAM13 shows

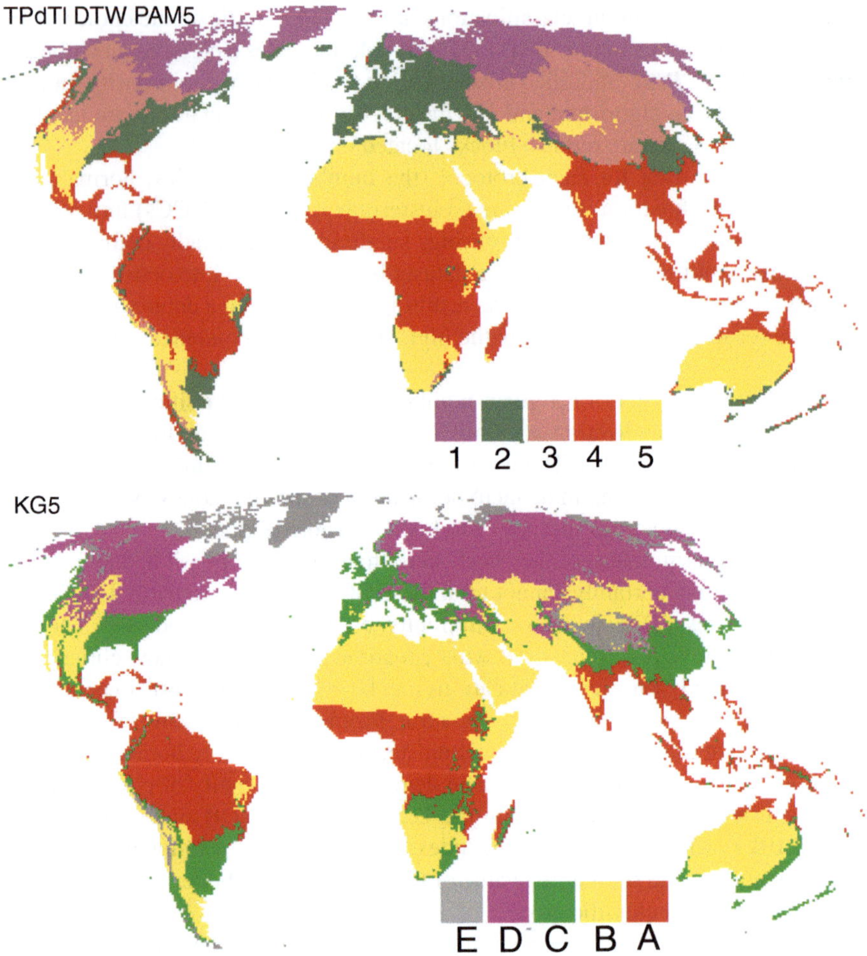

Fig. 5 A comparison of the clustering-based classification (TPdT1 DTW PAM5), having 5 climate types, with the KG5 classification. Each classification has its own legend to stress that cluster-derived climate types may have different meanings from KG climate type

13 unnamed climate types obtained via clustering. Of the 13 types resulting from clustering, 6 can be matched to the KGC climate types: 4→Af, 13→BW, 2→ET, 2→DF, 5→Dw, and 11→BS. The remaining clustering types cannot be matched to a KGC climate type.

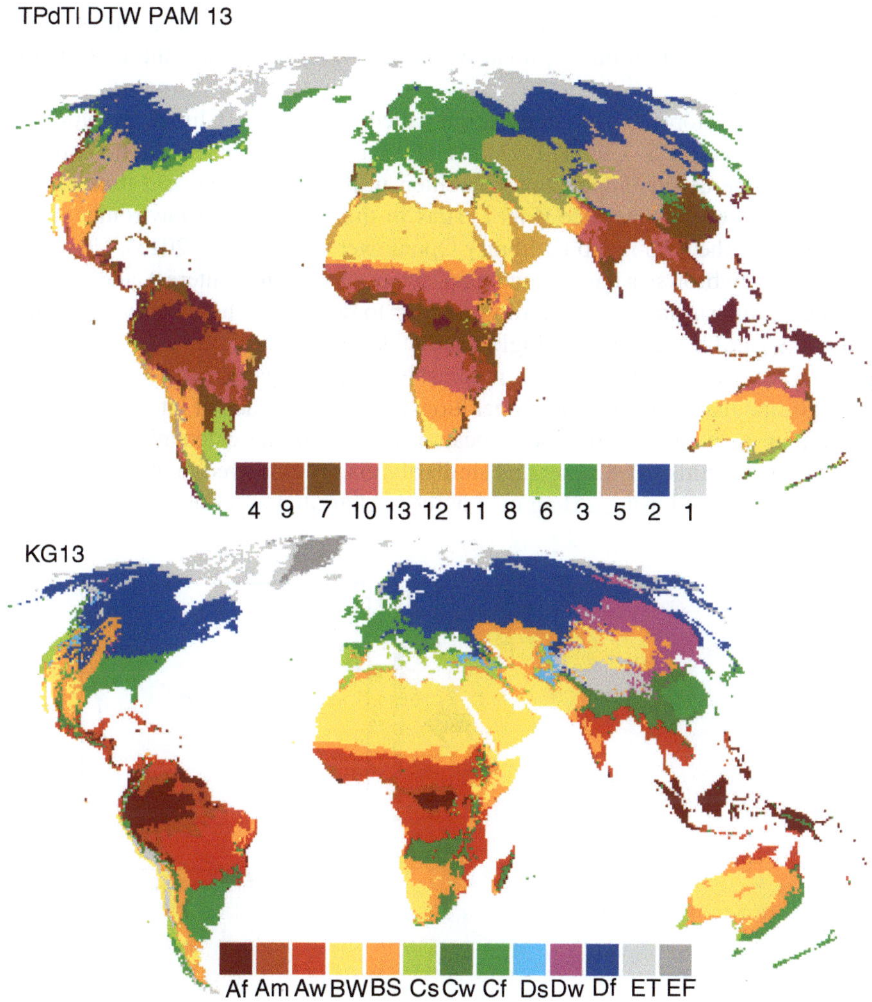

Fig. 6 A comparison of the clustering-based classification (TPdT1 DTW PAM5), having 13 climate types, with the KG13 classification. Each classification has its own legend to stress that cluster-derived climate types may have different meanings than KG climate types

4 Climate Search

CS is an interactive, online visualization tool for finding locations around the world having climate similar to a user-selected query. The main idea of a CS is to select a location, obtain a LC (represented by a time series) for this location, and ask an algorithm to find all locations around the world that have similar climates (on the basis of the DTW dissimilarity function). Thus, CS works on the principle of query-by-example. A CS uses three climate variables—T, P, dT—and the modified (l) normalization.

We have developed a web application called ClimateEx (Climate Explorer) to enable world-wide CS. ClimateEx is available at (http://sil.uc.edu). The core calculation engine of this application is written in the C language and uses parallel processing. The user interface (Fig. 7) runs in the web browser environment, and is powered by the web-mapping library OpenLayers 3 (Santiago 2015). ClimateEx allows a user to browse a background map, select a location of interest, and calculate a climate similarity map. It also allows a user to download the resulting similarity map in the GeoTIFF format. A single search takes about 40 s.

When working with ClimateEx, a typical work flow is as follows. A user browses a background map (pan and zoom operations) to locate a place of interest and clicks on the map to point to a precise a location that starts the calculations. Once the calculations are finished, a generated similarity map is displayed in the web browser that can be explored using a Bing map in the background as a reference. ClimateEx

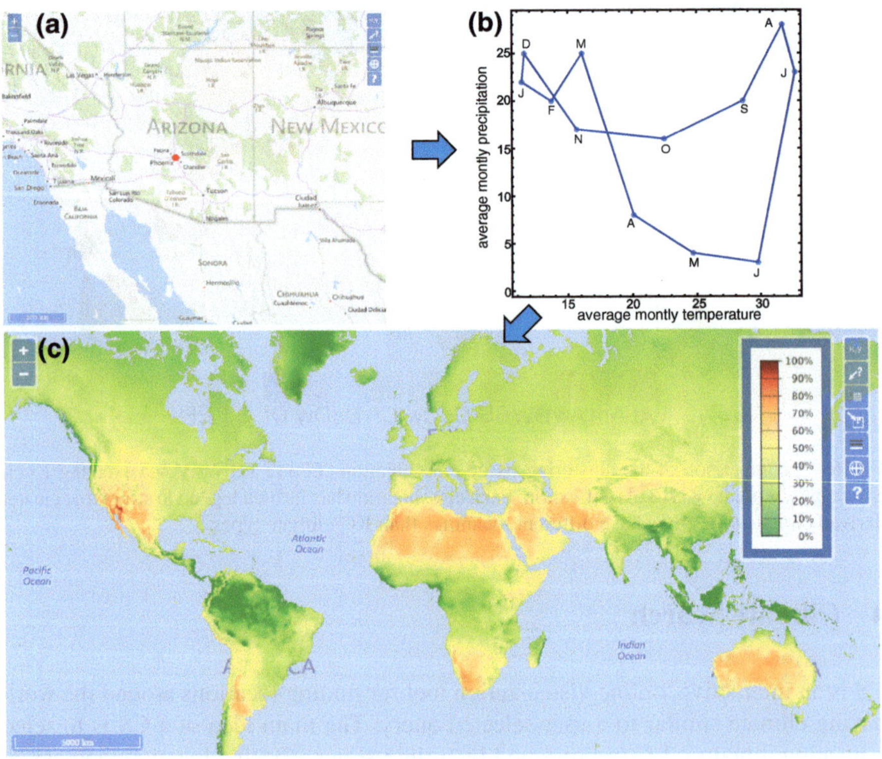

Fig. 7 A search for places with climate similar to that in Phoenix, Arizona. **a** Location of the query climate. **b** Circular climatogram of the query climate; months are labeled by their first letters. **c** A similarity map

can hold more than one similarity map at the same time to compare between different searches.

We demonstrate the utility of CS with two examples. The first example is the location of Phoenix, Arizona (Fig. 7), as a query produced a map of the world showing other locations having climates similar to that in the Phoenix area. A character of the LC in Phoenix (a query) is illustrated by a circular climatogram—a parametric graph with T on the x-axis and P on the y-axis. The closed polyline represents an annual cycle of these two parameters. The climatogram describes Phoenix climate as warm and dry, with the peak P in the summer months (monsoon). A red-to-green color gradient in the similarity map shows the degree of similarity to the query climate that goes large to small. The similarity map reveals that climate in Phoenix is similar (to various degrees) to LCs of North Africa, Australia's interior, the Arabian Peninsula, the Gobi desert, and the Kalahari desert, as well as a LC in Patagonia. These areas also are arid, and their annual cycle of T and P are similar to that in Phoenix except, for a possible time shift.

The second example is the location of Dallas, Texas (Fig. 8). The climatogram indicates that the LC in Dallas is warm and its P is, on average, about five times

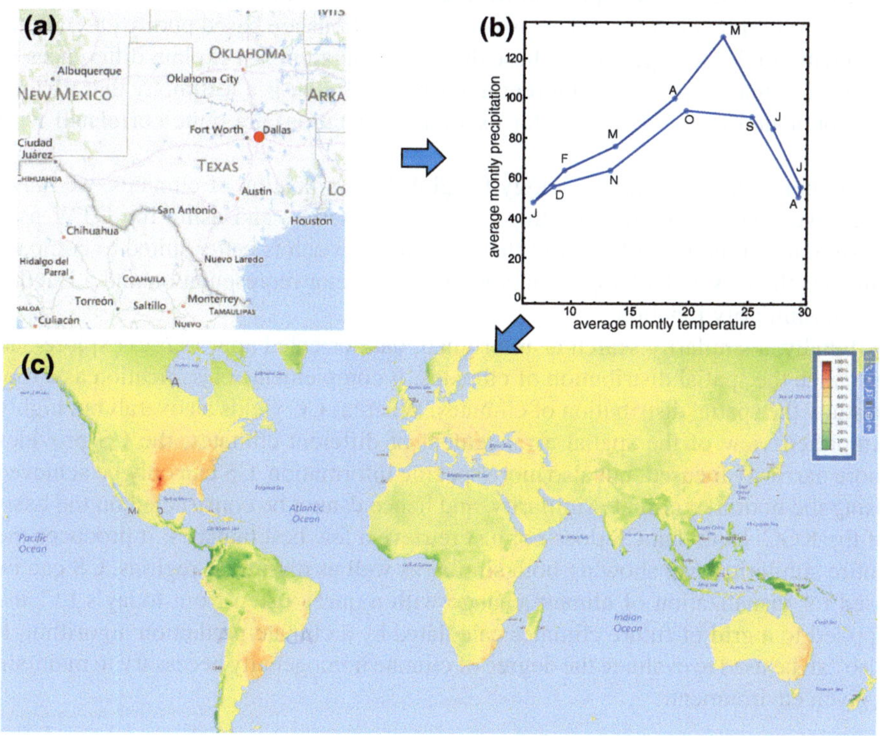

Fig. 8 A search for places with climate similar to that in Dallas, Texas. **a** Location of the query climate. **b** Circular climatogram of the query climate; months are labeled by their first letters. **c** A similarity map

higher then in Phoenix. A significant difference exists in the shape of the climatogram, indicating a different annual cycle of climate variables. The similarity map shows that climates similar to that in Dallas are found across the United States in the region that roughly coincides with "Tornado Alley" and also are found across South America in the region called "El Pasillo de los Tornados". Other regions in the world where similarity is 50 % (denoted by red in the map) also coincide with the occurrence of tornadoes. Users can compare the similarity map with the map of tornado regions at https://www.ncdc.noaa.gov/climate-information/extreme-events/us-tornado-climatology.

5 Conclusions

Our proposed data-oriented method of CC is a step forward in the effort to study spatial regionalization of world climates. There is no a single "best" classification of climate continuum. An advantage of our method over the KGC is that it is based on a data-driven methodology, and it produces classifications that are customized to a particular task. For example, the KGC uses vegetation as a proxy to discriminate between different climates, whereas our classifications are based purely on climatic data with no ties to vegetation. These differences in approach explain differences in maps of climate types. The different maps being similar is a testimony that various environmental factors (such as, for example, vegetation) are quite correlated with climate.

Within a clustering methodology for global classification of climates, we introduce two innovations: representing LCs as time series, and using the DTW as a dissimilarity function. We submit that such an approach is better suited to compare climates than a standard approach consisting of vector representation and a Euclidean dissimilarity function.

Finally, a similarity search is an original, data-oriented approach to explore and visualize the spatial distribution of climates. It complements classification as a tool to study the spatial distribution of climates. Whereas CC yields an overall but highly simplified view of the spatial arrangement of different climates, the CS provides more narrowly focused, but also more specific information. CS can only be achieved using the notion of climate similarity, and hence cannot be constructed on the basis of the KGC. Note that CS does not just retrieve a few best matches; it produces the entire similarity map showing both similar as well as dissimilar regions. CS can be used for visualization of climate change with a query taken from today's LC and applied to a grid of future climates calculated by a climate prediction algorithm. It also can be used to evaluate the degree of climate homogeneity necessary to maintain a given environment.

Acknowledgements This work was supported by the University of Cincinnati Space Exploration Institute, and by the National Aeronautics and Space Administration through grant NNX15AJ47G.

References

Berndt DJ, Clifford J (1994) Using dynamic time warping to find patterns in time series. KDD Workshop 10(16):359–370

Cannon AJ (2012) Köppen versus the computer: comparing Köppen-Geiger and multivariate regression tree climate classifications in terms of climate homogeneity. Hydrol Earth Syst Sci 16(1):217–229

Davies DL, Bouldin DW (1979) A cluster separation measure. IEEE Trans Pattern Anal Mach Intell 1(2):224–227

Hijmans RJ, Cameron SE, Parra JL, Jones P, Jarvis A (2005) Very high resolution interpolated climate surfaces for global land areas. Int J Climatol 25(15):1965–1978

Kaufman L, Rousseeuw P (1987) Clustering by means of medoids. In: Dodge Y (ed) Statistical data analysis based on the L1 norm and related methods. North-Holland, pp 405–416

Köppen W (1936) Das geographische system der klimate. In: Köppen W, Geiger R (eds) Handbuch der klimatologie. Gebrueder Borntraeger, Berlin, pp 1–44

Kottek M, Grieser J, Beck C, Rudolf B, Rubel F (2006) World map of the Köppen-Geiger climate classification updated. Meteorologische Zeitschrift 15(3):259–263

Metzger MJ, Bunce RGH, Jongman RHG, Sayre R, Trabucco A, Zomer R (2012) A high-resolution bioclimate map of the world: a unifying framework for global biodiversity research and monitoring. Glob Ecol Biogeogr 22(5):630–638

Rabiner L, Juand B (1993) Fundamentals of speech recognition. Prentice-Hall International Inc

Rosenberg A, Hirschberg J (2007) V-Measure: a conditional entropy-based external cluster evaluation measure. In: Joint conference on empirical methods in natural language processing and computational natural language learning, pp 410–420

Santiago A (2015) The book of OpenLayers 3. Theory and Practice, Leanpub, Victoria, BC

Stepinski T, Netzel P, Jasiewicz J (2014) Landex—a geoweb tool for query and retrieval of spatial patterns in land cover datasets. IEEE J Sel Top Appl Earth Obs Remote Sens 7(1):257–266

Usery EL, Seong J (2001) All equal-area map projections are created equal, but some are more equal than others. Cartogr Geogr Inf Sci 28(3):183–193

Ward JH (1963) Hierarchical grouping to optimize an objective function. J Am Stat Assoc 58:236–244

Wilkinson L, Friendly M (2009) The history of the cluster heat map. Am Stat 63(2):179–184

Zhang X, Yan X (2014) Spatiotemporal change in geographical distribution of global climate types in the context of climate warming. Clim Dyn 43(3–4):595–605

Zscheischler J, Mahecha MD, Harmeling S (2012) Climate classifications: the value of unsupervised clustering. Procedia Comput Sci 9:897–906

References

Attribute Portfolio Distance: A Dynamic Time Warping-Based Approach to Comparing and Detecting Common Spatiotemporal Patterns Among Multiattribute Data Portfolios

Jesse Piburn, Robert Stewart and April Morton

Abstract Frequently questions we ask cannot be answered by simply looking at one indicator. To answer the question asking which countries are similar to one another economically over the past 20 years is not just a matter of looking at trends in gross domestic product (GDP) or unemployment rates; "economically" encompasses much more than just one or two measures. In this chapter, we propose a method called attribute portfolio distance (APD) and a variant trend only APD (TO-APD) to address questions such as these. APD/TO-APD is a spatiotemporal extension of a data-mining algorithm called dynamic time warping used to measure the similarity between two univariate time series. We provide an example of this method by answering the question, Which countries are most similar to Ukraine economically from 1994–2013?

Keywords Dynamic time warping · Spatiotemporal · Similarity · High dimensional · Time series

J. Piburn (✉) · R. Stewart · A. Morton
Oak Ridge National Laboratory, 1 Bethel Valley Road, Oak Ridge, TN 37830, USA
e-mail: piburnjo@ornl.gov

R. Stewart
e-mail: stewartrn@ornl.gov

A. Morton
e-mail: mortonam@ornl.gov

R. Stewart
Department of Geography, University of Tennessee, Burchfiel Geography Building Room 304, Knoxville, TN 37996, USA

D.A. Griffith et al. (eds.), *Advances in Geocomputation*,
Advances in Geographic Information Science,
DOI 10.1007/978-3-319-22786-3_18

1 Introduction

Dynamic time warping (DTW) is a nonlinear data-mining algorithm used to measure the similarity between two time series by computing the minimum cumulative distance between them. The seminal paper on DTW by Sakoe and Chiba (1978) presents the algorithm as a means for automatic speech recognition, where the nonlinear time warping is able to handle robustly individual particularities associated with the speech patterns of different speakers, such as inflection, speed, and enunciation. DTW was first introduced to the knowledge discovery and data-mining community by Berndt and Clifford (1994). However, only after Keogh and Pazzani (2000) introduced a new algorithm for DTW, was it orders of magnitude faster, with no loss of accuracy, and did it become a viable approach for time series data mining. In the interim, DTW has seen continued refinement and variations for data mining (Keogh and Pazzani 2001; Rakthanmanon et al. 2013), and applications in several domains, including computer vision (Rath and Manmatha 2003) and musical information retrieval (Müller 2007). Although DTW has long seen application to time series, its transition into space-time has been negligible. Many in the geographic research community who are concerned with finding similar behaviors over both space and time use principal component analysis (PCA)-based methods. Demšar et al. (2012) furnish a recent and extensive review of the situation. These approaches often require assumptions that may or may not hold for spatiotemporal data, such as independence of observations and normality. Because of the potential for both spatial and temporal autocorrelation to exist, an approach that does not require these assumptions, such as DTW, may prove advantageous for certain applications. In this chapter, we propose a novel extension to DTW called *attribute portfolio distance* (APD), which estimates the spatiotemporal similarity of a portfolio (group) of attribute time series for geographic entities. Therefore, APD facilitates a holistic comparison of multiattributed temporal behaviors. For geocomputation, APD represents a novel response to highly generalized questions, such as what countries are most similar economically, where "economics" is defined by portfolio selection. We develop APD and a close variation known as trend only attribute portfolio distance (TO-APD). We apply both to a portfolio of World Bank economic data for European nations from 1994 to 2013 to show how APD can illuminate which nations are economically similar to the Ukraine when considering multiple attributes in space and time.

2 Dynamic Time Warping

Two vectors $\mathbf{x} = (x_1, x_2, \ldots, x_i, \ldots, x_n)$ and $\mathbf{y} = (y_1, y_2, \ldots, y_j, \ldots, y_m)$ can be arranged to form an $n \times m$ matrix, **C**, referred to as a cost matrix, where each (i,j) pair is the distance between x_i and y_j. Distance here is defined by a distance function, $\delta(.)$. Often this distance is Euclidean, but any appropriate measure of distance can be used. The goal is to find the warping path, $\mathbf{w} = (w_1, w_2, \ldots, w_k)$, that minimizes the distance between **x** and **y** in Eq. 1:

$$DTW(\mathbf{x}, \mathbf{y}) = arg\,min\,\mathbf{w} \sum_{k=1}^{p} \delta(w_k). \quad (1)$$

Finding the warping path that minimizes the distance between **x** and **y** is equivalent to finding the least cost path across the cost matrix, **C**, where $DTW(\mathbf{x}, \mathbf{y})$ is equal to the cumulative cost of the warping path (Fig. 1). For a more detailed description of DTW, see Berndt and Clifford (1994).

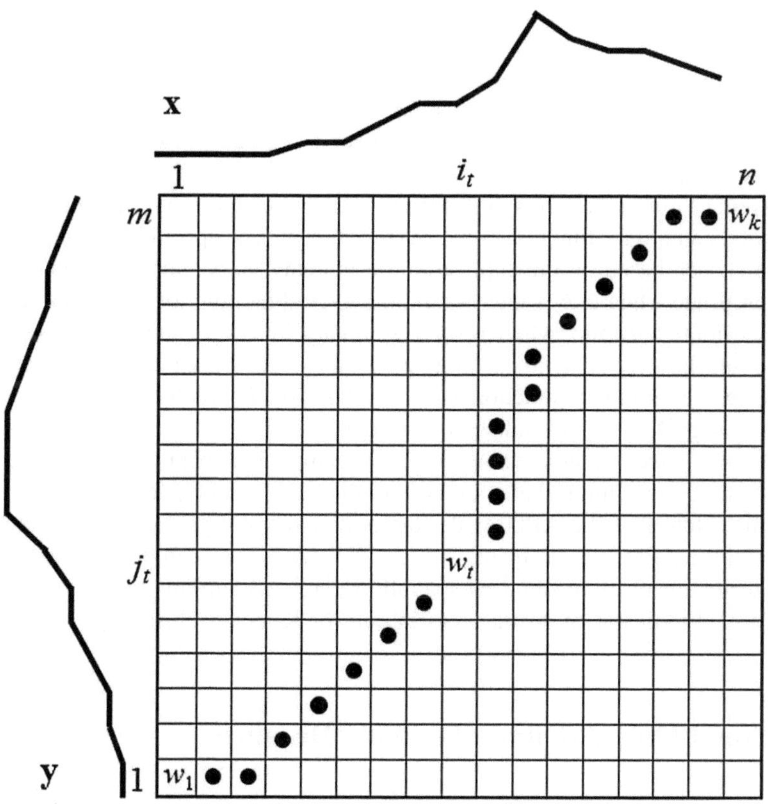

Fig. 1 Dynamic time warping

3 Attribute Portfolio Distance

Now that we have defined $DTW(.)$, we can extend it to the application of attribute portfolios. Given n countries and m attributes, an nx m distance matrix **D** can be formed, where each element (i,j) is equal to $DTW(\boldsymbol{\tau}_j, \mathbf{c}_{ij})$, where $\boldsymbol{\tau}_j$ is a vector corresponding to the values of attribute j for the chosen target country τ, and $\mathbf{c}_{ij}$ is a vector corresponding to the values of comparison country i for attribute j.

The resulting matrix, **D**, has country-centric rows and attribute-centric columns. At this point, each column is still in the units of the corresponding attribute. To allow for equal comparison across attributes, we must take an additional step before proceeding. To ensure attribute units are not influencing our distance measure, we now divide every element in a column by that column's corresponding root mean square (RMS), $\sqrt{\sum_{i=1}^{n} (x^2/n - 1)}$. The RMS is a measure of the magnitude of a varying quantity, and dividing each element in a column by its RMS effectively removes the attribute's original units, leaving only a measure of its variation, which allows each attribute to be equally compared regardless of its original unit of measurement.

To reduce our m measures of distance for each country to one summary measure, we apply a summary function, $\varphi(.)$, over each row. The specific summary function can be chosen based on the application of an analysis. A simple summation could be appropriate when one is interested in an overall measure of similarity across the entire attribute portfolio, whereas other summary functions, such as variance, allow one to see how consistent the similarities are across attributes in the portfolio. Applying the summary function $\varphi(.)$ row-wise on **D** results in a nx 1 vector,**d**, where $\mathbf{d}_i = \varphi(\mathbf{D}_{i,*})$ for all indices i from one to n.

We now have a single value for each comparison country, called the APD, representing how similar that country is to the target country for the given attribute portfolio. Note that the APD of the target country always equals zero, the lower bound of the APD measurement. These APD values now can be explored with the same techniques as any standard spatially referenced attribute, such as Moran's I, LISA (local indicator of spatial association), or K-means clustering.

4 Trend Only Attribute Portfolio Distance

The preceding APD values consider not only the temporal trend but also the magnitude of the trends. However, we can level the playing field and consider only the temporal trend by using TO-APD. It is not necessarily a modification of the APD methodology but rather a modification of the data that we use as input to

APD. To consider only the temporal trend of each attribute, we transform each country attribute vector, including the target country attributes, into standardized z-scores and perform APD on these transformed data. This transformation removes the differences in magnitude and allows for a comparison of just the temporal trends.

5 Application and Results

We apply these two methods by exploring the relationship between European countries and the Ukraine using a portfolio of selected economic indicators for the 20-year period of 1994–2013. The 34 comparison countries are listed in Table 1.

The selected attributes are a profile of each country's economic and population trends. These 23 attributes are listed in Table 2.

For our distance function $\delta(.)$, we use Euclidean distance, and our summary function $\varphi(.)$ is a simple summation.

Figure 2 portrays the results for our APD analysis, showing that Romania has the closest APD to the target country, the Ukraine, for our attribute portfolio, followed closely by Belarus and Lithuania. This means that, out of all the countries in our analysis, these three countries were the most similar in magnitude and temporal trend to the Ukraine for our attribute portfolio. Intuitively, this result makes sense given that these countries are similar in both the size and structure of their economies; that is, similar in both the magnitude and the temporal trend of the attributes in our portfolio. Also distinct is the division between east and west, with Russia being a large exception. This exception is likely due to Russia being much

Table 1 Countries being compared to the target country, the Ukraine

Comparison countries		
Albania	Germany	Portugal
Austria	Greece	Romania
Belarus	Hungary	Russian Federation
Belgium	Ireland	Serbia
Bosnia and Herzegovina	Italy	Slovak Republic
Bulgaria	Latvia	Slovenia
Croatia	Lithuania	Spain
Czech Republic	Macedonia, FYR	Sweden
Denmark	Moldova	Switzerland
Estonia	Netherlands	United Kingdom
Finland	Norway	
France	Poland	

Table 2 Names and sources of attributes included in the analysis

Attribute name	Source
Agriculture, value added (% of GDP)	World Bank
Birth rate, crude (per 1,000 people)	World Bank
Exports of goods and services (% of GDP)	World Bank
GDP (current US$)	World Bank
GDP growth (annual %)	World Bank
GDP per capita (current US$)	World Bank
GNI per capita, Atlas method (current US$)	World Bank
GNI per capita, PPP (current international $)	World Bank
GNI, Atlas method (current US$)	World Bank
GNI, PPP (current international $)	World Bank
Gross capital formation (% of GDP)	World Bank
Imports of goods and services (% of GDP)	World Bank
Industry, value added (% of GDP)	World Bank
Inflation, GDP deflator (annual %)	World Bank
Labor force, total	World Bank
Population ages 0–14 (% of total)	World Bank
Population ages 15–64 (% of total)	World Bank
Population, female (% of total)	World Bank
Population, total	World Bank
Services, etc., value added (% of GDP)	World Bank
Total reserves (include gold, current US$)	World Bank
Unemployment, total (% of total labor force) (modeled ILO estimate)	World Bank
Urban population	World Bank

GDP: Gross Domestic Product
GNI: Gross National Income
ILO: International Labour Organization
PPP: Purchasing Power Parity

larger in key attribute measures, such as GDP and population. Even if the Ukraine and Russia have the exact temporal trends for every attribute, Russia's much larger magnitude results in a large APD. To explore if this is the explanation, we use TO-APD to investigate the similarities of only the temporal trends. Figure 3 portrays the results.

When TO-APD analysis is applied to our attribute portfolio, with only the similarity of temporal trends being considered, Russia and Moldova emerge as the closest inTO-APD to the Ukraine. This means that, out of all the countries in our analysis, Russia and Moldova are the most similar in temporal trends to the Ukraine for our attribute portfolio. This finding highlights an important difference in APD and TO-APD: although neither Moldova nor Russia is similar to the Ukraine in the

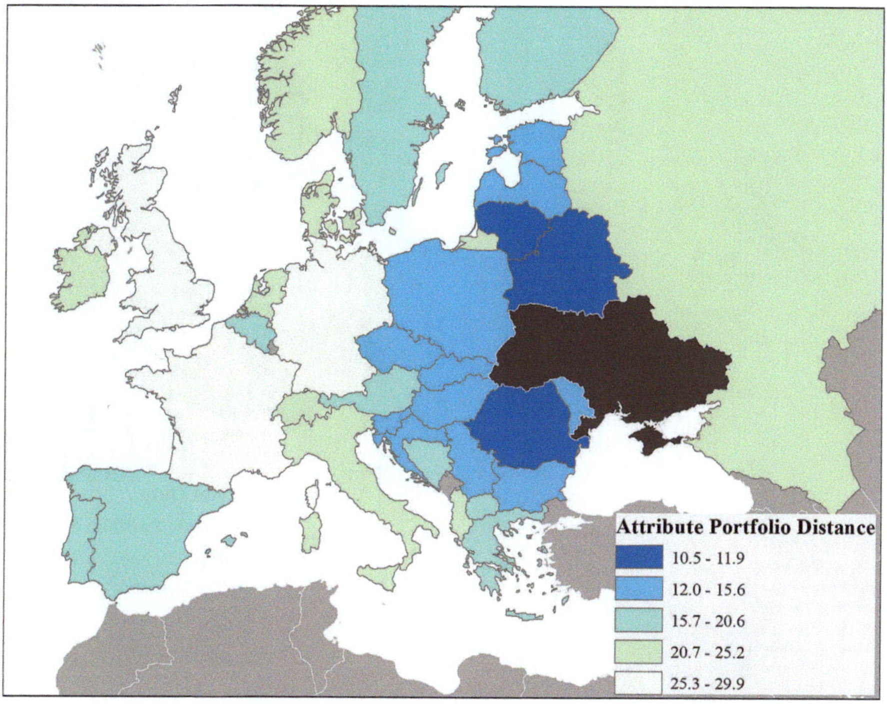

Fig. 2 Results of the APD analysis for a Ukrainian economic attribute portfolio

magnitude of its attributes, both are very close in their temporal trends. When considering only the temporal trends, the contrast between east and west appears even more pronounced than revealed by our APD results.

6 Summary

In this chapter, we introduce a spatiotemporal extension of the nonlinear data-mining algorithm DTW, which we call APD, and its close variant TO-APD. DTW provides a measure of overall similarity of two univariate temporal trends that is robust to individual particularities in each trend. We demonstrate a novel extension of this technique for high-dimensional spatiotemporal data. Both APD and TO-APD help modelers address questions such as Which countries are similar to one another economically? where economically is measured by a multitude of attributes, not just one, such as GDP. The results of APD/TO-APD are measures of distance that can be mapped and used as the input for further analyses, such as

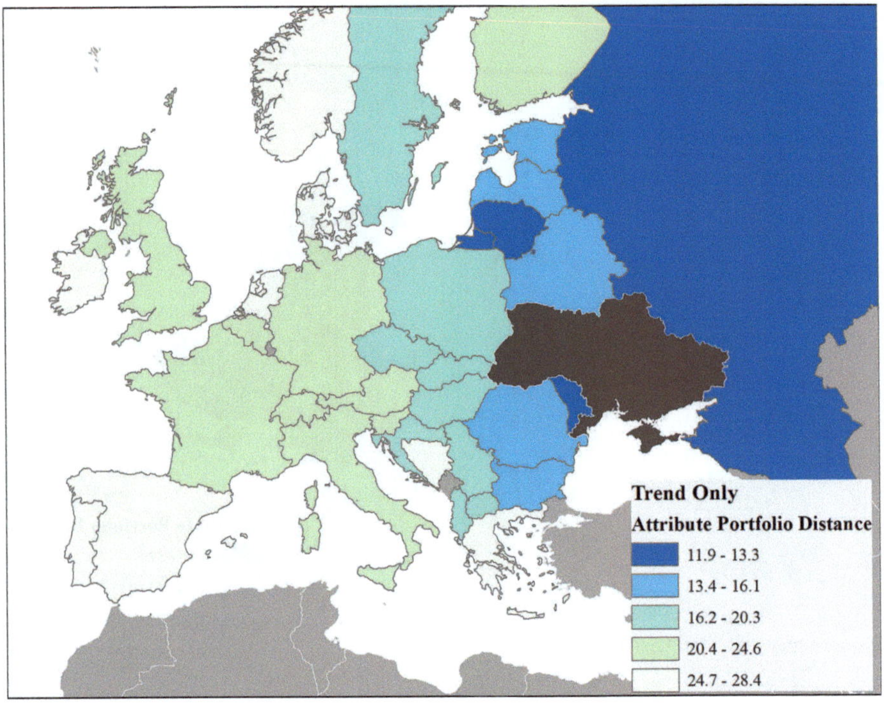

Fig. 3 Results of TO-APD analysis on a Ukrainian economic attribute portfolio

Moran's I or K-means clustering. We apply both APD and TO-APD to World Bank data to demonstrate how both are useful in detecting similar national behaviors for multiattributed portfolios. APD takes into account not only the trends but also the magnitude of those trends, and, as such, in our example, we found that Romania and Belarus are the most similar over time to the Ukraine in our economic portfolio analysis. When using TO-APD, which takes into consideration only the shape of the trend and not the magnitude of the values, we find that Russia and Moldova are the most similar economically to the Ukraine for our study period. This difference highlights important distinctions in the methods. Determining which method should be used of course begins with what question a researcher is addressing.

Acknowledgements This manuscript has been authored by employees of UT-Battelle, LLC, under contract DE-AC05-00OR22725 with the US Department of Energy. Accordingly, the United States Government retains and the publisher, by accepting the article for publication, acknowledges that the United States Government retains a nonexclusive, paid-up, irrevocable, worldwide license to publish or reproduce the published form of this manuscript, or allow others to do so, for United States Government purposes.

References

Berndt DJ, Clifford J (1994) Using dynamic time warping to find patterns in time series. In: KDD workshop, vol 10. Seattle, WA, pp 359–370

Demšar U, Harris P, Brunsdon C, Fotheringham AS, McLoone S (2012) Principal component analysis on spatial data: an overview. Ann Assoc Am Geogr 103(1):106–128. doi:10.1080/00045608.2012.689236

Keogh EJ, Pazzani MJ (2000) Scaling up dynamic time warping for datamining applications. In: Proceedings of the sixth ACM SIGKDD international conference on knowledge discovery and data mining. ACM, pp 285–289

Keogh EJ, Pazzani MJ (2001) Derivative dynamic time warping. In: SDM, vol 1. SIAM, pp 5–7

Müller M (2007) Information retrieval for music and motion, vol 2. Springer, Berlin

Rakthanmanon T, Campana B, Mueen A, Batista G, Westover B, Zhu Q et al (2013) Addressing big data time series: mining trillions of time series subsequences under dynamic time warping. ACM Trans knowl Disc Data (TKDD) 7(3):10

Rath TM, Manmatha R (2003) Word image matching using dynamic time warping. In: Proceedings of IEEE computer society conference on computer vision and pattern recognition, vol. 2. IEEE, pp II-521–II-527

Sakoe H, Chiba S (1978) Dynamic programming algorithm optimization for spoken word recognition. IEEE Trans Acoust Speech Sign Process 26(1):43–49

When Space Beats Time: A Proof of Concept with Hurricane Dean

Benoit Parmentier, Marco Millones, Daniel A. Griffith, Stuart E. Hamilton, Yongwan Chun and Sean McFall

Abstract In this research, we present an empirical case study to illustrate the new framework called "space beats time" (SBT). SBT is rooted in the expectation that predictions based on temporal autocorrelation typically outperform predictions based on spatial autocorrelation, except in the aftermath of abrupt disruptive events. Following such disruption scenarios, space is likely to outperform time, albeit often for a brief post event period. We illustrate the SBT concept by assessing the impact of Hurricane Dean on vegetation greenness using a remotely sensed spatiotemporal data series. We predict the normalized difference vegetation index (NDVI) using separate temporal-only and spatial-only models without the aid of covariates. We then compare each prediction model's performance before and after the hurricane event. Results suggest that SBT expected behaviors are valid in general terms but that some issues require attention. Our case study shows conspicuous SBT effects in the aftermath of the hurricane event in question, including increased performance in the geographic areas where the hurricane impact was more severe. In addition, we

B. Parmentier (✉)
Sustainability Solutions Initiative, Mitchell Center, University of Maine, Orono, USA
e-mail: benoit.parmentier@maine.edu

M. Millones
The College of William & Mary, Program of Public Policy, Williamsburg, USA
e-mail: mmmillones@wm.edu

M. Millones · D.A. Griffith · Y. Chun
School of Economic Political and Policy Sciences, The University of Texas at Dallas, Richardson, USA
e-mail: dagriffith@utdallas.edu

Y. Chun
e-mail: ywchun@utdallas.edu

S.E. Hamilton
Department of Geography and Geosciences, Salisbury University, Salisbury, USA
e-mail: sehamilton@salisbury.edu

S. McFall
Trout Unlimited Science Program, Emeryville, USA
e-mail: smcfall@tu.org

D.A. Griffith et al. (eds.), *Advances in Geocomputation*,
Advances in Geographic Information Science,
DOI 10.1007/978-3-319-22786-3_19

unexpectedly find that a more limited SBT pattern is present before the hurricane. This unanticipated pattern suggests that the presence of SBT features in an empirical study may vary, depending on the strength of a disruptive event as well as on the ability of a dataset and proxy variable to capture a disruptive event and its effects.

Keywords Spatiotemporal • Spatial statistics • Remote sensing • Natural disasters

1 Introduction

Many geospatial models combine both spatial and temporal components to improve their predictive capabilities. However, in the majority of cases, many researchers recognize that the temporal component outweighs the spatial component in terms of achieving predictive accuracy (Griffith 2010). While efficient and practical, combining spatial and temporal components does not always allow for the partitioning of the contribution of each component, or for the interpretation of shifts in their distinct individual contributions. Understanding when time or space is dominant becomes especially critical when observations are missing or scarce. For instance, many remote sensing products use gap-filling procedures that rely on the decomposition property by utilizing spatial or temporal information in lieu of the other when data are missing or unusable (Hall et al. 2010). In addition, Griffith (2013) and Griffith and Chun (2014) demonstrates the usefulness of spatiotemporal decomposition for the imputation of missing values and population forecasting. Beyond missing data imputation, understanding when spatial or temporal components are dominant may hint at changes in driving processes underlying the variability of social or natural systems.

When abrupt events, such as natural disasters or human interventions, occur, the temporal structure inherent in a system often is disrupted in such a manner that predictions using past trends from business-as-usual scenarios become inaccurate due to a reboot or reset in the system dynamics. Immediately after abrupt events, spatial models often outperform temporal models in their predictive ability during what we term the space beats time (SBT) window.

Discrete system disturbance events that result in SBT scenarios represent an opportunity to decouple spatial and temporal components, and to reconcile their relative dominance and trajectories. We examine the conditions under which spatial-only models become better predictors than temporal-only models in terms of model accuracy and performance. We also briefly outline a framework to analyze SBT using space-time data from a real-world event.

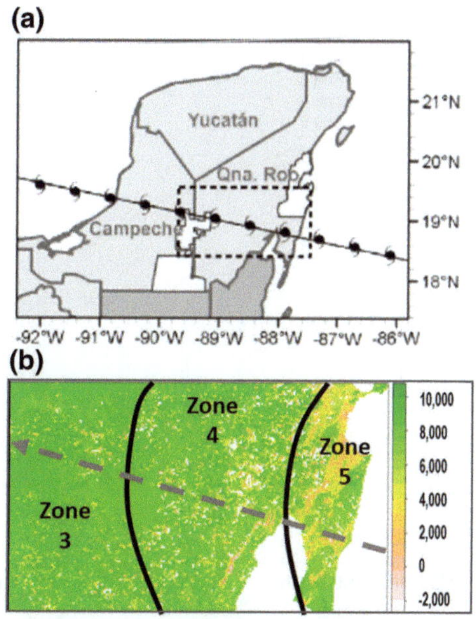

Fig. 1 Study area: **a** Study area location and hurricane path; **b** Saffir–Simpson (Schott et al 2012) magnitude zones overlaid on NDVI 16 days before Dean's landfall as background. Wind speeds diminished as the hurricane traveled inland. Zone 5 depicts the strongest winds. Greener values depict higher NDVI values (expressed in the −10,000, +10,000 range)

2 A Case Study: The Yucatan Peninsula—NDVI Before and After Hurricane Dean

In this case study, we illustrate the SBT concept by using models that predict NDVI before and after the arrival of Hurricane Dean. This hurricane made landfall along the southern Yucatán Peninsula on August 21, 2007, as a Category 5 hurricane affecting large swaths of forest cover. The study region occupies approximately 27,000 km^2 (Fig. 1) and has been documented as an area heavily impacted by Hurricane Dean (Rogan et al. 2011). Vegetation cover in the area is characterized by a mosaic of seasonally dry subtropical forests of varying stature and hydrophilic capacity, secondary successional vegetation, and agricultural landscapes (Vester et al. 2007).

3 Methods and Data

The general concept of the SBT framework is depicted in Fig. 3. The SBT methodology consists of four steps. First, we select the disruptive event (e.g., Hurricane Dean in T_0) and define its spatial and temporal characteristics (Figs. 1 and 2). Second, we compile a space-time dataset based on the measurement of a variable that acts as a proxy to describe the event of interest and the state of the system affected. We selected NDVI because it has spatial and temporal scales that

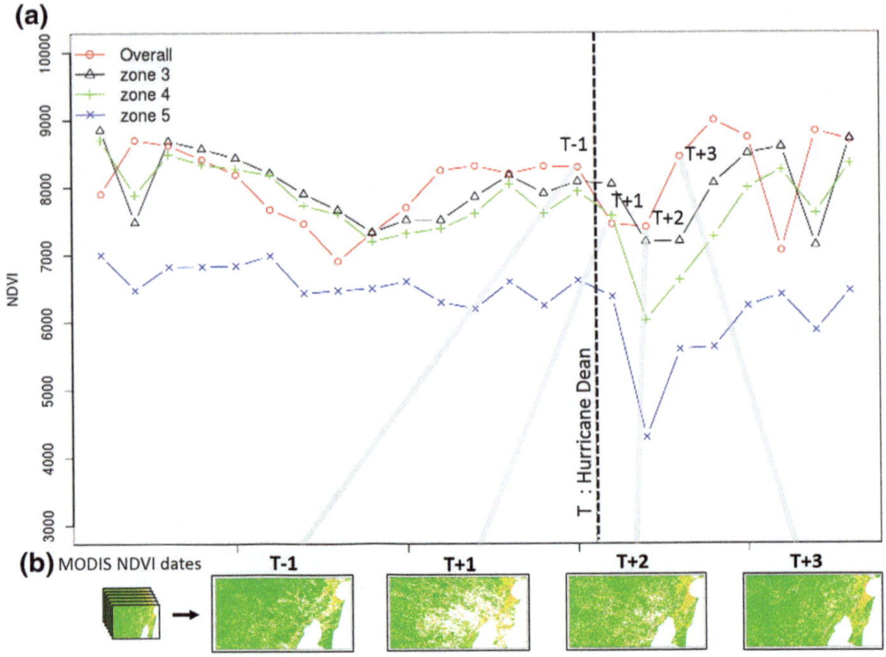

Fig. 2 MODIS NDVI time series data: **a** Temporal profile of the mean NDVI for the entire study area (*red circles*), zone 5 (*blue Xs*), zone 4 (*green plus signs*), and zone 3 (*black triangles*); and **b** NDVI images before (T − 1) and after (T + 1, T + 2, T + 3) the hurricane landing (T)

allow for the detection of the event and because it is known to be influenced by hurricanes. Third, we run prediction models with spatial and temporal components separately. Finally, we compare each model's performance using error metrics and interpret the results.

3.1 Data

To illustrate SBT, we use a measure of vegetation greenness, NDVI, derived from the moderate resolution imaging spectroradiometer (MODIS) sensor and its 1 km product (Huete et al. 2002). NDVI tiles matching the study area (H09V06 and H10V07) were mosaicked into a single raster stack for the 2001–2010 time period. We removed low-quality and cloud-affected observations using MODIS quality control flags. The processed dataset has 26,215 pixels per observation period, and 230 16-day time steps (Fig. 2b).

3.2 Methods: Temporal and Spatial Models

We use models that rely on neighboring observations in space and models that rely on neighboring observations in time to generate predictions. We use no covariates in the models because our aim is to demonstrate the usefulness of time or space used in isolation to predict NDVI values. For the temporal-only model, we use a variant of the autoregressive integrated moving average (ARIMA) model, which allows for the modeling of neighbors effects through autoregressive regression (AR) as well as temporal autocorrelation in the residuals and trends (Griffith 2010). For the spatial-only model, we use a spatial autoregressive model that includes the neighborhood effect via a lag variable (Anselin 2002). General model formulations are described in Eqs. (1) and (2).

$$y_{t+1} = \alpha y_t + \mathrm{e}, \tag{1}$$

where y is NDVI, t is a monthly time index, and e is the error.

$$y = \lambda W y + e, \tag{2}$$

where y is NDVI, and W is the spatial weight matrix, and e is error.

Processing and modeling were carried out in the open source computing R platform, using the raster and sp packages for data manipulation, spdep for the spatial autoregressive model, and the xts, zoo, and forecast packages for time series analyses (Bivand 2006; Hyndman and Khandakar 2007). We also examined maps of the residuals to assess visually where either model performs better.

3.3 Model Performance Assessment

To evaluate the accuracy of the predictions, we computed the mean absolute error (MAE) for both the temporal and the spatial models. MAE was chosen because of its ease of interpretation and model independence (Pontius et al. 2008; Willmott and Matsuura, 2005) We produced MAE temporal profiles to define the SBT window, which is the spatiotemporal window where spatial model error is substantially lower than temporal model error (Fig. 4a). Finally, we used wind speed zones (Fig. 2a) as proxies for potential impact zones to stratify model prediction performance with the expectation that SBT effect would be more evident where wind was stronger.

4 Results

We find that temporal models perform reasonably well before Hurricane Dean but poorly after this hurricane. Conversely, we find that the spatial models perform poorly before Hurricane Dean but well after the hurricane (Figs. 3 and 4b). This finding indicates the existence of a SBT window, although results do not exactly follow the idealized trajectory suggested in Fig. 1. Results suggest that the length of this window is short for this case study: within 32 days of the event, temporal models return to their pre-hurricane performance levels.

These findings suggest that the core of the SBT hypothesis holds for the case study but with various caveats. In addition, existing remaining challenges include to further generalize and improve the SBT methodology. Although we found evidence of a SBT effect, we observe that spatial models have slightly lower errors than temporal models even before Hurricane Dean. This unexpected pattern may reflect specific characteristics about the region and data used. The NDVI time series dataset contains considerable amounts of noise, with sizable date-to-date variation for each cell location. The high variability and noise make past observations poor predictors of future values. Therefore, temporal predictions display slightly higher

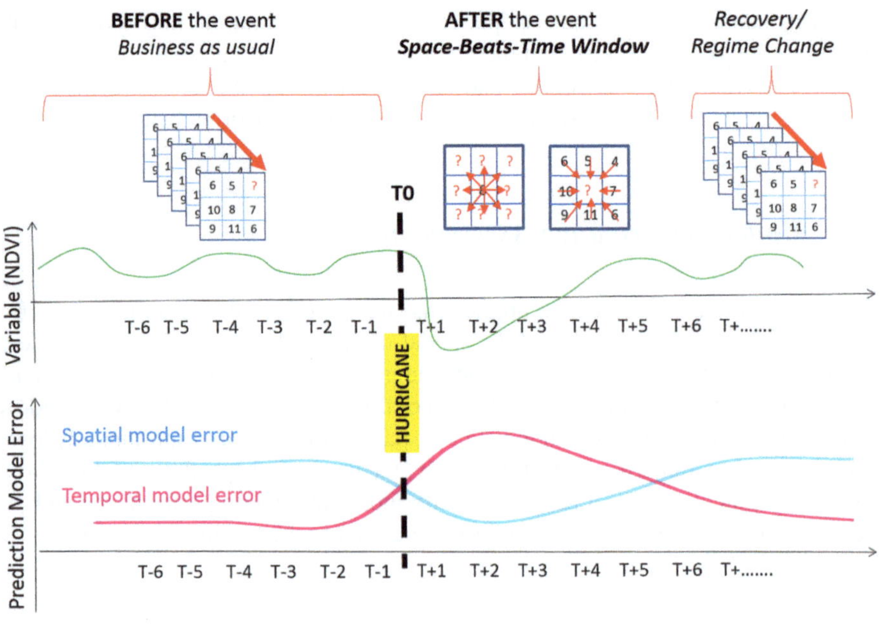

Fig. 3 A SBT event hypothetical trajectory based on a hurricane and NDVI. T labels indicate time steps with available observations before (T – 6 to T – 1) and after (T + 1 onward) after a disruptive event or intervention such as a hurricane (T0). SBT between T + 1 and T + 4, in terms of reducing model error, indicates a SBT window, after which the predictive power of time recovers

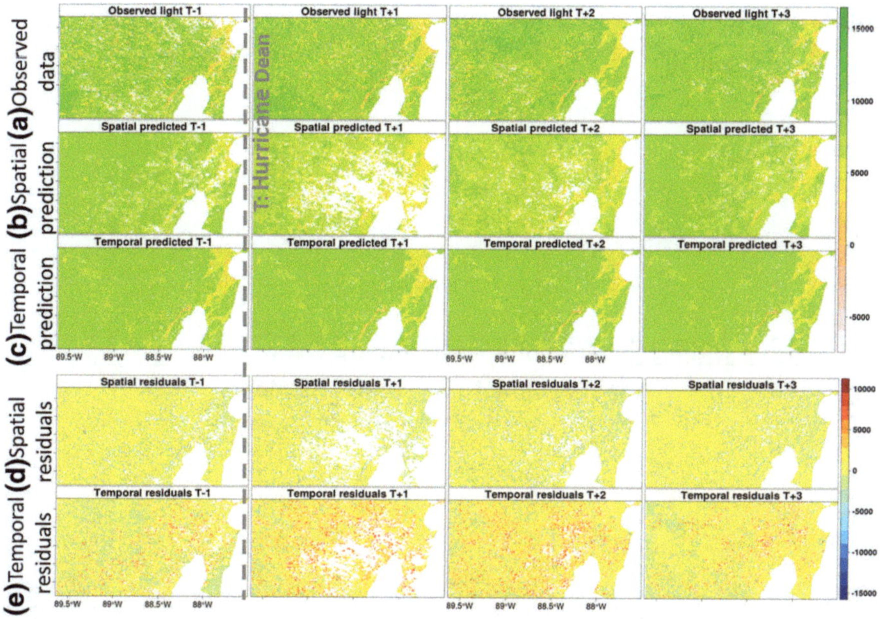

Fig. 4 Model predictions before (T − 1) and after the hurricane (T + 1, T + 2, T + 3). From *top* to *bottom*: **a** Observed NDVI data; **b** Spatial predictions; **c** Temporal predictions; **d** Spatial residuals; **e** Temporal residuals. Note that NDVI units are expressed in the –10,000 to 10,000 range

error levels than spatial predictions throughout the analyzed time period, even within the high wind zone closest to water (zone 5 in Fig. 5). Despite these unexpected results, we detect a visible SBT effect as illustrated by the large differences between MAE trajectories, particularly in wind zones 4 and 5 (Fig. 5).

In addition to data issues described, other potential reasons for the unexpected underperformance of the temporal model may lie in a mismatch between the 16-day sampling period of NDVI and the hurricane event under consideration, as well as in a mismatch between the measured variable and the phenomenon of interest. In this case study, NDVI recovered rapidly, but the 16-day measure failed to record changes in vegetation type because NDVI solely measures greenness. Hence, post-hurricane vegetation may differ substantially from the pre-storm vegetation, possibly obscuring the SBT window that occurred due to the sampling scale, although this appears not to be the case in this example. Thus, the recovery time and SBT detection are dependent on the magnitude of an event and on the sampling and measurement variable under consideration. Finally, the implementation of spatial models in available software currently is limited. Spatial autoregressive models typically are applied to socioeconomic datasets having relatively few observations stored in vector geographic information systems (GIS) format. In the future, spatial estimation methods should be improved to accommodate larger environmental

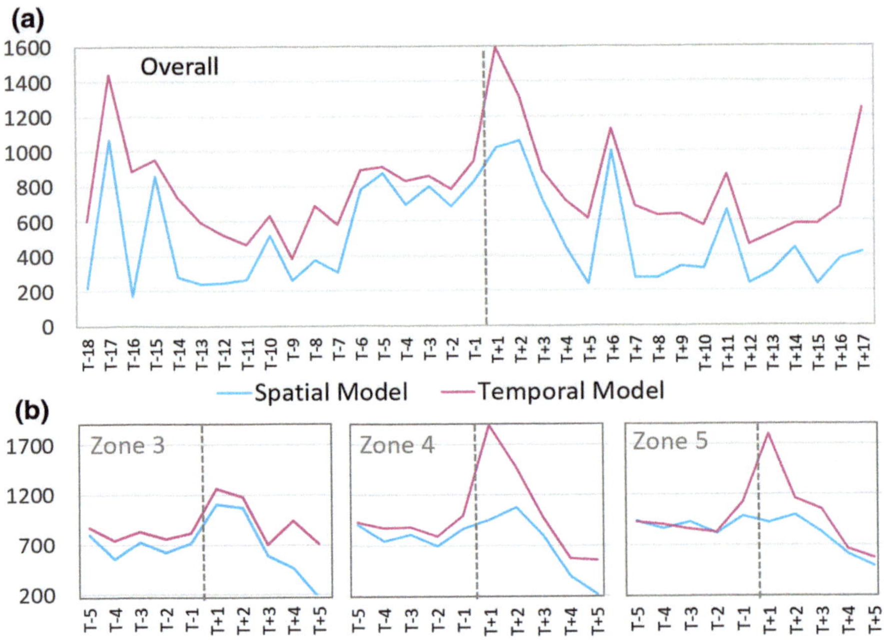

Fig. 5 Model performance based on MAE trajectories for the spatial (*blue*) and the temporal (*pink*) models with: **a** The overall MAE for the study area before and after the hurricane; **b** The MAE by hurricane wind zones before (T − 5 through T − 1) and after (T + 1 through T + 5) the event (*dashed black line*). The *horizontal axis* corresponds to time (16-day steps). The *vertical axis* corresponds to NDVI units

raster datasets. In the same manner, ARIMA time series modeling is a computationally complex and intensive task when performed on thousands of observations on desktop computer environments. To circumvent these issues, we parallelized the ARIMA processes to fit some 26,000 ARIMA models (one for each raster cell location) for every time step examined.

5 Conclusion and Discussion

Environmental and social systems present spatial and temporal variability that underlie particular processes unfolding in geographical areas. When abrupt events occur, the spatial and temporal components of variability may be affected in such a way that temporal model predictions using business-as-usual scenarios may fail because process drivers in the system may have changed. In this chapter, we present a new SBT framework, an approach designed to identify and exploit circumstances when spatial models are at their most useful in terms of prediction accuracy. We then demonstrate that spatial models outperform strongly temporal models immediately after a disruptive event in a period of time called the SBT window. In our case study, the SBT window is short. We posit that the length of the SBT window depends on the nature and magnitude of an event, as well as on the variable used to monitor the event. In particular, we find that NDVI recovers rapidly but that this recovery may hide changes in land cover and vary spatially, depending on the event and spatial heterogeneity of a landscape.

Future SBT research will incorporate additional case studies and variables that account for a variety of scenarios. In addition, we will assess SBT as a spatial data-mining tool aimed at the discovery of unknown historical events and as a tool to predict long-lag events that have yet to occur.

References

Anselin L (2002) Under the hood: issues in the specification and interpretation of spatial regression models. Agric Econ 27(3):247–267

Bivand R (2006) Implementing spatial data analysis software tools in R. Geogr Anal 38(1):23–40

Griffith DA (2010) Modeling spatio-temporal relationships: retrospect and prospect. J Geogr Syst 12(2):111–123

Griffith DA (2013) Estimating missing data values for georeferenced Poisson counts. Geogr Anal 45(3):259–284

Griffith DA, Chun Y (2014) An eigenvector spatial filtering contribution to short range regional population forecasting. Econ Bus Lett 3(4):208–217

Hall DK, Riggs GA, Foster JL, Kumar SV (2010) Development and evaluation of a cloud-gap-filled MODIS daily snow-cover product. Remote Sens Environ 114(3):496–503

Huete A, Didan K, Miura T, Rodriguez EP, Gao X, Ferreira LG (2002) Overview of the radiometric and biophysical performance of the MODIS vegetation indices. Remote Sens Environ 83(1):195–213

Hyndman RJ, Khandakar Y (2007) Automatic time series for forecasting: the forecast package for R. Department of Econometrics and Business Statistics, Monash University

Pontius RG Jr, Thontteh O, Chen H (2008) Components of information for multiple resolution comparison between maps that share a real variable. Environ Ecol Stat 15(2):111–142

Rogan J, Schneider L, Christman Z, Millones M, Lawrence D, Schmook B (2011) Hurricane disturbance mapping using MODIS EVI data in the southeastern Yucatán, México. Remote Sens Lett 2(3):259–267

Schott T, Landsea C, Hafele G, Lorens J, Taylor A, Thurm H, Ward B, Willis M, Zaleski W (2012) The Saffir-Simpson hurricane wind scale. National Weather Services, National Hurricane Centre, National Oceanic and Atmospheric Administration (NOAA) factsheet. URL: http://www.nhc.noaa.gov/pdf/sshws.pdf. Accessed 18 May 2016

Vester HF, Lawrence D, Eastman JR, Turner B, Calme S, Dickson R et al (2007) Land change in the southern Yucatán and Calakmul Biosphere Reserve: effects on habitat and biodiversity. Ecol Appl 17(4):989–1003

Willmott CJ, Matsuura K (2005) Advantages of the mean absolute error (MAE) over the root mean square error (RMSE) in assessing average model performance. Clim Res 30(1):79–82

Using Soft Computing Logic and the Logic Scoring of Preference Method for Agricultural Land Suitability Evaluation

Bryn Montgomery, Suzana Dragićević and Jozo Dujmović

Abstract A need exists for expanding agricultural lands due to increased demand for food production and security. Some regions can convert available land to agricultural land use. To evaluate available land for future agricultural production, geographic information systems (GIS) and GIS-based multicriteria evaluation (MCE) methods can be used to identify land suitability. This study proposes and implements the soft computing logic of the logic scoring of preference (LSP) method as an improved GIS-based MCE method for evaluating areas suitable for agriculture. Datasets from Boulder County, Colorado, USA, have been used to apply the GIS-based LSP method. The GIS-based MCE methods using additive scoring as simple aggregation has also been used to determine suitability and to compare with the LSP method. Results indicated that the LSP method can produce more refined agricultural land suitability maps. The proposed methodology can be used as an integral part of land use planning.

Keywords Soft computing logic · Logic scoring of preference · Multi-criteria evaluation · Geographic information systems · Agricultural land suitability evaluation

B. Montgomery · S. Dragićević (✉)
Geography Department, Simon Fraser University,
8888 University Drive, Burnaby, BC V5A 1S6, Canada
e-mail: suzanad@sfu.ca

B. Montgomery
e-mail: bmontgom@sfu.ca

J. Dujmović
Department of Computer Science, San Francisco State University,
1600, Holloway Avenue, San Francisco, CA 94132, USA
e-mail: jozo@sfsu.edu

D.A. Griffith et al. (eds.), *Advances in Geocomputation*,
Advances in Geographic Information Science,
DOI 10.1007/978-3-319-22786-3_20

1 Introduction

Multicriteria evaluation (MCE) is a well-known methodology for spatial decision making in the field of geography (Voogd 1983; Carver 1991; Jankowski 1995; Thill 1999; Malczewski 2004). Spatial decision making focuses primarily on urban land use (Wu 1998), environmental planning and management (Store and Kangas 2001), and agriculture (Ceballos-Silva and López-Blanco 2003). The most commonly used MCE methods supported by geographic information systems (GIS) software are simple additive scoring, multiattribute techniques, analytical hierarchy process (AHP), ordered weighted averaging (OWA), and outranking. The AHP and OWA methods are popular because of their ability to calculate weights and evaluate a range of decision-making alternatives (Saaty 1980; Yager 1988; Jiang and Eastman 2000; Boroushaki and Malczewski 2008). These methods are used in GIS and GIS-based software like IDRISI (Jiang and Eastman 2000); however, issues arise due to initial model assumptions and a lack of flexibility (Malczewski 2006; Dujmović et al. 2009). The commonly used GIS-based MCE methods usually rely on simple additive aggregation, which restricts the complete representation of human reasoning, to represent human decision-making processes. Through the use of hard and soft partial conjunction/disjunction, conjunctive/disjunctive partial absorption, and a range of logic conditions, the logic scoring of preference (LSP) method provides the necessary components to represent effectively human evaluation logic (Dujmović and De Tré 2011). Therefore, the main objective of this study is to integrate soft computing logic criteria, GIS, and the LSP method for application in agricultural land suitability evaluation.

2 Context of the Case Study

Due to increases in global population and its subsequent demand for food, allocating more land for increased food production is imperative. Evaluating the suitability of agricultural land is important in determining areas of future agricultural production. Because agricultural land suitability is influenced by a combination of socioeconomic and environmental factors, methods are needed to address a large range of criteria. Previous studies have used AHP and OWA methods with a relatively small number of criteria to evaluate land suitability (Hill et al. 2005; Chen et al. 2010; Akinci et al. 2013). Previous studies using MCE methods have limited the ability to incorporate a sufficient number of input criteria and completely represent observed human decision-making logic. A need exists for methods that can permit a detailed evaluation based on justifiable criteria in order to identify the suitability of land for conversion to agricultural use. For this reason, the soft computing logic and GIS-based LSP method is proposed to evaluate agricultural land suitability. A combination of socioeconomic and environmental datasets from Boulder County, Colorado, USA, as raster GIS layers at 50 m spatial resolution, have been used to demonstrate the applicability of LSP-based evaluation criteria. Regional land use and accessibility data were obtained from the Boulder County Department of Geographic Information Systems (Boulder County and Colorado 2015). Additional agricultural land use, economic, topological, and soil datasets were obtained from a variety of United States Department of Agriculture (USDA) databases and tools (NRCS 2009; USDA 2014a, 2014b).

3 The Logic Scoring of Preference Method

The LSP method is based on soft computing evaluation logic and is used for evaluation of any (typically large) number of input attributes. It provides logic operators that are observed in human reasoning (Dujmović et al. 2009). LSP is applied in computer science but recently has been linked with GIS and applied to geographic applications such as urban points of interest (Dujmović and Scheer 2010), residential home selection (Dujmović and De Tré 2011), and residential land use (Hatch et al. 2014).

Land suitability is not an objectively measurable physical property but a perception generated in a human mind after a careful expert analysis of the degree of satisfaction of stakeholder requirements (Shindler and Cramer 1999). The LSP method used in this study seeks to quantify this process without making it "objective." By definition, each evaluation is subjective; that is, based on human expertise. One assumption is that the parameters of LSP models (e.g., selection of attributes, the weights that describe degrees of importance, logic conditions) are selected by expert teams with the intention to reflect as precisely as possible the goals and interests of stakeholders. Quantification based on justifiable criteria is a way to minimize the impact of political interests that sometimes dominate decision making in this area.

The LSP method comprises three main components: an attribute tree, elementary attribute criteria, and an LSP aggregation structure. The attribute tree (Fig. 1) is created by hierarchical decomposition of suitability categories (Dujmović et al. 2010). For each suitability attribute, an elementary attribute criterion needs to be created that specifies individual requirements for that specific attribute reflecting stakeholder or decision-maker preferences (Hatch et al. 2014). Fuzzy suitability functions are used to standardize attribute criteria as well as to determine the level of satisfaction for each attribute criterion. As a result, elementary criteria generate attribute preference scores representing the degree of satisfaction of attribute criteria. For example, if the angle of slope is α and the suitability $S(\alpha) \in [0, 1]$ satisfies the conditions $S(\alpha) = 1,\ 0 \leq \alpha \leq \alpha_{\min}$ and $S(\alpha) = 0,\ \alpha \geq \alpha_{\max}$, then a simple elementary attribute criterion for slope could be $S(\alpha) = \min\{1, \max[0, (\alpha_{\max} - \alpha)/(\alpha_{\max} - \alpha_{\min})]\}$. All suitability scores are normalized to the interval [0, 1], where zero is unacceptable and one is perfect.

If a mandatory requirement is not satisfied, the resulting overall suitability is zero. For each point in an evaluated area, attribute suitability degrees are stepwise

Fig. 1 A sample attribute tree for agricultural land suitability; elementary attributes are classified as mandatory (+) or optional (−) based upon their need to be satisfied in an evaluation process

1. Agricultural Land Suitability
   - 1.1 Land Capability
     - 1.1.1 Slope (+)
     - 1.1.2 Elevation (-)
     - 1.1.3 Aspect (-)
     - 1.1.4 Soil Texture (+)
     - 1.1.5 Organic Matter (+)
     - 1.1.6 Depth to Restrictive Layer (+)
     - 1.1.7 Available Water (+)
     - 1.1.8 Drainage Class (+)
     - 1.1.9 Bulk Density (+)
   - 1.2 Climate
     - 1.2.1 Precipitation (+)
     - 1.2.2 Temperature (+)
     - 1.2.3 Frost Free Days (+)
     - 1.2.4 Water Retention (-)
     - 1.2.5 Flooding (-)
   - 1.3 Accessibility
     - 1.3.1 Location of Highly Capable Soils (+)
     - 1.3.2 Distance to Water for Irrigation (-)
     - 1.3.3 Distance to Open Space (-)
     - 1.3.4 Distance to Major Roads (+)
     - 1.3.5 Distance to Local Roads (+)
     - 1.3.6 Distance to Urban Areas (+)
     - 1.3.7 Distance to Markets (+)
   - 1.4 Management
     - 1.4.1 Designated Open Space (+)
     - 1.4.2 Zoning (+)
     - 1.4.3 Crop Type (+)
     - 1.4.4 Farm Product Consumption (+)
     - 1.4.5 Vacant Land (-)
   - 1.5 Economics
     - 1.5.1 Cash Crops (+)
     - 1.5.2 Annual Income (+)
     - 1.5.3 Price of Land (+)
     - 1.5.4 Economic Hazards (-)
     - 1.5.5 Land Renting (-)

aggregated using LSP aggregators until an overall suitability degree in the analyzed point is computed and the final suitability map is generated. The goal of LSP aggregators is to model logic relationships between suitability of attributes according to those relationships that are observable in intuitive human evaluation reasoning. Human reasoning is not based on only one aggregator, the simple arithmetic mean, which is frequently used in the context of AHP and some forms of OWA. Table 1 summarizes selected types of logic aggregators that are visible in human reasoning and are used in our criteria based on the LSP method.

A sample aggregation structure has been used for aggregating agricultural land suitability attributes, as shown in Fig. 2. The logic properties of all the aggregators are described in Table 1, and the mathematical implementation details can be found in Dujmović (2007). The aggregators denoted C-+, CA, C+-, C+, and C++ belong

Table 1 Logic properties of selected LSP aggregation operators

Name of aggregator	Logic properties and implementation of aggregator
Neutrality (A)	The weighted arithmetic mean of inputs $x_1, \ldots, x_n$. Fixed and balanced simultaneity and substitutability requirements. Low and high inputs have an equal opportunity to affect the output. This criterion is not satisfied only if all inputs are not satisfied. The criterion is completely satisfied only if all inputs are completely satisfied. Neither the mandatory/sufficient requirements, nor the adjustable degree of simultaneity/substitutability can be modeled using this aggregator. Implemented using normalized weights $W_1, \ldots, W_n$ that denote relative importance, as follows: $y = W_1x_1 + \ldots + W_nx_n,\ 0 \le x_i \le 1,\ 0 < W_i < 1,\ i = 1, \ldots, n,\ \ W_1 + \ldots + W_n = 1$
Soft partial conjunction (SPC)	Modeling the requirement for an adjustable low to medium degree of simultaneity that does not support mandatory requirements. All inputs affect output. Low input values have a stronger influence on the output than do high input values. To satisfy this criterion completely, all inputs must be completely satisfied. The criterion is not satisfied only if all inputs are not satisfied. Implemented using the weighted power mean, as follows: $y = (W_1x_1^r + \ldots + W_nx_n^r)^{1/r},\ 0 < r < 1,\ 0 \le x_i \le 1,\ 0 < W_i < 1,\ W_1 + \ldots + W_n = 1$
Hard partial conjunction (HPC)	Modeling the requirement for an adjustable high degree of simultaneity that supports mandatory requirements. Only one completely unsatisfied input is sufficient to completely not satisfy the entire criterion; so, it is mandatory to at least partially satisfy all inputs. If no input is completely unsatisfied, then all inputs affect the output. Low input values have a significantly stronger influence on the output than do high input values. To satisfy this criterion completely, all inputs must be completely satisfied. Implemented using the weighted power mean with negative exponents, as follows: $y = (W_1x_1^r + \ldots + W_nx_n^r)^{1/r},\ -\infty < r < 0,\ 0 \le x_i \le 1,\ 0 < W_i < 1$. The degree of simultaneity is adjusted by selecting the appropriate value of r
Conjunctive partial absorption (CPA)	Output depends on two asymmetric inputs: the mandatory input x and the optional input y. If the mandatory input is completely unsatisfied and has a zero value, then the output also is zero. If the mandatory input is positive, and the optional input is zero, then the output is positive. For a partially satisfied mandatory input, a higher/lower optional input can increase/decrease the output value with respect to the mandatory input for an adjustable reward/penalty. CPA (x, y) is implemented as a combination of the arithmetic mean A(x,y) and the HPC(x, y) as follows: CPA (x, y) = HPC$(x$, A$(x, y))$. So, CPA $(0, y) = 0$. Assuming $0 < x < 1$, we have CPA $(x, 0) = x$-*penalty* > 0, and CPA $(x,1) = x +$ *reward* < 1

to the hard partial conjunction (HPC) type with an increasing degree of simultaneity. A denotes the simple weighted arithmetic mean, and the compound aggregator consisting of a combination of A and CA aggregators is an implementation conjunctive partial absorption (CPA).

Other MCE methods do not support the nine types of logic aggregators that are observable in human reasoning and are therefore necessary if suitability is modeled in a way that is consistent with human decision making. These capabilities make the LSP method more effective than other MCE methods.

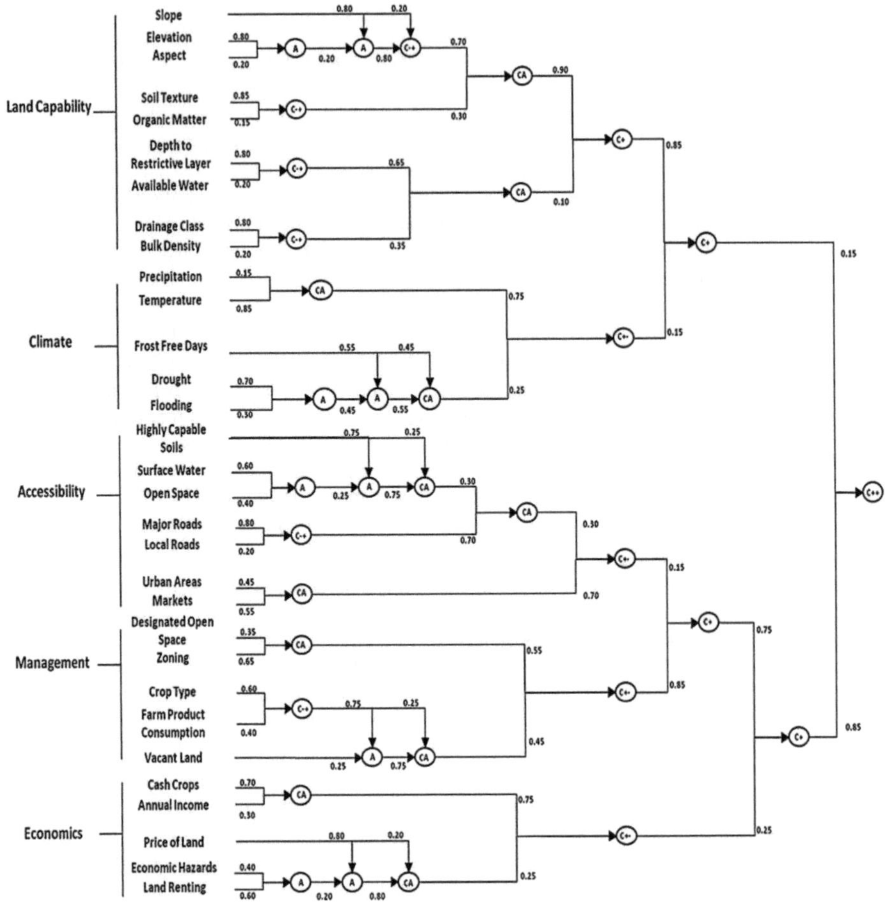

Fig. 2 The LSP aggregation structure

For all aggregators, a user must select two fundamental parameters: (1) formal logic parameters (degrees of simultaneity or substitutability) and (2) semantic parameters (degrees of importance expressed using weights). In this study, three types of aggregators have been used: A, HPC, and CPA, and were implemented using a weighted power means. The weights and the exponent *r*, introduced in Table 1, can be determined from a desired training set using the preferential neural network training tool ANSY[1] (Dujmović 1991).

The implementation of CPA using a weighted power means is shown in Fig. 2. The first aggregator is the arithmetic mean, and the second aggregator is a HPC.

---

[1]ANSY denotes ANalysis and SYnthesis of aggregation operators. This acronym is used for numerical analysis and synthesis of partial conjunction, partial disjunction, conjunctive partial absorption, and disjunctive partial absorption.

Table 2 The implementation of CPA using weighted power means

CPA based on weighted power means∇ = partial disjunction Δ = partial conjunction	x = mandatory input y = desired/optional input 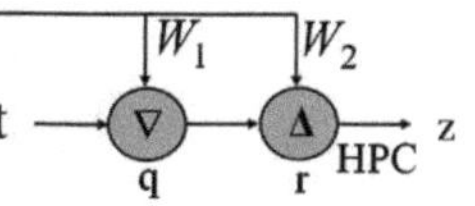 $z=\{(1-W_2)[W_1x^q+(1-W_1)y^q]^{r/q}+W_2x^r\}^{1/r}$, $0<W_1<1,\quad 0<W_2<1,\ 0\le x\le 1,\quad 0\le y\le 1,$ $-\infty\le q\le+\infty,\quad -\infty\le r\le+\infty,\quad 0\le z\le 1$
CPA derived from the desired values of mean penalty and mean reward	x = mandatory input y = desired input 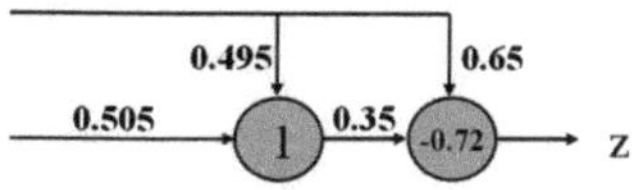 $z(x,y)=[0.35(0.495x+0.505y)^{-0.72}+0.65x^{-0.72}]^{-1/0.72}$ Penalty: $P=z(x,y)-z(x,0)$; Reward: $R=z(x,1)-z(x,y)$ Mean penalty = 25%, Mean reward = 10%

The HPC acts as an AND gate: if the mandatory input is zero, the output is unconditionally also zero. However, if the mandatory input is positive and the optional input is zero, then the output is positive. As shown in Tables 1 and 2, the output is reduced (with respect to the mandatory input) for a desired value of a penalty. Similarly, if an optional input has the highest value one, then the output is increased (with respect to the mandatory input) for a desired value of a reward.

All CPA parameters can be derived from a training set using ANSY. However, in practice, for convenience, users select desired values of a mean penalty and a mean reward (Table 2); then, all parameters of CPA can be easily found using tables (Dujmović 1979) or other software tools.

4 LSP Land Suitability Maps

In order to evaluate agricultural land suitability, an attribute tree was designed, followed by elementary criteria for evaluating agricultural land suitability. Additionally, the LSP aggregation structure (Fig. 2) was developed using the five categories presented in Fig. 1: land capability, climate, accessibility, management, and economics. Weights and logic aggregators were adjusted to reflect agronomic requirements. Figure 3 presents the resulting agricultural suitability map. The presented results show areas with different values of suitability for agricultural land use and production. The areas of excellent suitability were determined to be in close proximity to municipalities and corresponding to farmers' access to urban markets.

The LSP method is nonlinear and strictly follows observable properties of human evaluation reasoning. This is not the case with methods that use additive scoring or

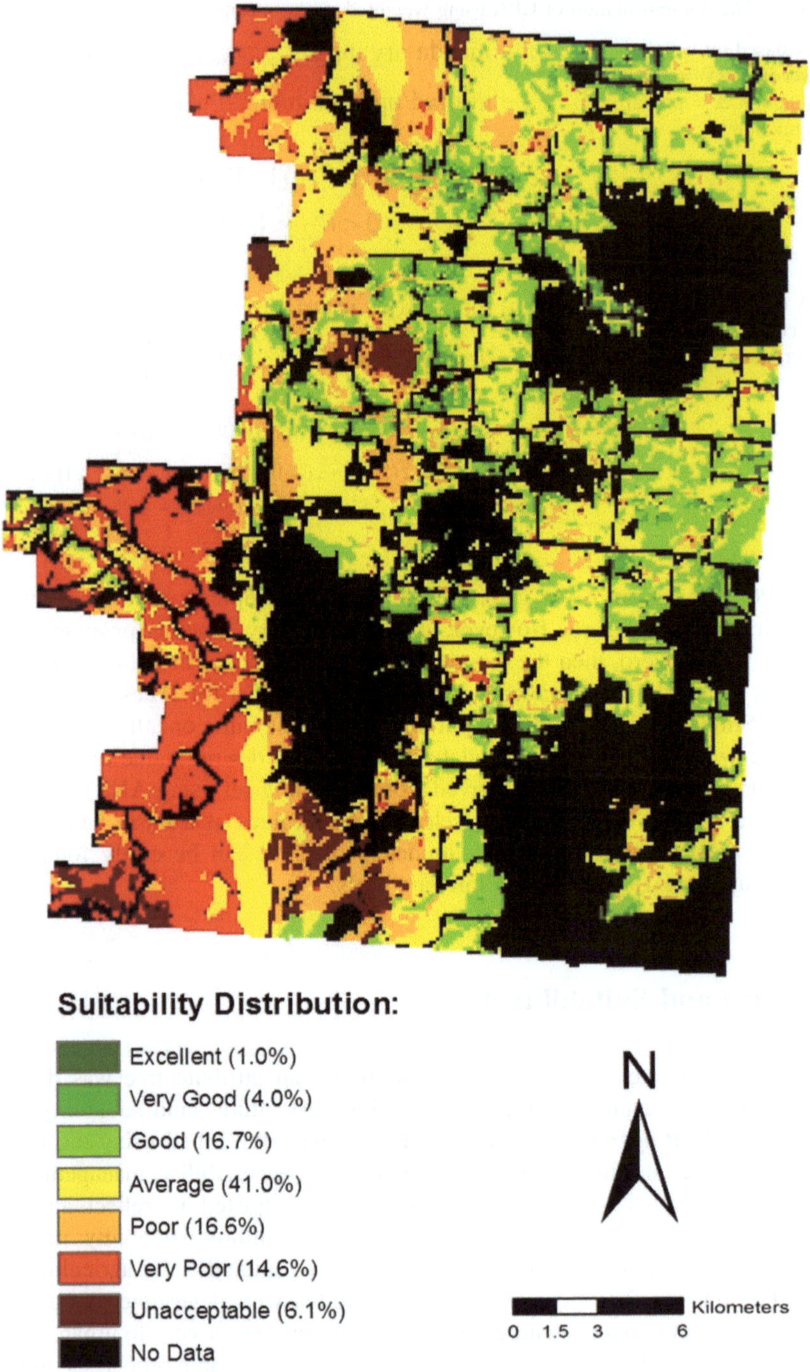

Fig. 3 The resulting LSP suitability map for agricultural land use

Table 3 Characteristic differences between LSP and SAS methods

Suitability degree	Method	
	LSP (%)	SAS (%)
Excellent	1	3.9
Very good	4	16.4
Unacceptable	6.1	0.8

do not support all nine types of aggregators that are used in human evaluation logic. The AHP- and OWA-based methods do not support the nine necessary aggregators and consequently may generate results that are inconsistent with human evaluation reasoning. To illustrate problems that are generated by methods that do not support aggregators used in human evaluation logic, an additional suitability map has been generated using the simple additive scoring (SAS) method and then compared with the LSP method. Table 3 summarizes typical differences between these two methods in the areas of extreme values of suitability.

The differences are obvious and consist of unjustified, exceedingly optimistic results from the SAS. On the one hand, the LSP method used conjunctive criteria and found that only 5 % of the analyzed area is very good or excellent for agriculture. On the other hand, the SAS model found that more than 20 % of area is very suitable for agriculture. That is a fivefold increase. The other side of the suitability spectrum shows similar anomalies. Because of mandatory requirements that agricultural production must satisfy (e.g., physical characteristics of land such as slope, soil texture, availability of water), the LSP model revealed that 6.1 % of terrain is fully unacceptable for agricultural production. However, in the case of SAS, the lack of support for mandatory requirements and HPC caused the reduction of unacceptable areas to only 0.8 %, which is more than a sevenfold decrease. In fact, SAS methods can produce zero overall suitability only if all inputs are zero. A small subset of positive inputs can produce a positive overall suitability, and that is the reason why, in the case of SAS, unacceptable results are so rare. Consequently, SAS and other methods that use additive aggregation regularly are prone to produce false positive results.

5 Conclusion

The primary advantage of the LSP approach is its flexibility in modeling logic conditions such as simultaneity, substitutability, mandatory and sufficient requirements, and asymmetric mandatory/optional and sufficient/optional aggregation. These features are visible in human intuitive reasoning but are not available in the case of additive methods. If additive aggregation is used in conjunction with AHP or OWA, then these methods produce unrealistic results and are inferior compared to the expressive power of the LSP approach.

The overall results obtained indicate that the LSP method is an improved MCE approach by its ability to incorporate a large number of inputs and precisely reflect the goals and interests specified by stakeholders. In particular, this study presents significant advantages of the LSP method compared to all versions of additive aggregation methods. The resulting LSP suitability maps offer results that can be used by various planners, land evaluators, farmers, investors, managers, governmental officials, decision analysts, and stakeholders, making the LSP method an integral part of land use management and planning decision procedures.

Acknowledgements This study was fully funded by a Natural Sciences and Engineering Research Council (NSERC) of Canada Discovery Grant awarded to the second author. The authors are thankful for valuable comments from two anonymous reviewers.

References

Akinci H, Özalp AY, Turgut B (2013) Agricultural land use suitability analysis using GIS and AHP technique. Comput Electron Agric 97:71–82

Boroushaki S, Malczewski J (2008) Implementing an extension of the analytical hierarchy process using ordered weighted averaging operators with fuzzy quantifiers in ArcGIS. Comput Geosci 34(4):399–410

County B, Colorado (2015) Geographic information systems (GIS) downloadable data. http://www.bouldercounty.org/gov/data/pages/gisdldata.aspx. Accessed July 2015

Carver S (1991) Integrating multi-criteria evaluation with geographical information systems. Int J Geogr Inf Syst 5(3):321–339

Ceballos-Silva A, López-Blanco J (2003) Delineation of suitable areas for crops using a multi-criteria evaluation approach and land use/cover mapping: a case study in Central Mexico. Agric Syst 77(2):117–136

Chen Y, Khan S, Paydar Z (2010) To retire or expand? A fuzzy GIS-based spatial multicriteria evaluation framework for irrigated agriculture. Irrig Drain 59(2):174–188

Dujmović JJ (1979) Partial absorption function. J Univ Belgrade, EE Dept Ser Math Phys 659:156–163

Dujmović JJ (1991) Preferential neural networks. In: Antognetti P, Milutinović V (eds) Neural networks: concepts, applications, and implementations, vol 2. NJ, Prentice-Hall, Englewood Cliffs, pp 155–206

Dujmović JJ (2007) Continuous preference logic for system evaluation. IEEE Trans Fuzzy Syst 15 (6):1082–1099

Dujmović JJ, De Tré G (2011) Multicriteria methods and logic aggregation in suitability maps. Int J Intell Syst 26(10):971–1001

Dujmović JJ, De Tré G, Dragićević S (2009) Comparison of multicriteria methods for land-use suitability assessment. In: Proceedings of the joint 2009 IFSA World Congress/EUSFLAT conference, Lisbon, Portugal, pp 1404–1409, 20–24 Jul 2009

Dujmović JJ, De Tré G, Weghe N (2010) LSP suitability maps. Soft Comput 14(5):421–434

Dujmović JJ, Scheer D (2010) Logic aggregation of suitability maps. In: 2010 IEEE international conference on fuzzy systems (FUZZ), Barcelona, July 18–23, pp 1–8

Hatch K, Dragićević S, Dujmović JJ (2014) Logic scoring of preference and spatial multicriteria evaluation for urban residential land use analysis. In: Duckham M et al (eds) GIScience 2014. Lecture notes in computer science, vol 8728, Springer, Switzerland, pp 64–80

Hill MJ, Braaten R, Veitch SM, Lees BG, Sharma S (2005) Multi-criteria decision analysis in spatial decision support: the ASSESS analytic hierarchy process and the role of quantitative methods and spatially explicit analysis. Environ Model Softw 20(7):955–976

Jankowski P (1995) Integrating geographical information systems and multiple criteria decision-making methods. Int J Geog Inf Syst 9(3):251–273
Jiang H, Eastman JR (2000) Application of fuzzy measures in multi-criteria evaluation in GIS. Int J Geogr Inf Sci 14(2):173–184
Malczewski J (2004) GIS-based land-use suitability analysis: a critical overview. Prog Plann 62 (1):3–65
Malczewski J (2006) GIS-based multicriteria decision analysis: a survey of the literature. Int J Geogr Inf Sci 20(7):703–726
Natural Resources Conservation Service (NRCS) (2009) Web soil survey. http://websoilsurvey.sc.egov.usda.gov/App/WebSoilSurvey.aspx. Accessed July 2015
Saaty TL (1980) The analytical hierarchy process. McGraw Hill, New York
Shindler B, Cramer L (1999) Shifting public values for forest management: making sense of wicked problems. West J Appl For 14(1):28–34
Store R, Kangas J (2001) Integrating spatial multi-criteria evaluation and expert knowledge for GIS-based habitat suitability modelling. Landscape Urban Plan 55(2):79–93
Thill JC (1999) Multicriteria decision-making and analysis: a geographic information sciences approach. Ashgate, New York
United States Department of Agriculture (USDA) (2014a) 2012 Census of agriculture: Colorado state and county data. In: 2012 Census of agriculture, vol 1, Chap 1: States level data. http://www.agcensus.usda.gov/Publications/2012/Full_Report/Volume_1,_Chapter_1_State_Level/Colorado/. Accessed July 2014
United States Department of Agriculture (USDA) (2014b) CropScape. http://nassgeodata.gmu.edu/CropScape/. Accessed July 2014
Voogd H (1983) Integrating multi-criteria evaluation with geographical information systems. Taylor & Francis, London
Wu F (1998) SimLand: a prototype to simulate land conversion through the integrated GIS and CA with AHP-derived transition rules. Int J Geogr Inf Sci 12(1):63–82
Yager R (1988) On ordered weighted averaging aggregation operators in multicriteria decision making. IEEE Trans Syst Man Cybern 18(1):183–190

Surgical Phase Recognition using Movement Data from Video Imagery and Location Sensor Data

Atsushi Nara, Chris Allen and Kiyoshi Izumi

Abstract The automatic recognition of surgical phases has strong potential to help medical staff understand individual and group patterns, optimize work flows, and identify potential work flow risks that lead to adverse medical events in an operating room. In this chapter, we investigate the performance of context recognition on the movement of operating room staff throughout their work environment, which was measured by imaging and tracking. We employed an optical flow algorithm and trajectory clustering techniques to extract movement characteristics of surgical staff from video imagery and time-stamped location data collected by an ultrasonic location aware system, respectively. Then we applied a Support Vector Machine to time-stamped location data, optical flow estimates, trajectory clusters, and combinations of these three data to examine the intraoperative context recognition rate. Our results show that the integration of both video imagery and location sensor data improves context awareness of neurosurgical operations.

Keywords Location sensor data · Context awareness · Moving object · Optical flow · Trajectory clustering

---

A. Nara (✉) · C. Allen
Department of Geography, San Diego State University,
5500 Campanile Dr, San Diego, CA 92182-4493, USA
e-mail: anara@mail.sdsu.edu

C. Allen
e-mail: allen14@rohan.sdsu.edu

K. Izumi
Graduate School of Engineering, University of Tokyo, 7-3-1 Bunkyo-ku, Hongo
Tokyo 113-8656, Japan
e-mail: izumi-sec@socsim.org

D.A. Griffith et al. (eds.), *Advances in Geocomputation*,
Advances in Geographic Information Science,
DOI 10.1007/978-3-319-22786-3_21

1 Introduction

Advances in location sensing and computing technologies enable automatic tracking of moving objects at a high level of detail in space and time. Context awareness from such moving object data is one of the key research challenges in data mining and ubiquitous computing. Activity recognition and situation awareness associated with locations, time, and moving objects facilitate the interaction between users and computing system, which ultimately supports decision making in applications such as transportation (Andrienko et al. 2011), video surveillance (Rougier et al. 2011), and offender monitoring systems (Yuan and Nara 2015). This study examines context awareness in the operating room (OR) environment because a better understanding of human activity during surgery can help improve patient treatment and increase hospital efficiency; for example, automatic surgical phase recognition supports dynamic scheduling and resource allocations (Sutherland and van den Heuvel 2006), and work flow analysis (Padoy et al. 2012) aids work flow optimization and standardization.

To achieve context recognition of intraoperative activities, various monitoring approaches have been proposed: a patient's vital signs (Xiao et al. 2005), instrument signals (Padoy et al. 2007), a surgeon's elbow and wrist movements using two video cameras (Ohnuma et al. 2006), eye-gaze tracking data (James et al. 2007), and standardized free-hand movement by a Kinect sensor (Yoshimitsu et al. 2014). While previous research demonstrates that various sensors can recognize activities, work flows, and phases during an operation, most of them ignore the comparative study of sensor technologies and their context recognition performances.

This chapter investigates the performance of context recognition from moving object data collected by video imaging and sensor-based tracking. We used three measurements to estimate the amount of movement in an OR during a neurosurgical operation: (1) tag moving distance, (2) optical flow, and (3) trajectory clusters. We collected time-stamped location data of tag-attached surgical staff by an ultrasonic location aware system to estimate a tag moving distance. We used trajectory clustering techniques to further extract movement characteristics from the time-stamped location data. We employed an optical flow algorithm to measure movements captured from video imagery. Then, we applied a support vector machine to three measures and their combinations to compare and examine the context recognition rate. The context recognition in this study refers to the recognition of intraoperative surgical phases, such as preparation, craniotomy, tumor resection, and magnetic resonance (MR) imaging.

2 Data Collection

In this study, we employ two data collection methodologies, video imaging and active location sensing by an ultrasonic location aware system, for capturing the amount of intraoperative movement.

2.1 *Video Imagery*

We recorded single-channel intraoperative video imagery by a camera mounted on the wall near the entrance of an OR to record the surgical field and staff (Fig. 1). The sequence of captured image frames is used to measure the number of motions during an operation by optical flow.

2.2 *Ultrasonic Location Aware System*

The ultrasonic location aware system (Fig. 2) consists of ultrasonic tags (transmitters), receivers, and four control units. The receivers collect ultrasonic pulses emitted from multiple tags. Four control units identify each tag's identification and detect associated three-dimensional (3D) positions in an OR. To estimate a location, the system records the time-of-flight, which is the travel time of the signal from transmission to reception. Based on more than three time-of-flight results, the system computes a 3D position using the trilateration method based on a robust estimation algorithm known as random sample consensus (RANSAC) (Fischler and Bolles 1981). Table 1 summarizes the system specification in a typical environment.

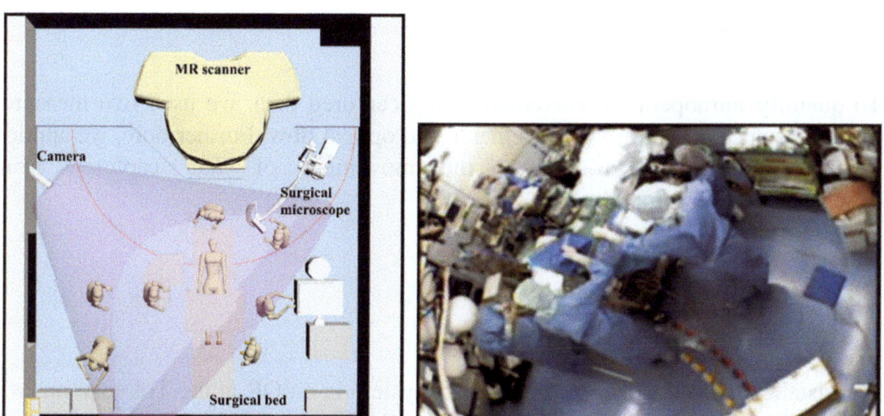

Fig. 1 OR layout (*left*) and camera view (*right*)

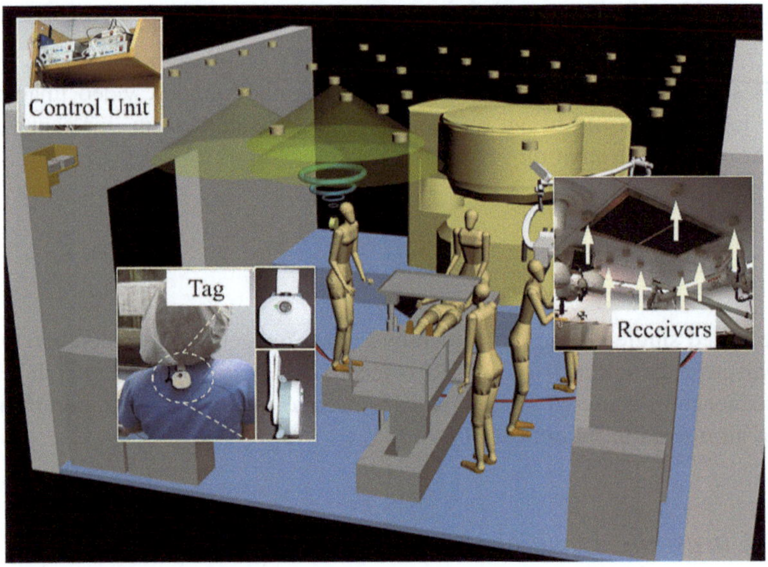

Fig. 2 Ultrasonic location aware system

Table 1 Specifications of the ultrasonic location aware system

Frequency of ultrasound	40 kHz
Position estimation error	Less than 80 mm
Sampling frequency	Up to 50 Hz
Measurement range (distance)	Vertical distance from a reader ≈7 m
Measurement range (angle)	Vertical angle from a reader ≈100°
Maximum number of tags	2,048

3 Methods

To quantify intraoperative movement using captured data, we used two measurements, distance based on tag movements and optical flow. Furthermore, we applied trajectory clustering to analyze the movement of data captured from distanced-based tag movement.

3.1 *Tag Movements*

We installed the ultrasonic location aware system in an OR at the Tokyo Women's Medical University (TWMU), Tokyo, Japan (Nara et al. 2011). The room is 5.8 m (width) × 4.8 m (depth) × 2.9 m (height) in size (Fig. 2). We deployed 33 ultrasonic receivers on the ceiling and set four control units on the wall near the room entrance. The wearable ultrasonic tag is 44 mm (width) × 75 mm (height) × 24 mm (depth) in size, and 40 g (tag: 30 g, battery: 10 g) in weight. For the purpose of minimum disturbance during a surgical operation, a single tag was hooked on surgical clothes around the nape of a surgical staff's neck. We have also verified that the system does not conflict with other surgical devices, including MR scanners.

At TWMU, a typical neurosurgical operating team includes surgeons, anesthetists, engineers, scrub nurses, and assistant nurses. We collected tag movement data from all but the engineers, who are not continuously present in an OR during neurosurgical operations. In this study, we used movement data from ten neurosurgery cases collected between 2009 and 2010, where the average operative time for all ten cases was 7 h 40 min and the standard deviation was 1 h 10 min.

3.2 Optical Flow

Optical flow measures attempt to track the movement of individual features from one frame to the next and produce a set of motion vectors that describe the direction and magnitude of these movements. The most straightforward of these algorithms uses the concept of block matching, which divides a frame into subblocks and searches for corresponding subblocks in a second frame. To search for candidate subblocks, block matching techniques often rely upon metrics such as the root mean squared error (RMSE) or the sum of absolute differences (SAD). Block matching has been effectively applied to many computer vision problems, such as segmenting moving objects (Bradski and Davis 2002) and measuring cyclical motion in artery walls (Golemati et al. 2003).

In this study, we generated frame images using one-second sampling, which is consistent with the sampling frequency of the tag sensor data. Subsequently, we calculated optical flow vectors between two consecutive frames. Due to run-time performance considerations, we used the Lucas–Kanade method (Lucas and Kanade 1981), which is optimized for real-time analysis. The magnitudes for each of these vectors was summed in order to derive a global measure of activity during a given time interval. Because our camera was stationary, we did not need to account for any movement of the camera when calculating optical flow. This global approach diverges from most studies that use optical flow to track individual objects; here we are concerned only with aggregate motion within a scene.

3.3 Trajectory Clustering

The ultrasonic location aware system collects a set of trajectories from multiple surgical staff {trajectory set: $T = T_1, T_2, T_3, \ldots, T_i$, where i denotes the number of surgical staff} during a surgical operation. Each trajectory comprises a sequence of four-dimensional points {{$T_i = p_1, p_2, p_3, \ldots, p_j$, where j denotes the number of points in trajectory i}, {$p_j = x, y, z, t$}}. To extract intraoperative movement characteristics of surgical staff, we employed trajectory data-mining techniques, which include two procedures: trajectory partitioning and trajectory clustering (Nara et al. 2011). The trajectory partitioning process partitioned an entire trajectory of each surgical staff into trajectory partitions (subtrajectories). By grouping trajectory partitions for each surgical role, the unsupervised clustering process describes surgical events and procedures that have similar trajectory patterns.

For each trajectory partition, we obtain multidimensional vectors to characterize the partition trajectory. The vector values include total distance (x-y axes), distance between start and end nodes (x-y axes), total distance (z axis), and time duration (Nara et al. 2011). Then, k-means cluster analysis is used with standardized values of these vectors. To estimate the quality of clusters for determining the number of cluster k in k-means automatically, we apply the IGRC (information gain ratio for cluster) index (Yoshida et al. 2006).

4 Results

Figure 3 shows the relationship between optical flow and tag moving distances during one neurosurgical operation. The thick red line represents the amount of optical flow, whereas the thick black line represents the total tag moving distance. Figure 3 shows a strong correlation between two values, suggesting that movements obtained from video imagery and the ultrasonic location aware system describe similar movement behaviors. The blue stacked vertical bars represent the

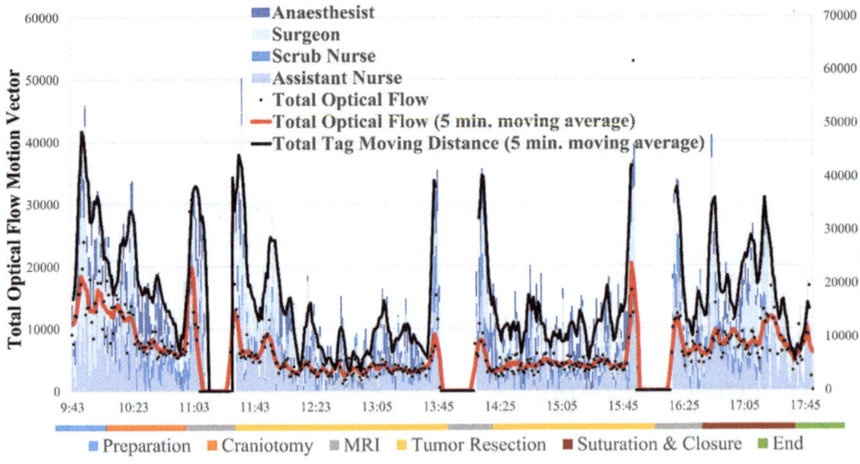

Fig. 3 The relationship between optical flow and tag movements

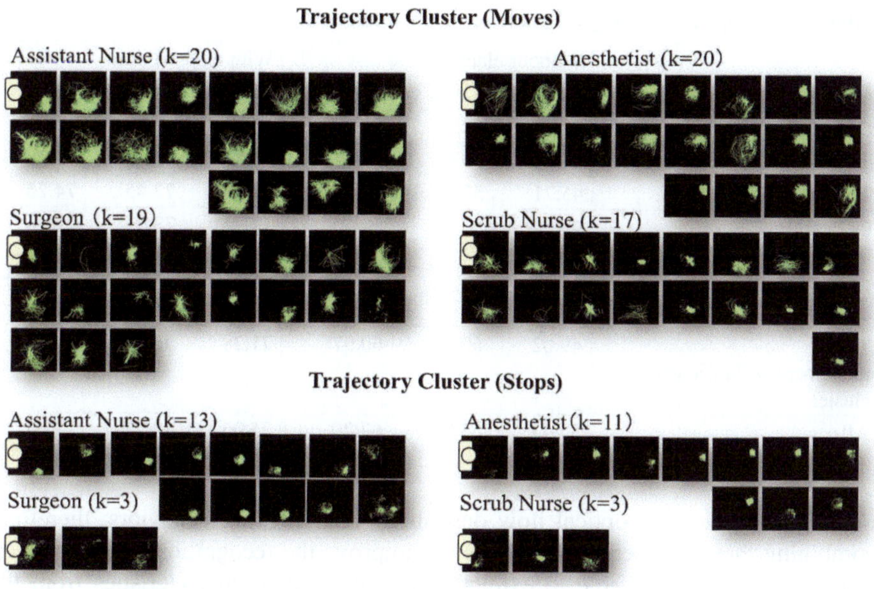

Fig. 4 Trajectory clustering results

amount of moving distance by each role (i.e., surgeons, anesthetists, scrub nurses, assistant nurses).

Figure 4 portrays the result of trajectory clustering, which uncovers groups of similar moving and stopping behaviors. For each staff's role, we quantified moving behaviors and created time-sequence vectors by counting the total amount of duration represented for each trajectory cluster for a specified time window. In this study, we selected 1-min, 5-min, and 10-min time windows. To compare and integrate data for three movement characteristics, we applied a time window averaging method to the results of optical flow and tag moving distances and created time-sequence vectors using the same time window size used for trajectory clustering. Finally, we applied a support vector machine to tag moving distances, optical flow, and trajectory clustering, and combinations of these three groups of data to evaluate the performance of the intraoperative context recognition using tenfold cross-validation. Table 2 summarizes the results of the context recognition rate by the three measures and their combinations.

5 Discussion

Our results show that the integration of both video imagery and location sensor data improves context awareness of neurosurgical operations, particularly when utilizing trajectory clustering outcomes (numbers in red in Table 2). As shown on Fig. 3, tag

Table 2 A comparison of surgical phase recognition rates (%)

	Without elapsed time			With elapsed time		
Moving window average	1 min.	5 min.	10 min.	1 min.	5 min.	10 min.
Optical flow	40.08	48.37	49.09	67.14	68.57	70.82
Tag movement	34.78	36.94	36.42	61.83	60.82	63.38
Trajectory cluster	45.71	57.86	58.15	69.46	72.45	74.25
Optical flow + Tag movement	49.56	50.41	50.50	72.44	73.57	72.44
Optical flow + Trajectory cluster	50.30	62.55	65.19	71.24	76.43	77.46
Tag movement + Trajectory cluster	49.02	60.41	60.97	71.28	74.59	73.84
All	55.43	65.92	67.20	75.05	78.27	77.67

moving distances and optical flow describe similar moving behaviors; therefore, combining these two sets of data does not improve the recognition rate. However, trajectory clustering is able to extract unique movement characteristics from the same data and thus provides a more comprehensive description of surgical phases.

Finally, further work is required to use these methods in a real-world scenario. In this study, we processed data and computed variables offline to extract movement characteristics in a parallel way using multiple central processing units (CPUs). To implement the approach in a real-time clinical scenario, further development and implementation of online classification algorithms and computational optimization procedures are necessary.

Acknowledgements This study was partially funded by NEDO (New Energy and Industrial Technology Development Organization) through the Intelligent Surgical Instruments Project. We thank Drs. Hiroshi Iseki, Takashi Suzuki, Takashi Maruyama, and Masahiko Tanaka, and other staff of the Tokyo Women's Medical University for participating in data collection.

References

Andrienko G, Andrienko N, Heurich M (2011) An event-based conceptual model for context-aware movement analysis. Int J Geogr Inf Sci 25(9):1347–1370

Bradski GR, James WD (2002) Motion segmentation and pose recognition with motion history gradients. Mach Vis Appl 13(3):174–184

Fischler MA, Bolles RC (1981) Random sample consensus: a paradigm for model fitting with applications to image analysis and automated cartography. Commun ACM 24(6):381–395

Golemati S, Sassano A, Lever MJ, Bharath AA, Dhanjil S, Nicolaides AN (2003) Carotid artery wall motion estimated from B-mode ultrasound using region tracking and block matching. Ultrasound Med Biol 29:387–399

James A, Vieira D, Lo B, Darzi A, Yang, GZ (2007) Eye-gaze driven surgical workflow segmentation. In: Ayache N, Ourselin S, Maeder A (eds) Medical image computing and computer-assisted intervention—MICCAI 2007, Brisbane, Australia, 29 Oct–2 Nov 2007. Lecture notes in computer science, vol 4792. Springer, Berlin, pp 110–117

Lucas BD, Kanade T (1981) An iterative image registration technique with an application to stereo vision. In: Proceedings of the 7th international joint conference on artificial intelligence (IJCAI), Vancouver, BC, 24–28 Aug 1981, pp 674–679

Nara A, Izumi K, Iseki H, Suzuki T, Nambu K, Sakurai Y (2011) Surgical workflow monitoring based on trajectory data mining. In: Onada T, Bekki D, McCready E (eds), New frontiers in artificial intelligence. JSAI-isAI 2010 workshops, Tokyo, Japan, 18–19Nov 2010. Lecture notes in computer science (Lecture note in artificial intelligence), vol. 6797. Springer, Berlin Heidelberg, pp 283–291

Ohnuma K, Masamune K, Yoshimitsu K, Sadahiro T, Vain J, Fukui Y, Miyawaki F (2006) Timed-automata-based model for laparoscopic surgery and intraoperative motion recognition of a surgeon as the interface connecting the surgical and the real operating room. Int J Comput Assist Radiol Surg 1:442–445

Padoy N, Blum T, Ahmadi SA, Feussner H, Berger MO, Navab N (2012) Statistical modeling and recognition of surgical workflow. Med Image Anal 16(3):632–641

Padoy N, Blum T, Essa I, Feussner H, Berger MO, Navab N (2007) A boosted segmentation method for surgical workflow analysis. In: Ayache N, Ourselin S, Maeder A (eds), Medical image computing and computer-assisted intervention—MICCAI 2007, Brisbane, Australia, 29 Oct–2 Nov 2007. Lecture notes in computer science, vol 4792. Springer, Berlin, pp 102–109

Rougier C, Meunier J, St-Arnaud A, Rousseau J (2011) Robust video surveillance for fall detection based on human shape deformation. IEEE Trans Circuits Syst Video Technol 21(5):611–622

Sutherland J, van den Heuvel WJ (2006) Towards an intelligent hospital environment: adaptive workflow in the OR of the future. In: Proceedings of the 39th annual Hawaii international conference on system sciences 2006 (HICSS), Kauai, HI, 4–7 Jan 2006, vol 5, p 100b. doi:10.1109/HICSS.2006.494

Xiao Y, Hu P, Hu H, Ho D, Dexter F, Mackenzie CF, Seagull FJ, Dutton RP (2005) An algorithm for processing vital sign monitoring data to remotely identify operating room occupancy in real-time. Anesth Analg 101(3):823–829

Yoshida T, Shoda R, Motoda H (2006) Graph clustering based on structural similarity of fragments. Lect Notes Comput Sci 3847:97–114

Yoshimitsu K, Muragaki Y, Maruyama T, Yamato M, Iseki H (2014) Development and initial clinical testing of "OPECT": An innovative device for fully intangible control of the intraoperative image-displaying monitor by the surgeon. Neurosurgery 10:46–50

Yuan M, Nara A (2015) Space-time analytics of tracks for the understanding of patterns of life. In: Kwan MP, Richardson D, Wang D, Zhou C (eds) Space-time integration in geography and GIScience. Springer, Netherlands, Dordrecht, pp 373–398

Part III
Spatial Statistical and Geostatistical Modeling

Statistical modeling and computation have contributed continuously to advances in geocomputation. Spatial statistics and geostatistics especially are widely utilized in statistical modeling of georeferenced data. Conference papers in this section exemplify this contribution in theoretical and practical aspects of data analysis. Morris et al. extend spatial autocorrelation into qualitative data analysis. Specifically, they investigate how spatial autocorrelation affects and enhances qualitative data sampling. Luo et al. examine two popular spatial autocorrelation indices, Moran's I and Geary's c. They compare the statistical efficiency and power of those indices for various spatial tessellations and sample sizes.

Two papers address variogram estimation for spatial processes. Pu and Tiefelsdorf propose a modified Box–Cox transformation to reduce bias in variance estimation in clustered geographic samples. Kim and Zhu propose a filter-based nonparametric variogram estimation method to estimate variance of a nonstationary geographic process.

Two papers utilize the eigenvector spatial filtering (ESF) technique. Sinha et al. examine how the coefficients of eigenvectors that are selected in an ESF model are distributed and report that this distribution closely conforms to a gamma distribution. Lee et al. investigate how location and measurement errors impact parameter and variance estimates of regression models. They utilize ESF to account for spatial autocorrelation in the regression models.

Three other papers analyze georeferenced data using spatial and geostatistical models. Brown and Oleson develop a spatiotemporal epidemiology model in a Bayesian context using libSpatialSEIR, an R-package module. Stoudt estimates the distribution of uranium across the coterminous USA using geostatistical models. Finally, Sinha extends ESF to account for spatial autocorrelation in a multinomial logit model to analyze land use change.

Respondent-Driven Sampling and Spatial Autocorrelation

E. Scott Morris, Vaishnavi Thakar and Daniel A. Griffith

Abstract Respondent-driven sampling (RDS) is a type of sampling method used to survey rare and hard-to-reach populations. RDS was developed to address the issue of bias associated with snowball sampling in qualitative research. Although, RDS has evolved by addressing major issues involved with the snowball sampling method, the issue of how the presence of spatial autocorrelation (SA) affects RDS had not been studied. SA refers to the clustering of similar attribute values in geographic space. Quantitative studies show that the presence of positive SA leads to an underestimation of the appropriate sample size. If RDS is not affected by SA, then the samples are expected to be dispersed in geographic space and not clustered around a sampling seed that initiates a sequence of respondents. This chapter presents impacts of SA on RDS when a social network displays a geographic pattern. The geographic distribution of the samples and associated socioeconomic and demographic variables are analyzed with respect to sequences of respondents. Social network RDS data for Rio de Janeiro, Brazil, are analyzed. Previous research indicates that, in these social network RDS data, samples are clustered around their initial seeds and do not spread out in geographic space as the sequence of respondents progresses. This tendency may result in increased sampling variance, which raises a concern about appropriate sample size determination in RDS.

Keywords Respondent-driven sampling · Snowball sampling · Spatial autocorrelation · Social network

E.S. Morris · V. Thakar · D.A. Griffith (✉)
University of Texas at Dallas, 800W. Campbell Road,
Richardson, TX 75080, USA
e-mail: dagriffith@utdallas.edu

E.S. Morris
e-mail: exm123030@utdallas.edu

V. Thakar
e-mail: vxt110130@utdallas.edu

D.A. Griffith et al. (eds.), *Advances in Geocomputation*,
Advances in Geographic Information Science,
DOI 10.1007/978-3-319-22786-3_22

1 Introduction

Snowball sampling, introduced by Goodman in 1961, is a survey strategy commencing by selecting an initial group of participants known as seeds. Once surveyed, each seed recommends potential respondents (nodes) with shared connections (edges) on the basis of a research topic and whom each referrer believes is likely also to participate. This process proceeds in a similar fashion over a series of waves, and the nodes and edges define a social network. Respondent-driven sampling (RDS), developed by Heckathorn in 1997, is a formalized method, based on the snowball strategy, that compensates for the non-random process of data collection. Previous studies concluded that little bias exists among RDS results compared to simple random sampling (SRS). However, subsequent research indicates the presence of a variance inflation factor (VIF) and increased design effect among underlying attributes of the members of RDS networks versus SRS. The result of this effect is the tendency to underestimate the appropriate sample size among RDS surveys. A prominent contributor to the VIF may be positive spatial autocorrelation (SA) attributed to the geographic configuration of the targeted population. If a social network displays a geographic pattern, the variance for a RDS is likely to be impacted by SA (Rudolph et al. 2015). To analyze this impact, a simulation can be designed based upon real-world data from a RDS survey conducted with heavy drug users in Rio de Janeiro, Brazil. Figure 1 portrays the location of the study area. This simulation proceeds through the obtained social network based upon empirical probabilities determining the number of nodes for each subsequent wave. The purpose of this chapter is to establish a basis for designing this type of simulation experiment and to demonstrate that the VIF attributable to SA is an outcome.

2 Data

Two datasets are utilized for the purpose of spatial analysis of the impact of SA upon RDS. The first set includes the connectivity structure defining the social network. The second set entails the demographic variables selected to describe the general population within the spatial region of interest.

2.1 Network

The social network analyzed in this research is from an RDS study conducted in Rio de Janeiro in 2009. The network consists of 611 heavy drug users defined as having injected illegal narcotics in the last six months, and/or using illicit drugs, other than marijuana or hashish, at least 25 days in the last six months. Respondents

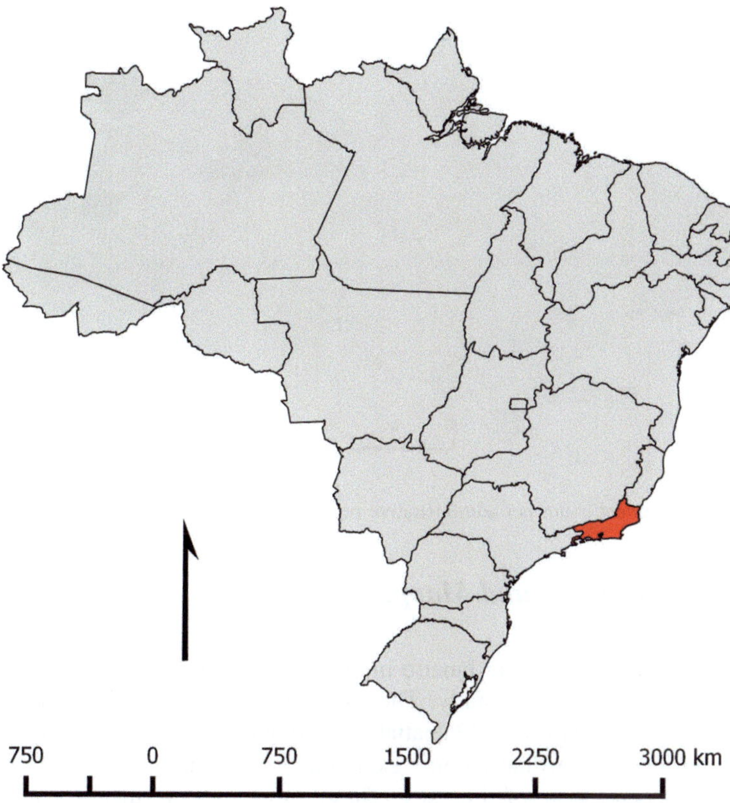

Fig. 1 Rio de Janeiro study area within Brazil

are over the age of 18 and meet the protocol of the study. The original study utilized RDS as a technique for surveying hard-to-reach populations, specifically, those affected by HIV transmission associated with heavy drug use (Toledo et al. 2011). Network data are configured in two tables. First, the node data consist of anonymous respondent identification numbers (ID) and their corresponding administrative regions (AR) of residence; 140 of the respondents' locations are unknown. Figure 2 portrays the number of respondents by AR.

2.2 *Demographics*

Demographic data were obtained from the Instituto Brasileiro de Geografia e Estatística (IBGE) from the 2010 census online at http://www.ibge.gov.br/home. Four attribute variables were selected assuming that they demonstrate correlation with heavy drug use and reflect at least a moderate degree of SA.

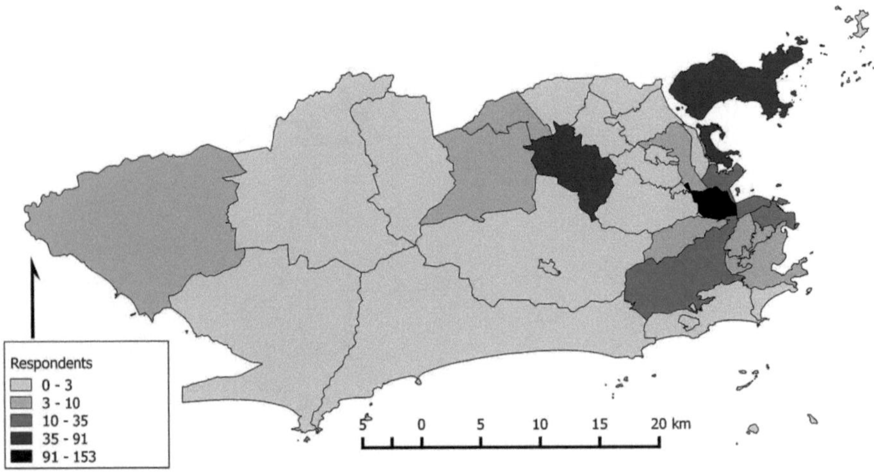

Fig. 2 RDS respondent count per administrative region of Rio de Janeiro

2.3 *Transformation and Mapping*

Thirty-three ARs in the Rio de Janeiro municipality in the state of Rio de Janeiro, Brazil, were selected for this study. The following socioeconomic variables were obtained at the AR level for all 33 spatial units: population density, median income including no income, median income excluding no income, percentage of unemployment, and percentage of illiteracy. These variables were mapped in order to visualize their spatial distributions. Maps of the socioeconomic variables for the administrative regions in the Rio de Janeiro municipality are provided in the Appendix.

2.4 *Spatial Autocorrelation*

Tobler's first law of geography states that "Everything is related to everything else, but near things are more related than distant things" (Tobler 1970, 236). Hubert et al. (1981, 224) define SA as "Given a set *S* containing *n* geographical units, the relationship between some variable observed in each of the *n* localities and a measure of geographical proximity defined for all $n(n - 1)$ pairs chosen from *S*." Standard inferential statistics assumes complete randomness (of observations, which is referred to as an independent random process [IRP] or complete spatial randomness [CSR]). Spatial data, however, violate this assumption due to the presence of SA. Positive SA causes variance inflation (Griffith 2005). Hence, spatial statistics measures are used to quantify the degree of self-correlatedness of a variable as a function of nearness. A spatial weights matrix is needed to measure

SA, which gives the information about relative location of pairs of adjacent neighboring locations (binary rook, queen neighbors; first-, second-order neighbors) or all other locations (distance based—Euclidian, rectilinear, or network).

Two widely used indices of SA are provided by Moran (1948) and Geary (1954). The global Moran's I tests for spatial randomness (null hypothesis) and detects the nature (positive or negative) as well as degree of SA. The Moran's I values range from roughly –1 (high negative SA; dissimilar values cluster on a map) to 1 (high positive SA; similar values cluster on a map)—this lower bound actually can be between −1 and –0.5, whereas this upper bound can range from 0.8 to more than 1.3—Moran's I value denoting no SA is $-1/(n-1)$, which is slightly less than zero. The exact extreme values are functions of eigenvalues of the spatial weights matrix.

The global Geary's c (null hypothesis of no SA) values range from roughly zero (extreme positive SA), to one (no SA), to 2+ (extreme negative SA). The exact extreme values are functions of eigenvalues of the spatial weights matrix.

Moran's I and Geary's c for variable y are given by

$$I = \frac{n}{\sum_{i=1}^{n} \sum_{j=1}^{n} w_{ij}} \frac{\sum_{i=1}^{n} \sum_{j=1}^{n} w_{ij}(y_i - \bar{y})(y_j - \bar{y})}{\sum_{i=1}^{n} (y_i - \bar{y})^2}$$

$$c = \frac{n-1}{2(\sum_{i=1}^{n} \sum_{j=1}^{n} w_{ij})} \frac{\sum_{i=1}^{n} \sum_{j=1}^{n} w_{ij}(y_i - y_j)^2}{\sum_{i=1}^{n} (y_i - \bar{y})^2},$$

where

$\bar{y}$ denotes the mean of variable y;
y_i denotes the variable value at a particular location *i*; and
w_{ij} denotes the spatial weight between locations *i* and *j* (a spatial weights matrix, **W**, cell entry).

Previous quantitative geographic research documents that human attributes tend to display moderate positive SA. This idea was captured by 2005 Nobel Prize–winning economist Thomas Schelling (1969, 1971) in his model of segregation. Keeping this observation in mind, the Moran and Geary SA indices were used to measure the nature and degree of SA for the selected Rio de Janeiro socioeconomic variables.

SA indices need a spatial weights matrix that defines polygons sharing common boundaries as neighbors. The Rio study area polygon shapefile includes two disconnected islands, which were manually edited and connected to the mainland before creating the spatial weights matrix. This was done by observing the bridges connecting the islands with the mainland, so that each of the spatial units (polygons) shares at least one common boundary or edge. A row standardized spatial weights matrix was created using the modified polygon shapefile with connected islands.

A Box–Cox power transformation (log) was performed on population density to make its frequency distribution more bell-shaped. Table 1 summarizes the Moran

Table 1 Summary of Moran coefficient and Geary ratio values for socioeconomic variables in the Rio de Janeiro municipality

Variable name	Moran coefficient	Z-scores (Moran's I)	Geary ratio	Z-scores (Geary's c)
Population density	0.23	2.2135	0.56	–3.2990
Median income excluded	0.47	4.5227	0.71	–1.7969
Median income included	0.56	5.2740	0.55	–2.8990
Percentage of unemployment	0.37	3.3796	0.65	–2.4992
Percentage of illiteracy	0.33	3.2141	0.68	–2.1011

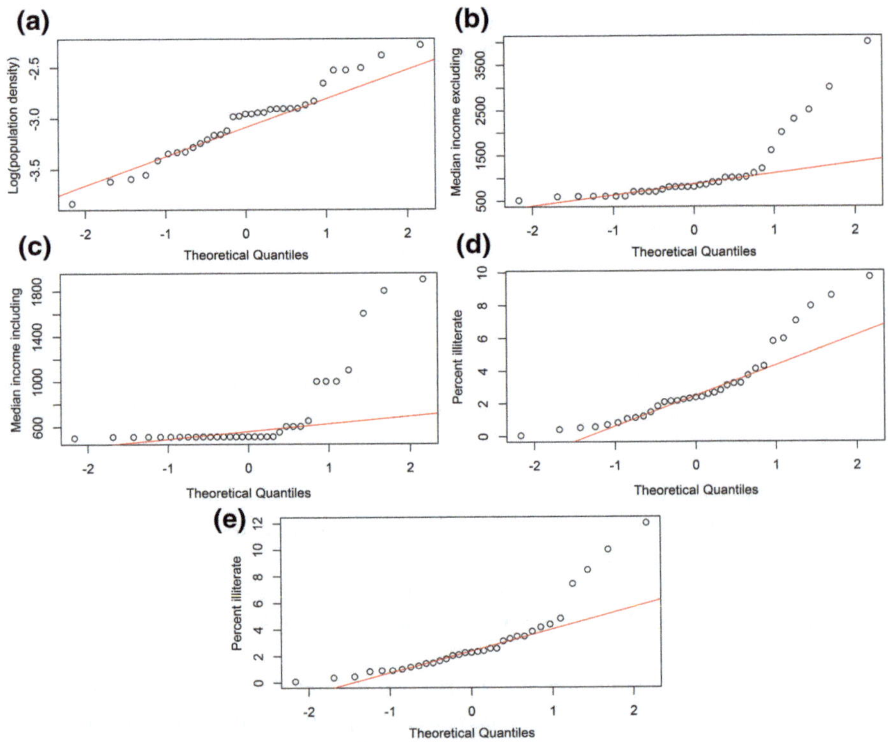

Fig. 3 Normal quantile plots for selected socioeconomic variables: **a** Population density (transformed), **b** Median income excluding no income, **c** Median income including no income, **d** Percentage of unemployment, **e** Percentage of illiteracy

coefficient and Geary ratio values of SA for each socioeconomic variable; all five variables contain weak to moderate positive SA. Figure 3 displays the normal quantile plots for each variable.

Table 2 Number of chains and subchains by length

Nodes	2	3	4	5	6	7	8	9	10	11	12	Total
Chains	5	13	21	24	42	44	31	80	53	46	18	377
Subchains	2,656	2,279	1,907	1,548	1,210	896	624	396	199	82	18	11,815

3 Methodology

Simulations were conducted to demonstrate the impact of variance inflation in the presence of SA and network autocorrelation. SA measures are based upon the geographic connectivity of the region, while network autocorrelation is measured with the social connectivity of the target population.

3.1 Network Chains

A network is developed by respondents recommending subsequent participants. The process is initiated by the selection of initial participants, or seeds, from whom the social network emanates. The network analyzed in this study originated with six seeds, one of which yielded no referrals. Based upon the referral connectivity, the network comprises 377 chains. Each chain is defined by a seed—a never-referred respondent—and an end—a respondent with no referrals. Chain lengths range from 2 to 12 respondents (Table 2).

3.2 Simulation Design

An analysis of the network can be conducted with simulations based upon the network connectivity. Starting points and chain lengths could be selected, and subchains could be traced based upon underlying empirical probabilities. Therefore, the 377 complete chains are partitioned into subchains, providing a total number of 11,815 possible chains. For example, one chain of length 5 consists of two subchains of length 4, three of length 3, and four of length 2. The prevalence of missing locational data introduces complications and constraints to such a simulation experiment.

4 Anticipated Results

The anticipated result of this research is that the geographic distribution of a social network displays SA. In turn, this feature inflates the sampling variance. Inflation occurs because of two factors: (1) the sampling probabilities no longer are equal;

and (2) covariance between individuals no longer is zero. This covariation is a function of the structure of a social network coupled with SA in its geographic landscape, which are correlated. One way to capture this latter effect is to couple a social network with its corresponding spatial weights matrix, conceptualizing the social network as being articulated first. Griffith (2005) outlines the VIF attributable to SA. Extending this specification and considering only the case of positive SA,

$$1 \leq \mathrm{VIF} \leq \mathrm{TR}(\mathbf{V}_s^{-1})\mathrm{TR}(\mathbf{V}_N^{-1})/n^2,$$

where n is the number of observations (i.e., individuals or areal units), $\mathbf{V}_s$ denotes the spatial autoregressive variance component (e.g., $(\mathbf{I}-\rho_s\mathbf{W})^T(\mathbf{I}-\rho_s\mathbf{W})$, where ρ_s is the spatial autoregressive parameter), and $\mathbf{V}_N$ denotes the network autoregressive variance component. This specification indicates that network autocorrelation inflates variance beyond what SA does, and vice versa. It also could be modified by including a geographic aggregation matrix, which would smooth the SA effects.

Acknowledgements This research was conducted with support from the US National Science Foundation, Grant BCS-1262717; any opinions, findings, and conclusions or recommendations expressed in this material are those of the authors and do not necessarily reflect the views of the National Science Foundation. We thank Dr. Francisco Inacio Bastos of the Oswaldo Cruz Foundation for providing the data facilitating this research and Dr. Parmanand Sinha for his initial work on retrieving the social network data.

Appendix

See Figs. 4, 5, 6, 7 and 8.

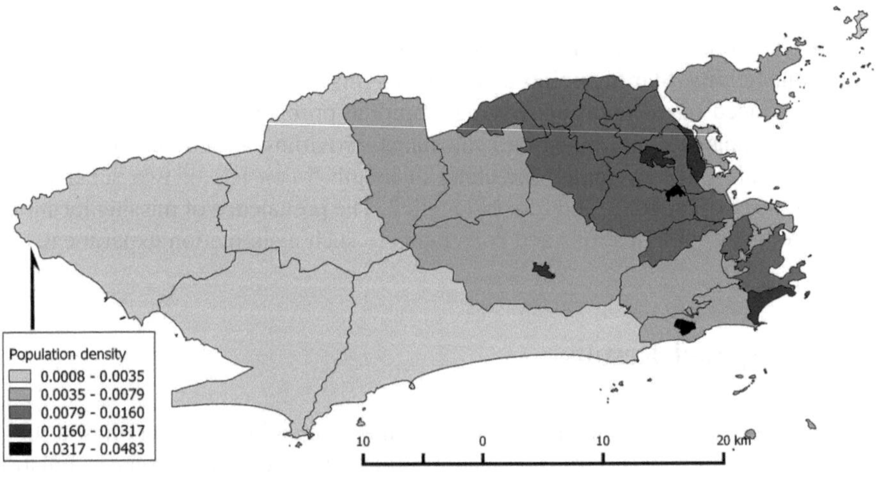

Fig. 4 Log of population density by administrative region

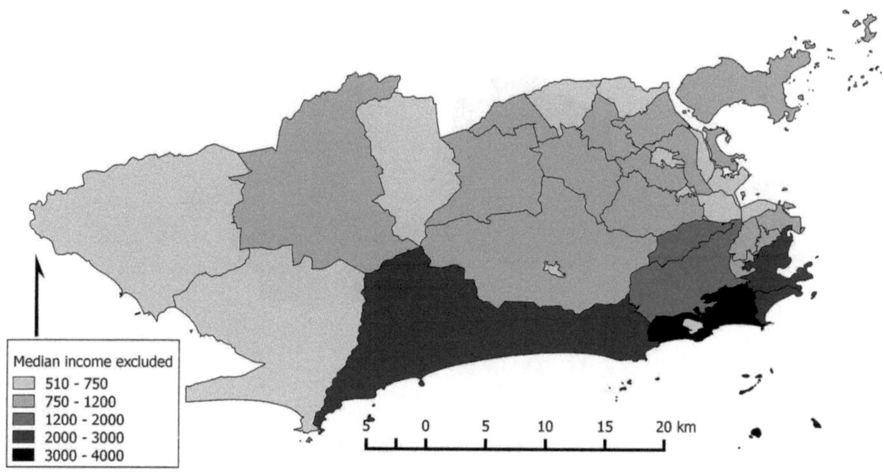

Fig. 5 Median income by administrative region excluding responses of no income

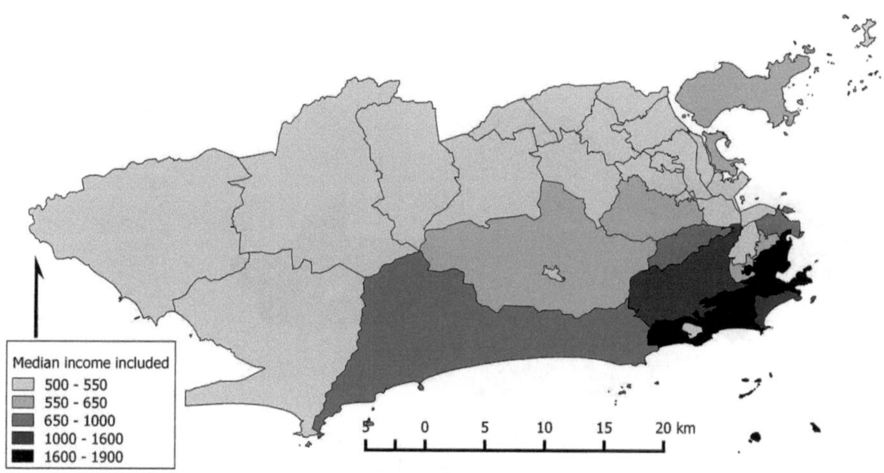

Fig. 6 Median income by administrative region including responses of no income

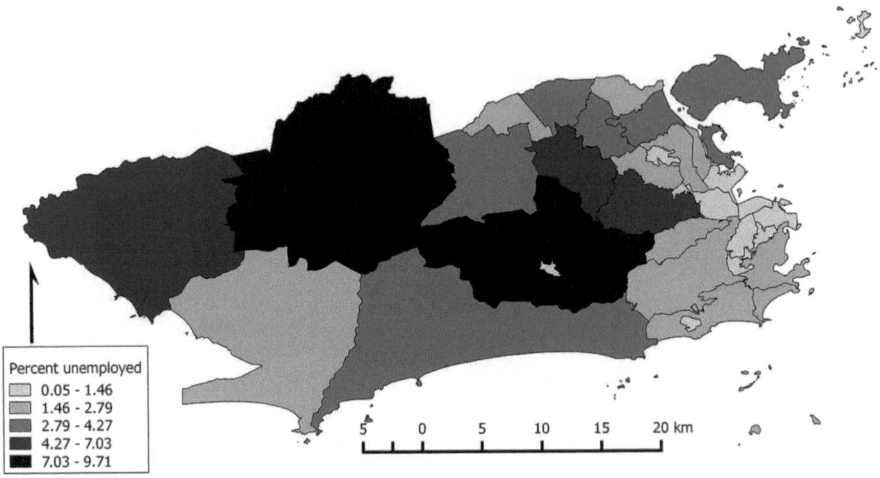

Fig. 7 Percentage of unemployment by administrative region

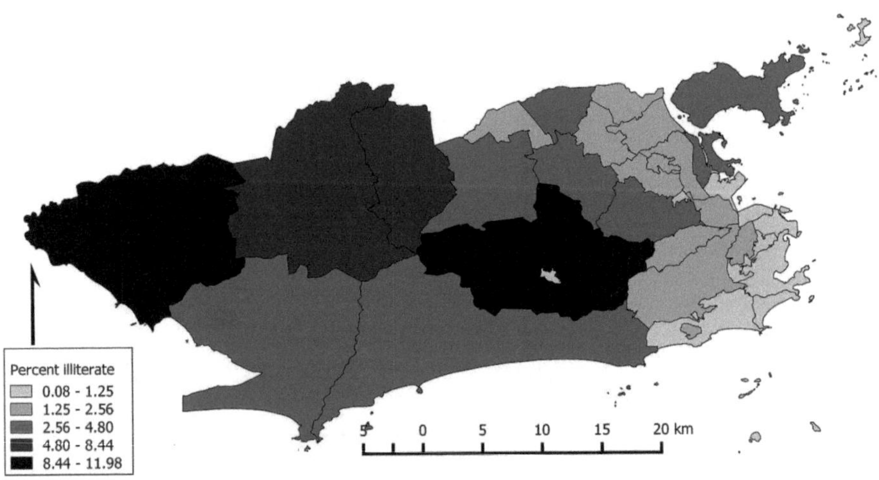

Fig. 8 Percentage of illiteracy by administrative region

References

Geary R (1954) The contiguity ratio and statistical mapping. Inc Stat 5:115–141

Goodman L (1961) Snowball sampling. Annals of Mathematical Statist 32:148–170

Griffith D (2005) Effective geographic sample size in the presence of spatial autocorrelation. Ann Assoc Am Geogr 95:740–760

Heckathorn DD (1997) Respondent-driven sampling: a new approach to the study of hidden populations. Soc Probl 44(2):174–199

Hubert LJ, Golledge RG, Costanzo CM (1981) Generalized procedures for evaluating spatial autocorrelation. Geogr Anal 13(3):224–233

Moran P (1948) The interpretation of statistical maps. J Roy Stat Soc B 10:243–251

Rudolph A, Young A, Lewis C (2015) Assessing the geographic coverage and spatial clustering of illicit drug users recruited through respondent-driven sampling in New York City. J Urban Health 92:352–378

Schelling TC (1969) Models of segregation. Am Econ Rev 59(2):488–493

Schelling TC (1971) Dynamic models of segregation. J Math Sociol 1(2):143–186

Tobler WR (1970) A computer movie simulating urban growth in the Detroit region. Econ Geog 46(2):234–240

Toledo L, Codeço CT, Bertoni N, Albuquerque E, Malta M, Bastos FI (2011) Putting respondent-driven sampling on the map: insights from Rio de Janeiro. Brazil. Acquir Immune Defic Syndr 57(3):S136–S143

The Moran Coefficient and the Geary Ratio: Some Mathematical and Numerical Comparisons

Qing Luo, Daniel A. Griffith and Huayi Wu

Abstract This chapter discusses relationships between the Moran coefficient (MC) and the Geary ratio (GR) under different distributional assumptions (normal, uniform, beta, and exponential) and selected geographic neighborhood definitions (linear, square rook, hexagon, square queen, maximum planar, maximum hexagon, and a constant number of neighbors). It focuses on comparisons of efficiency and power for the MC and the GR. Its results should inform features of spatial data analysis.

Keywords Moran coefficient · Geary ratio · Efficiency · Power · Geographic configuration

1 Introduction

The Moran coefficient (MC) (Moran 1950) and Geary ratio (GR) (Geary 1954) are statistics used to quantify the nature and degree of spatial autocorrelation. Cliff and Ord (1973, 1981) establish their asymptotic normal sampling distribution properties and the power superiority of the MC versus the GR for only a few types of selected small surface partitionings. Tiefelsdorf and Boots (1995) derive the exact distribution of the MC for small samples, which is a seminal work establishing the novel eigenvector spatial filtering spatial statistics methodology (Griffith 1996). This

Q. Luo (✉) · H. Wu
Wuhan University, 129 Luoyu Road, Hongshan District,
Wuhan 430079, Hubei, China
e-mail: luoqing11@whu.edu.cn

H. Wu
e-mail: wuhuayi@whu.edu.cn

Q. Luo · D.A. Griffith
The University of Texas at Dallas, 800W. Campbell Rd, Richardson,
TX 75080-3021, USA
e-mail: dagriffith@utdallas.edu

D.A. Griffith et al. (eds.), *Advances in Geocomputation*,
Advances in Geographic Information Science,
DOI 10.1007/978-3-319-22786-3_23

chapter explores relationships between the MC and GR for a wide range of surface partitionings across sizes that expand to infinity, derives their approximate variances under different distributional assumptions, analyzes their statistical efficiency, presents large sample power comparisons for them, and documents some comparative features.

2 The Relationship Between the MC and GR

Let X be the georeferenced variable of interest distributed over a tessellation. Its observations are $x_1, x_2, \cdots, x_n$. The average of these observations is denoted by

$$\overline{x} = \sum_{i=1}^{n} x_i \,/n. \tag{1}$$

For regular surface partitionings, $n = P \times Q$, where P and Q respectively are the number of rows and columns. Let $C = (c_{ij})_{n \times n}$ be a surface partitioning's connectivity matrix, where $c_{ij} = 1$ if i and j are adjacent (i.e., neighbors), and zero otherwise; matrix $\boldsymbol{C}$ is symmetric.

The sample MC and GR of variable X are defined by

$$MC = n \sum_{i=1}^{n} \sum_{j=1}^{n} c_{ij}(x_i - \overline{x})(x_j - \overline{x}) / \sum_{i=1}^{n} \sum_{j=1}^{n} c_{ij} \sum_{i=1}^{n} (x_i - \overline{x})^2, \tag{2}$$

and

$$GR = (n-1) \sum_{i=1}^{n} \sum_{j=1}^{n} c_{ij}(x_i - x_j)^2 / 2 \sum_{i=1}^{n} \sum_{j=1}^{n} c_{ij} \sum_{i=1}^{n} (x_i - \overline{x})^2. \tag{3}$$

The GR can be rewritten as (Griffith 1987, 44)

$$\frac{n-1}{2 \sum\limits_{i=1}^{n} \sum\limits_{j=1}^{n} c_{ij}} \frac{2 \sum\limits_{i=1}^{n} (x_i - \overline{x})^2 \left(\sum\limits_{j=1}^{n} c_{ij} \right)}{\sum\limits_{i=1}^{n} (x_i - \overline{x})^2} - \frac{n-1}{n} MC. \tag{4}$$

Proof Substituting Eq. (2) into Eq. (4) yields

$$GR = (n-1) \left[2 \sum_{i=1}^{n} (x_i - \overline{x})^2 \left(\sum_{j=1}^{n} c_{ij} \right) - 2 \sum_{i=1}^{n} \sum_{j=1}^{n} c_{ij}(x_i - \overline{x})(x_j - \overline{x}) \right] / \left[2 \sum_{i=1}^{n} \sum_{j=1}^{n} c_{ij} \sum_{i=1}^{n} (x_i - \overline{x})^2 \right]. \tag{5}$$

Substituting $(x_i - x_j)^2 = \left[(x_i - \bar{x}) - (x_j - \bar{x})\right]^2$ and utilizing the symmetry of matrix $\boldsymbol{C}$ yield GR = Eq. (4). ■

3 Derivation of the MC and GR Asymptotic Variances

The exact variances of these two statistics (Cliff and Ord 1973), where subscript N denotes normality and R denotes randomization, are

$$Var_N(MC) = \frac{n^2 S_1 - nS_2 + 3{S_0}^2}{(n-1)(n+1){S_0}^2} - \frac{1}{(n-1)^2}, \tag{6}$$

$$Var_R(MC) = \frac{n\left[(n^2 - 3n + 3)S_1 - nS_2 + 3{S_0}^2\right] - b_2\left[(n^2 - n)S_1 - 2nS_2 + 6{S_0}^2\right]}{(n-1)(n-2)(n-3){S_0}^2} - \frac{1}{(n-1)^2}, \tag{7}$$

$$Var_N(GR) = \left[(2S_1 + S_2)(n-1) - 4{S_0}^2\right] / \left[2(n+1){S_0}^2\right], \tag{8}$$

and

$$Var_R(GR) = \frac{(n-1)S_1[n^2 - 3n + 3 - (n-1)b_2] - \frac{1}{4}(n-1)S_2[n^2 + 3n - 6 - (n^2 - n + 2)b_2]}{n(n-2)(n-3){S_0}^2} + \frac{{S_0}^2\left[n^2 - 3 - (n-1)^2 b_2\right]}{n(n-2)(n-3){S_0}^2}, \tag{9}$$

where $S_0 = \sum_{i=1}^{n}\sum_{j=1}^{n} c_{ij}, S_1 = \sum_{i=1}^{n}\sum_{j=1}^{n}(c_{ij} + c_{ji}), S_2 = \sum_{i=1}^{n}\left(\sum_{j=1}^{n}(c_{ij} + c_{ji})\right)^2$, and for $z_i = x_i - \bar{x}$, $b_2 = \frac{1}{n}\sum_{i=1}^{n} z_i^4 / \left(\frac{1}{n}\sum_{i=1}^{n} z_i^2\right)^2$ is kurtosis. Because matrix $\boldsymbol{C}$ is symmetric, $S_1 = 2\sum_{i=1}^{n}\sum_{j=1}^{n} c_{ij} = 2S_0$, and $S_2 = 4\sum_{i=1}^{n}\left(\sum_{j=1}^{n} c_{ij}\right)^2$.

Griffith (2010) proposes simplifying Eqs. (6)–(9) through asymptotics, assuming a normal distribution, producing

$$Var_A(MC) = 2/\sum_{i=1}^{n}\sum_{j=1}^{n} c_{ij} = 2/S_0, \tag{10}$$

and

$$Var_A(GR) = 2/\sum_{i=1}^{n}\sum_{j=1}^{n} c_{ij} + 2\sum_{i=1}^{n}\left(\sum_{j=1}^{n} c_{ij}\right)^2 \Big/ \left(\sum_{i=1}^{n}\sum_{j=1}^{n} c_{ij}\right)^2 = \frac{2}{S_0} + \frac{S_2}{2S_0{}^2} = \frac{2S_0 + S_2}{2S_0{}^2}, \quad (11)$$

where the subscript A denotes asymptotic.

The asymptotic variance for the MC is insensitive to the normality and randomization assumptions.

Theorem 1 $\lim_{n\to\infty} Var_N(MC) = Var_A(MC)$.

Proof $\lim_{n\to\infty} Var_N(MC) = \lim_{n\to\infty}[Eq.(6)] = 2/S_0 = Var_A(MC)$. ■

Theorem 2 $\lim_{n\to\infty} Var_R(MC) = Var_A(MC)$.

Proof $\lim_{n\to\infty} Var_R(MC) = \lim_{n\to\infty}[Eq.(7)] = 2/S_0 = Var_A(MC)$. ■

In contrast, the asymptotic variance of the GR is sensitive to the normality and randomization assumptions.

Theorem 3 $\lim_{n\to\infty} Var_N(GR) = Var_A(GR)$.

Proof $\lim_{n\to\infty} Var_N(GR) = \lim_{n\to\infty}[Eq.(8)] = (2/S_0) + \left(S_2/2S_0{}^2\right) = Var_A(GR)$. ■

Theorem 4 $\lim_{n\to\infty} Var_R(GR)$ *depends on* b_2*, the kurtosis of a distribution.*

Proof $\lim_{n\to\infty} Var_R(GR) = \lim_{n\to\infty}[Eq.(9)] = (2/S_0) + \left[(b_2 - 1)S_2\ /4S_0{}^2\right]$. ■

For the normal, uniform, beta ($\alpha = \beta = 0.5$), and exponential distributions, the b_2 values are 3, 9/5, 3/2, and 9, respectively, yielding asymptotic variances for GR of

$$Var_{AN}(GR) = 2/S_0 + S_2/2S_0{}^2, \quad (12)$$

$$Var_{AU}(GR) = 2/S_0 + S_2\ /5S_0{}^2, \quad (13)$$

$$Var_{AB}(GR) = 2/S_0 + S_2\ /8S_0{}^2, (\alpha = \beta = 0.5), \quad (14)$$

and

$$Var_{AE}(GR) = 2/S_0 + 2S_2\ /S_0{}^2, \quad (15)$$

where the subscripts *AN, AU, AB*, and *AE* respectively denote asymptotic variance for the normal, uniform, beta, and exponential distributions. Equation (12) coincides with Griffith's (2010) result.

4 Efficiency Analysis

A statistic with a smaller variance is more efficient. The variance ratio of the MC and the GR may be defined by

$$r_{exact} = Var_{exact}(MC)/Var_{exact}(GR), \tag{16}$$

where subscript *exact* denotes the exact MC and GR variances, given by Eqs. (6) and (8), or (7) and (9).

The following asymptotic variances also are of interest:

$$r = Var_A(MC)/Var_{A*}(GR) = 2/S_0/S, \tag{17}$$

where $A*$ denotes *AN, AU, AB,* or *AE*, and *S* denotes Eqs. (12), (13), (14), or (15). If $r < 1$, then the MC is more efficient than the GR.

Table 1 presents five different geographic configurations (linear, square rook, square queen, hexagon, and maximum planar connectivity) rendering different connectivity matrices (different summations of neighbors are listed beside their respective configurations) that illuminate the range of possible geographic situations. Analysis here focuses on their respective variance ratios, which converge to one in the limit for all but the maximum planar connectivity.

In Table 1, (1a)–(4a) describe some regular surface partitionings, but because the "preponderance of surface partitionings encountered in georeferenced data analyses are completely irregular" (Griffith and Sone 1995, p. 170), (5a) is introduced. This partitioning can be viewed as a "realization" of a planar graph that has the maximum number of edges 3(n − 2), where n is the number of nodes. Its corresponding n-by-n adjacency matrix **C** has 6(n − 2) ones (every connection is counted twice), and the largest principle eigenvalue for n > 11. In this "realization," n equals Q + 2 (i.e. P-by-(Q + 2) becomes 1-by-(Q + 2)); two units are adjacent to all Q + 1 other units. Table 2 also introduces a new maximum hexagon connectivity (1a–2a) case, whose purpose is to better illustrate the role of (5a). The number of units in both configurations can be expressed as P-by-(Q + 2). In addition, some of the configurations in Table 1 can be generalized to three dimensions (3D). For example, the two ends of a linear landscape (Table 1 (1a)) can be connected so that it becomes a closed circle (which also can be 2D); two pairs of the opposite ends of a square partitioning can be connected so that it becomes a torus. Table 2 portrays these 3D cases (3a–4a); their neighbor sums are $\sum_{i=1}^{n}\sum_{j=1}^{n} c_{ij} = kn$, $\sum_{i=1}^{n}\left(\sum_{j=1}^{n} c_{ij}\right)^2 = k^2 n$, where k is a constant (2, 4, or 8).

Table 1 Areal unit configuration cases and their neighbor sums

Geographic configuration	Neighbor sums
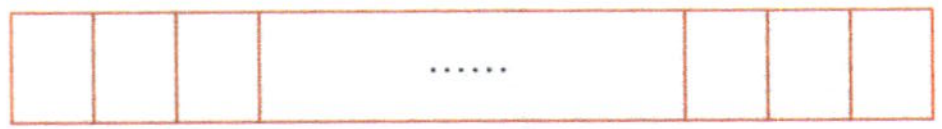 (1a) Linear	$\sum_{i=1}^{n}\sum_{j=1}^{n} c_{ij} = 2(n-1)$, $\sum_{i=1}^{n}\left(\sum_{j=1}^{n} c_{ij}\right)^2 = 2(2n-3)$ (1b)
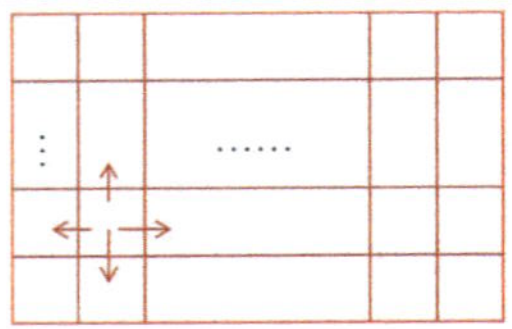 (2a) Square rook	$\sum_{i=1}^{n}\sum_{j=1}^{n} c_{ij} = 2(2PQ-P-Q)$ $\sum_{i=1}^{n}\left(\sum_{j=1}^{n} c_{ij}\right)^2 = 2(8PQ-7P-7Q+4)$ $(n=P\times Q)$ (2b)
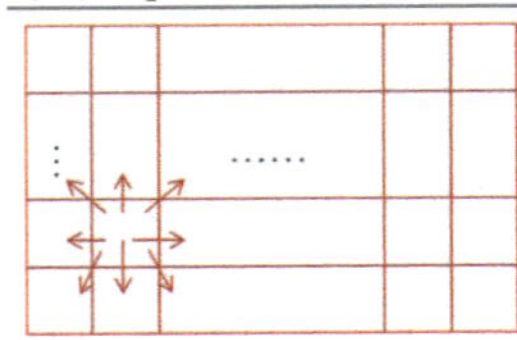 (3a) Square queen	$\sum_{i=1}^{n}\sum_{j=1}^{n} c_{ij} = 2(4PQ-3P-3Q+2)$ $\sum_{i=1}^{n}\left(\sum_{j=1}^{n} c_{ij}\right)^2 = 2(32PQ-39P-39Q+46)$ $(n=P\times Q)$ (3b)
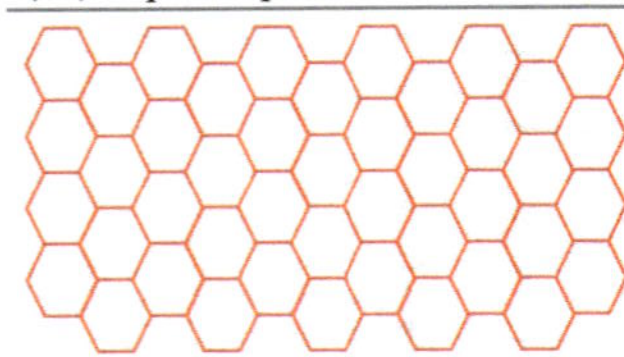 (4a) Hexagon	$\sum_{i=1}^{n}\sum_{j=1}^{n} c_{ij} = 2(3PQ-2P-2Q+1)$ $\sum_{i=1}^{n}\left(\sum_{j=1}^{n} c_{ij}\right)^2 = 2(18PQ-20P-19Q+19)$ $(n=P\times Q)$ (4b)
 (5a) Maximum planar connectivity	$\sum_{i=1}^{n}\sum_{j=1}^{n} c_{ij} = 6(n-2)$ $(n=Q+2)$ $\sum_{i=1}^{n}\left(\sum_{j=1}^{n} c_{ij}\right)^2 = 2(n^2+6n-22)$ (5b)

Table 2 Other connectivity cases and their neighbor sums

Geographic configuration	Neighbor sums
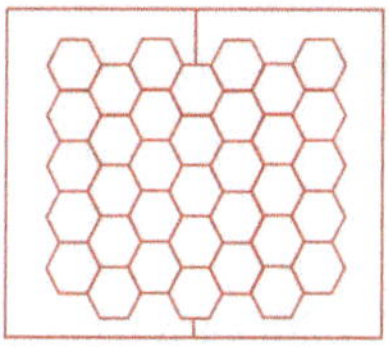 (1a) Maximum hexagon connectivity, odd Q	$\sum_{i=1}^{n}\sum_{j=1}^{n} c_{ij} = 6PQ \ (n = P \times Q + 2)$ $\sum_{i=1}^{n}\left(\sum_{j=1}^{n} c_{ij}\right)^2 = 2(P^2 + Q^2 + 20PQ - 11P - 10Q + 6)$ (1b)
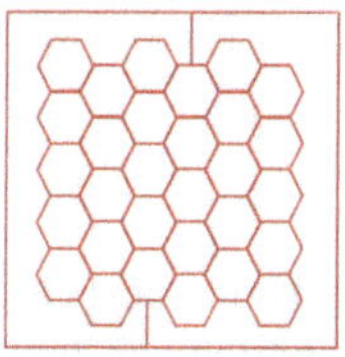 (2a) Maximum hexagon connectivity, even Q	$\sum_{i=1}^{n}\sum_{j=1}^{n} c_{ij} = 6PQ \ (n = P \times Q + 2)$ $\sum_{i=1}^{n}\left(\sum_{j=1}^{n} c_{ij}\right)^2 = 2(P^2 + Q^2 + 20PQ - 11P - 10Q + 8)$ (2b)
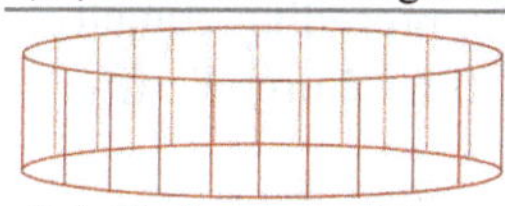 (3a) Circle connectivity	$\sum_{i=1}^{n}\sum_{j=1}^{n} c_{ij} = 2n, \ \sum_{i=1}^{n}\left(\sum_{j=1}^{n} c_{ij}\right)^2 = 4n$ $(n = P \times Q)$ (3b)
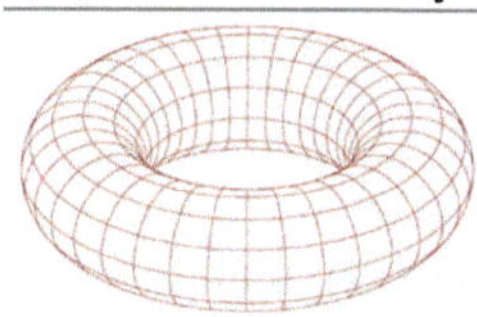 (4a) Torus connectivity	$\sum_{i=1}^{n}\sum_{j=1}^{n} c_{ij} = 4n, \ \sum_{i=1}^{n}\left(\sum_{j=1}^{n} c_{ij}\right)^2 = 16n$ (rook) $\sum_{i=1}^{n}\sum_{j=1}^{n} c_{ij} = 8n, \ \sum_{i=1}^{n}\left(\sum_{j=1}^{n} c_{ij}\right)^2 = 64n$ (queen) $(n = P \times Q)$ (4b)

4.1 Normal Variance Ratios

Substituting Eqs. (10) and (12) into Eq. (17) yields

$$[Var_A(MC)\,/\,Var_{AN}(GR)]_* = S_0\,/\,(S_0 + S_2\,/4), \tag{18}$$

where * denotes linear (*L*), square rook (*SR*), square queen (*SQ*), hexagon (*H*), maximum planar (*MP*), maximum hexagon (*MH*), and constant neighbors (*CN*) connectivity. Substituting the corresponding S_0 and S_2 values (see Tables 1 (1b–5b) and 2 (1b–4b)) into Eq. (18) yields the following asymptotic variance ratios between the MC and GR:

$$[Var_A(MC)/Var_{AN}(GR)]_L = (n-1)/(3n-4),$$
$$[Var_A(MC)/Var_{AN}(GR)]_{SR} = (2PQ-P-Q)/(10PQ-8P-8Q+4),$$
$$[Var_A(MC)/Var_{AN}(GR)]_{SQ} = (4PQ-3P-3Q+2)/(36PQ-42P-42Q+48),$$
$$[Var_A(MC)/Var_{AN}(GR)]_H = (3PQ-2P-2Q+1)/(21PQ-22P-21Q+20),$$
$$[Var_A(MC)/Var_{AN}(GR)]_{ML} = 3(n-2)/\left(n^2-9n-28\right),$$
$$[Var_A(MC)/Var_{AN}(GR)]_{MH} = 6PQ/\left(2P^2+2Q^2+46PQ-22P-20Q+a\right),$$

and

$$[Var_A(MC)/Var_{AN}(GR)]_{CN} = 1/(k+1).$$

For the *MH* ratio, the a in the denominator is a constant, which is 12 for odd Q or 16 for even Q. For the *CN* ratio, k is the number of neighbor sums, which can be 2, 4, or 8. Letting $n \to +\infty (n = PQ)$ furnishes the limits for the preceding equations. These results are summarized in the "AVR" subrow of the "Normal" row in Table 3.

4.2 Uniform Variance Ratios

Repeating the steps in Sect. 4.1, but with Eq. (13) rather than (12), yields

$$[Var_A(MC)/Var_{AU}(GR)]_* = S_0/(S_0+S_2/10), \tag{19}$$

and hence the asymptotic variance ratios become

$$[Var_A(MC)/Var_{AU}(GR)]_L = 5(n-1)/(9n-11),$$
$$[Var_A(MC)/Var_{AU}(GR)]_{SR} = (10PQ-5P-5Q)/(26PQ-19P-19Q+8),$$
$$[Var_A(MC)/Var_{AU}(GR)]_{SQ} = (20PQ-15P-15Q+10)/(84PQ-93P-93Q+102),$$
$$[Var_A(MC)/Var_{AU}(GR)]_H = (15PQ-10P-10Q+5)/(51PQ-50P-48Q+43),$$
$$[Var_A(MC)/Var_{AU}(GR)]_{ML} = 15(n-2)/\left(2n^2+27n-74\right),$$
$$[Var_A(MC)/Var_{AU}(GR)]_{MH} = 30PQ/\left(4P^2+4Q^2+110PQ-44P-40Q+2a\right),$$

and

$$[Var_A(MC)/Var_{AU}(GR)]_{CN} = 5/(2k+5).$$

The terms a and k have the same values as their counterparts in Sect. 4.1. The limits of these preceding equations are summarized in the "AVR" subrow of the "Uniform" row in Table 3.

Table 3 Asymptotic-to-exact variance ratios and adjustment factors

Landscape / Distribution		Linear	Square rook	Hexagon	Square queen	Maximum planar	Maximum hexagon	Constant neighbors
Normal	AVR	1/3	1/5	1/7	1/9	0	3/25	1/(k + 1)
	EVR	1	1	1	1	0	3/7	1
	AVMC	1	1	1	1	1/3	1	1
	AVGR	1/3	1/5	1/7	1/9	1	7/25	1/(k + 1)
Uniform	AVR	5/9	5/13	5/17	5/21	0	15/59	5/(5 + 2k)
	EVR	1	1	1	1	0	0.6522	1
	AVMC	1	1	1	1	1/3	1	1
	AVGR	5/9	5/13	5/17	5/21	1	1/2.565	5/(5 + 2k)
Beta (a = b = 0.5)	AVR	2/3	1/2	2/5	1/3	0	6/17	4/(4 + k)
	EVR	1	1	1	1	0	3/4	1
	AVMC	1	1	1	1	1/3	1	1
	AVGR	2/3	1/2	2/5	1/3	1	8/17	4/(4 + k)
Exponential	AVR	1/9	1/17	1/25	1/33	0	3/91	1/(1 + 4k)
	EVR	1	1	1	1	0	0.1579	1
	AVMC	1	1	1	1	1/3	1	1
	AVGR	1/9	1/17	1/25	1/33	1	1/4.79	1/(1 + 4k)

4.3 Beta Variance Ratios

Repeating the steps in Sect. 4.1, but with Eq. (14) rather than (12), yields

$$[Var_A(MC)/Var_{AB}(GR)]_* = S_0/(S_0 + S_2/16), \tag{20}$$

and hence the asymptotic variance ratios become

$$[Var_A(MC)/Var_{AB}(GR)]_L = 4(n-1)/(6n-7),$$
$$[Var_A(MC)/Var_{AB}(GR)]_{SR} = (8PQ - 4P - 4Q)/(16PQ - 11P - 11Q + 4),$$
$$[Var_A(MC)/Var_{AB}(GR)]_{SQ} = (16PQ - 12P - 12Q + 8)/(48PQ - 51P - 51Q + 54),$$
$$[Var_A(MC)/Var_{AB}(GR)]_H = (12PQ - 8P - 8Q + 4)/(30PQ - 28P - 27Q + 23),$$
$$[Var_A(MC)/Var_{AB}(GR)]_{ML} = 12(n-2)/\left(n^2 + 18n - 46\right),$$
$$[Var_A(MC)/Var_{AB}(GR)]_{MH} = 24PQ/\left(2P^2 + 2Q^2 + 64PQ - 22P - 20Q + a\right),$$

and

$$[Var_A(MC)/Var_{AB}(GR)]_{CN} = 4/(k+4).$$

The terms a and k have the same values as their counterparts in Sect. 4.1. The limits of these preceding equations are summarized in the "AVR" subrow of the "Beta (a = b = 0.5)" row in Table 3.

4.4 Exponential Variance Ratios

Repeating the steps in Sect. 4.1, but with Eq. (15) rather than (12), yields

$$[Var_A(MC)/Var_{AE}(GR)]_* = S_0/(S_0 + S_2), \tag{21}$$

and hence the asymptotic variance ratios become

$$[Var_A(MC)/Var_{AE}(GR)]_L = (n-1)/(9n-13),$$
$$[Var_A(MC)/Var_{AE}(GR)]_{SR} = (2PQ - P - Q)/(34PQ - 29P - 29Q + 16),$$
$$[Var_A(MC)/Var_{AE}(GR)]_{SQ} = (4PQ - 3P - 3Q + 2)/(132PQ - 159P - 159Q + 186),$$
$$[Var_A(MC)/Var_{AE}(GR)]_H = (3PQ - 2P - 2Q + 1)/(75PQ - 82P - 78Q + 77),$$
$$[Var_A(MC)/Var_{AE}(GR)]_{ML} = 3(n-2)/\left(4n^2 + 27n - 94\right),$$
$$[Var_A(MC)/Var_{AE}(GR)]_{MH} = 6PQ/\left(8P^2 + 8Q^2 + 166PQ - 88P - 80Q + 4a\right),$$

and

$$[Var_A(MC)/Var_{AE}(GR)]_{CN} = 1/(4k+1).$$

The terms a and k have the same values as their counterparts in Sect. 4.1. The limits of these preceding equations are summarized in the "AVR" subrow of the "Exponential" row in Table 3.

Table 3 contains four major rows, each containing four subrows. For example, "Normal" denotes the normal distribution; that is to say, all values are calculated assuming a normal distribution. "AVR" denotes the limit of the asymptotic variance ratio between the MC and the GR, and "EVR" denotes the limit of the exact variance ratio between the MC and the GR. "AVMC" and "AVGR" are adjustment factors for the asymptotic variance versus the exact variance of the MC and the GR, respectively. Except for the maximum planar and hexagon cases, all exact variance ratios are one, whereas their asymptotic counterparts are not; hence, asymptotic ratios need to be adjusted. Furthermore, calculating ratios of asymptotic and exact variances of the MC and the GR reveals that the GR's asymptotic ratios need to be adjusted (this is because all of the MC ratios are one). This result furnishes quantitative evidence that the GR, unlike the MC, is far more sensitive to the underlying frequency distribution of an attribute variable. For the maximum planar case, the asymptotic variance and exact variance ratios are zero, indicating that in either case the MC is more efficient than the GR; but in this particular case, the asymptotic variance of the MC should be adjusted by multiplying it by 1/3, whereas the GR does not need to be adjusted. For the maximum hexagon case, the MC is more efficient than the GR in both the asymptotic and the exact cases; GR's asymptotic variance needs to be adjusted.

Table 4 presents the minimum sample size for which the asymptotic variances approximate their exact variance counterparts reasonably well. Columns 2 and 3 are different approximation metrics, where **Abs(Asy/ex-1) <= 0.025** denotes $\left|\frac{V_{asyptotic}}{V_{exact}} - 1\right| \le 0.025$, and **Abs(Ex-asy) <= 0.01** denotes $\left|V_{exact} - V_{asymptotic}\right| \le 0.01$. For example, 23 is the smallest sample size (n) satisfying the inequality

Table 4 Minimum sample sizes for the MC and the GR

Landscape	Abs(Asy/ex-1) <= 0.025		Abs(Ex-asy) <= 0.01	
	MC	GR	MC	GR
Linear	42	56	23	27
Square rook	88	12	36	11
Square queen	161	7	37	93
Hexagon	121	7	34	72
Maximum planar	15	403	10	333
Maximum hexagon	157	14	34	5438
Circle	43	83	23	35
Torus rook	84	124	29	36
Torus queen	167	207	37	41

$\left|V_{exact} - V_{asymptotic}\right| \leq 0.01$. These smallest sample sizes indicate that the MC asymptotic variance furnishes a useful result for relatively small sample sizes.

4.5 Variance Ratio Convergence

Figure 1 portrays exact variance ratio curves as well as values for 184 specimen irregular surface partitions. Figure 1a–d are ideal increasing sample size ratio

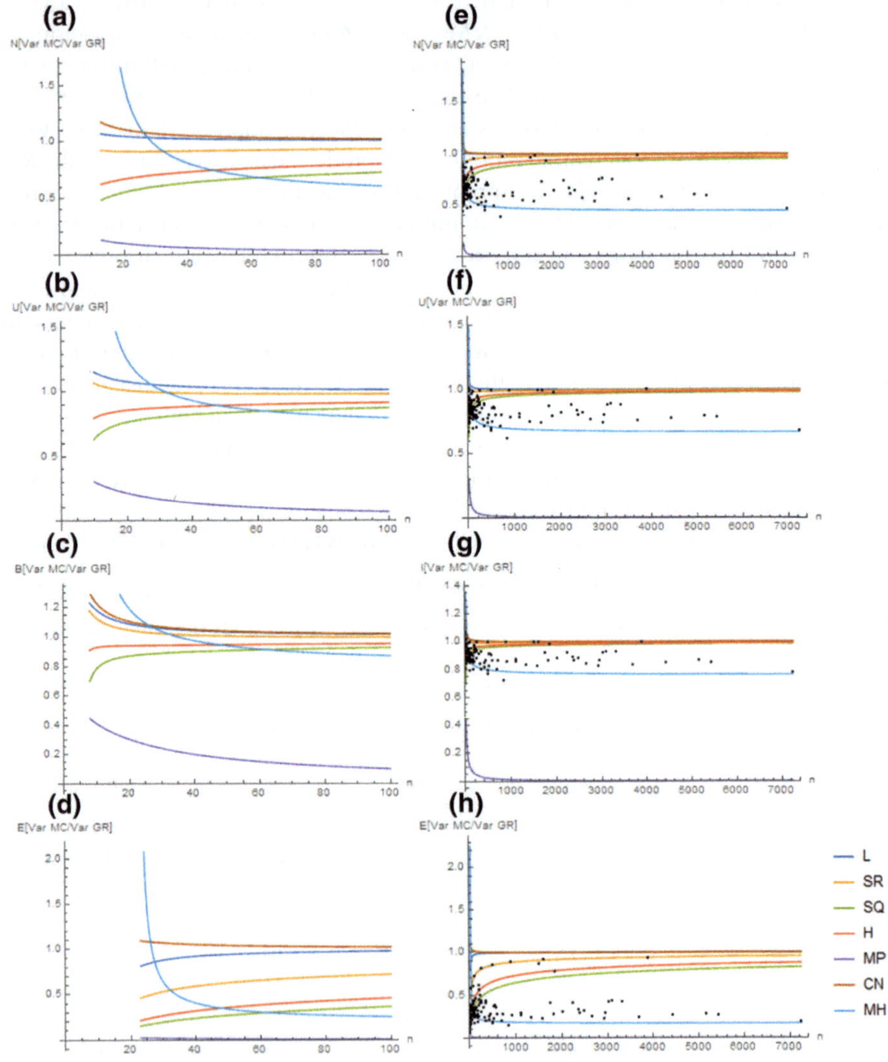

Fig. 1 Exact variance ratio curves

trajectories for normal, uniform, beta, and exponential random variables, respectively, which respectively depict convergence in the interval [13, 100], [10, 100], [8, 100], [23, 100]. Figure 1e–h are the same curves extended to n = 7,250, with the 184 specimen surface values superimposed (black dots).

5 A Power Comparison

Cliff and Ord (1973) conducted limited simulation experiments comparing the power of the MC and GR, concluding that the MC is more powerful. One remaining unanswered question asks what happens across a wider range of sample sizes and attribute variable types than they inspected.

5.1 Establishing Statistical Power

The power of a test is $1-\beta$, where β is the probability of committing a Type II error (i.e., failing to reject the null hypothesis when it is false) for a given significance level α, which is the probability of committing a Type I error (i.e., rejecting the null hypothesis when it is true). If $1-\beta_{MC} > 1-\beta_{GR}$ (i.e., the Type II error probability for the MC is less than the Type II error probability for the GR), then the MC is more powerful than the GR.

Step 1 Let the null and alternative hypotheses be

H_0: *No spatial autocorrelation*; H_1: *Nonzero spatial autocorrelation.* Suppose $\alpha = 0.05$.

Step 2a Consider the Moran test assuming normality such that

$$MC \sim N\left(-\frac{1}{n-1}, Var_N(MC)\right), \tag{22}$$

where Eq. (6) gives $Var_N(MC)$. The critical value (CV) is $z_{MC} = [MC + 1/(n-1)] / \sqrt{Var_N(MC)}$. If $|z_{MC}| < 1.96$, then the statistical decision is to fail to reject H_O.

Step 2b Parallel results for the GR include

$$GR \sim N(1, Var_N(GR)), \tag{23}$$

where Eq. (8) gives $Var_N(GR)$, and $z_{GR} = (GR-1)/\sqrt{Var_N(GR)}$. If $|z_{GR}| < 1.96$, then the statistical decision is to fail to reject H_O.

Step 3a If the statistical decision is to fail to reject H_O when it is not true, a Type II error occurs. The CV under the true sampling distribution is given by

$$MC \sim N(a, Var_N(MC))(a \in [-1, 1], a \neq -1/(n-1)). \tag{24}$$

A way of gaining understanding of the range of a (from –1 to 1, a rook connectivity is assumed here) is to notice that MC is a function of the eigenvalues of $\left(I - \frac{11^T}{n}\right)C\left(I - \frac{11^T}{n}\right)$. The CV under the null distribution (Eq. (22)) is

$$z_{CV} = \pm 1.96\sqrt{Var_N(MC)} - 1/(n-1). \tag{25}$$

The CV under the true distribution (Eq. (24)) is $z_{ts} = (z_{CV} - a) / \sqrt{Var_N(MC)}$. Substituting Eq. (25) into this equation yields $z_{ts} = \pm 1.96 - \frac{1}{\sqrt{Var_N(MC)}}\left(\frac{1}{n-1} + a\right)$, where the subscript ts denotes the standard CV under the true distribution. Given $z_{\alpha/2} = 1.96 - \frac{1}{\sqrt{Var_N(MC)}}\left(\frac{1}{n-1} + a\right)$ and $-z_{\alpha/2} = -1.96 - \frac{1}{\sqrt{Var_N(MC)}}\left(\frac{1}{n-1} + a\right)$, the power of the MC is

$$1 - \beta_{MC} = 1 - \mathrm{P}(x \leq z_{\alpha/2}) + \mathrm{P}(x \leq -z_{\alpha/2}). \tag{26}$$

Step3b For the GR, the CV under the true distribution is given by

$$GR \sim N(b, Var_N(GR)), (b \in [0, 2], b \neq 1), \tag{27}$$

[0, 2] is the range of b (a rook connectivity is assumed here), where [0, 1] stands for positive spatial autocorrelation, with GR value approaching zero, positive spatial autocorrelation becoming stronger and stronger; 1 stands for zero spatial autocorrelation, and [1, 2] stands for negative spatial autocorrelation, with GR value approaching 2, negative spatial autocorrelation becoming stronger and stronger. The CV under the null distribution (Eq. (23)) is

$$z'_{cv} = \pm 1.96\sqrt{Var_N(GR)} + 1. \tag{28}$$

The CV under the true distribution is $z'_{ts} = \left(z'_{CV} - b\right) / \sqrt{Var_N(GR)}$. Substituting Eq. (28) into this formula yields Eq. (29):

$$z'_{ts} = \pm 1.96 + \frac{1}{\sqrt{Var_N(GR)}}(1 - b), \tag{29}$$

Defining $z'_{\alpha/2}$ and $-z'_{\alpha/2}$ in the same way as $z_{\alpha/2}$ and $-z_{\alpha/2}$, the power of the GR is

$$1 - \beta_{GR} = 1 - \mathrm{P}\left(x \leq z'_{\alpha/2}\right) + \mathrm{P}\left(x \leq -z'_{\alpha/2}\right). \tag{30}$$

Figures 2 and 3 present selected power curves plotted with their respective minimum sample sizes. Figure 2 reveals that the MC is uniformly more powerful

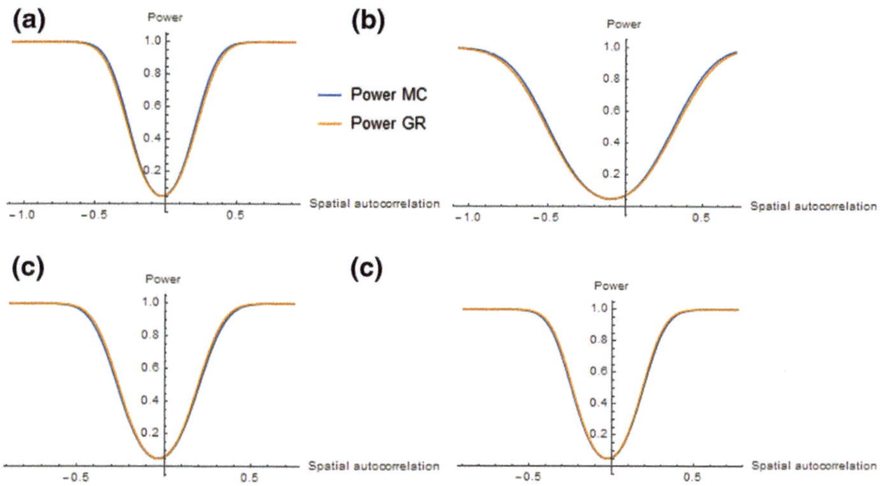

Fig. 2 Planar and torus rook square surface partitionings. **a** *Top left*: square rook, P = 6, Q = 6; **b** *Top right*: square rook, P = 3, Q = 4; **c** *Bottom left*: torus rook, P = 5, Q = 6; **d** *Bottom right*: torus rook, P = 6, Q = 6

than the GR for the rook configuration (for Fig. 2a, b, the blue curve is slightly above the orange curve), whereas the GR is uniformly more powerful than the MC for the torus rook connectivity case (for Fig. 2c, d, the orange curve is slightly above the blue curve). Figure 3 shows the hexagon and maximum hexagon cases: the MC is more powerful than the GR for positive spatial autocorrelation (the blue curve is above the orange curve) but not always for negative spatial autocorrelation (for Fig. 2c, the blue curve is below the orange curve).

5.2 *Theoretical Evaluation*

Equation (2) can be rewritten using matrix notation such that $\left(I - \frac{11^T}{n}\right)C\left(I - \frac{11^T}{n}\right)$ is in the numerator of the MC, where **1** is an n-by-1 vector of ones, and T denotes the matrix transpose operation. The eigenvalues of this matrix times $\frac{n}{1^T C1}$ furnish the complete set of distinct MC values for a geographic landscape, with the extreme values establishing the minimum and maximum possible MC values. Corresponding GR values can be calculated with the eigenvectors of this matrix. According to Eq. (4), the MC is negatively correlated with the GR; the relationship between MC and GR in this context is given by $GR = a + b(MC - MC_{\min})^c$. For linear (circle) adjacency and a square tessellation with rook (torus rook) adjacency, the estimates of this equation's parameters are $\widehat{a} = 2$, $\widehat{b} = -1$, and $\widehat{c} = 1$. For a queen

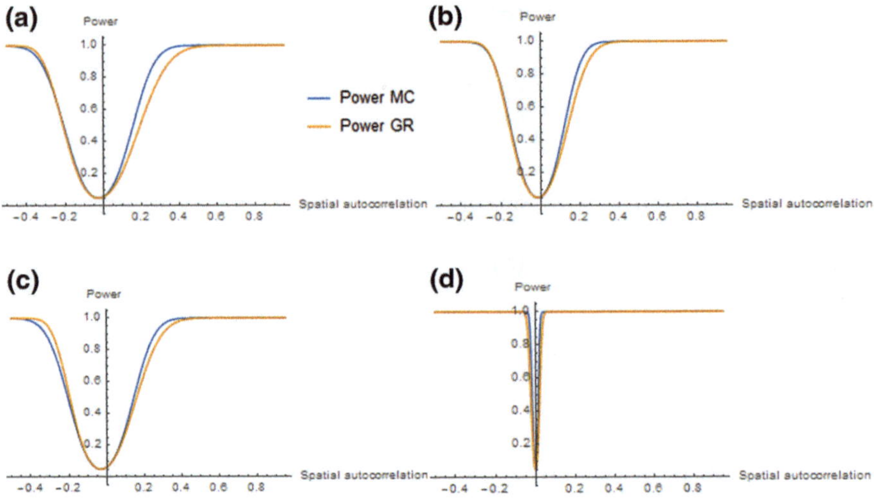

Fig. 3 Hexagon and maximum hexagon surface partitionings. **a** *Top left*: hexagon, P = 5, Q = 7; **b** *Top right*: hexagon, P = 8, Q = 9; **c** *Bottom left*: maximum hexagon, P = 5, Q = 7; **d** *Bottom right*: maximum hexagon, P = 73, Q = 74

(torus queen) adjacency, $\widehat{a} = 1.5$, $\widehat{b} = -1$, and $\widehat{c} = 1$. For a P-by-Q hexagonal tessellation, $\widehat{a} = 1.5$,

$$\widehat{b} = -0.99063 - 0.78935(1/P + 1/Q)^{0.87125} + 0.00205P/Q,$$

and

$$\widehat{c} = 1.05828 - 0.87248(1/P + 1/Q)^{0.57358} + 0.00385P/Q.$$

6 Conclusions

The MC may not be uniformly more powerful than the GR for all sample sizes and geographic configurations. For a regular square tessellation, it is uniformly more powerful than the GR, but for a hexagonal tessellation, it is more powerful than the GR only for positive spatial autocorrelation. The MC asymptotic variances are reliable for most modern day sample sizes, which are greater than 100, and preferable because its smallest sample sizes for approximating its exact variance are relatively smaller than those for the GR. The GR asymptotic variance varies with geographic configuration as well as attribute variable type, both of which determine efficiency when geographic sample size goes to infinity. Finally, the relationship between the MC and the GR appears to vary across areal unit configuration types.

References

Cliff A, Ord J (1973) Spatial autocorrelation. Pion, London
Cliff A, Ord J (1981) Spatial process. Pion, London
Geary R (1954) The contiguity ratio and statistical mapping. The incorporated statistician 5 (3):115–145
Griffith D (1987) Spatial autocorrelation: a primer. AAG, Pennsylvania
Griffith D (1996) Spatial autocorrelation and eigenfunctions of the geographic weights matrix accompanying geo-referenced data. The Can Geogr 40(4):351–367
Griffith D (2010) The Moran coefficient for non-normal data. J Stat Plann Infer 140:2980–2990
Griffith D, Sone A (1995) Trade-offs associated with normalizing constant computational simplifications for estimating spatial statistical models. J Stat Comput Simul 51(2–4):165–183
Moran P (1950) Notes on continuous stochastic phenomena. Biometrika 37:17–23
Tiefelsdorf M, Boots B (1995) The exact distribution of Moran's I. Environ Plann A 27:985–999

A Variance-Stabilizing Transformation to Mitigate Biased Variogram Estimation in Heterogeneous Surfaces with Clustered Samples

Xiaojun Pu and Michael Tiefelsdorf

Abstract Due to the inherent variance heterogeneity in clustered preferential sampling, the underlying variogram cannot be estimated directly. A variance-stabilizing declustering method is proposed here using a modified Box–Cox transformation. In contrast to the traditional Box–Cox transformation that aims at achieving normally distributed data, its modified version has the objective to match the variance in clustered sample observations to the variance of the remaining more dispersed background sample observations. The proposed approach leads to predictions with lower standard errors than alternative proposed methods.

Keywords Clustered preferential sampling • Variance-stabilizing transformation • Variogram estimation • Kriging prediction

1 Introduction

Geostatistical techniques, especially kriging, are widely used to predict natural features with a continuous spatial distribution. The performance of kriging critically relies on (1) the identification of any underlying spatial trend to capture the variation in the expected values, (2) the variogram estimation, which is supposed to capture the spatial autocorrelation structure within the unknown population values (Richmond 2002; Kovitz and Christakos 2004), and (3) the spatial structure of the sample locations among which kriging interpolation is performed. The sampling locations for the variogram estimation and those for the subsequent surface prediction do not necessarily need to be identical. The most efficient sampling procedure for variogram estimation requires capturing reliably the semivariogram at all relevant intersample distances, whereas for the purpose of prediction, evenly

X. Pu (✉) · M. Tiefelsdorf
The University of Texas at Dallas, 800 W. Campbell Rd, Richardson, TX, USA
e-mail: xxp102020@utdallas.edu

M. Tiefelsdorf
e-mail: tiefelsdorf@utdallas.edu

D.A. Griffith et al. (eds.), *Advances in Geocomputation*,
Advances in Geographic Information Science,
DOI 10.1007/978-3-319-22786-3_24

distributed sample locations across an entire study area are preferred. However, for both variogram estimation and prediction, clustered preferential sampling may occur due to external factors such as financial limitations and hostile environmental conditions (Olea 2007; Menezes et al. 2008). Sample observations coming from a heterogeneous population when the underlying spatial trend has been ignored, such as in clustered preferential sampling, lead to compromised variogram estimates. Usually, the variability within a cluster is substantially larger at shorter distances than that for the remainder of the sample points in the less variable region of a study area. In particular, at short distances, this local variance heterogeneity can lead to unrepresentative joint variogram estimation. In clustered preferential sampling, the status of whether a sample observation belongs to a cluster or the remainder of the study area is well identified during the initial sample collection. However, when predicting data values between sample locations, this status generally is unknown.

To avoid the problem caused by clustered preferential sampling, a number of declustering methods have been proposed to improve variogram estimation. Two major branches exist to control the induced bias in the variogram estimation: the calibration of weighted variograms (Bourgault 1997; Richmond 2002; Kovitz and Christakos 2004; Menezes et al. 2008) and the subsampling approach (Olea 2007). Most researchers adopt the weighted variogram approach, such as using the ratio of correlation matrix determinants (Bourgault 1997), the two-point declustering method based on cells or clusters (Richmond 2002), declustering weights based on zones of proximity (Kovitz and Christakos 2004), or the robust kernel variogram estimator (Menezes et al. 2008). This chapter employs the two-point declustering method (Richmond 2002) for comparison purposes to evaluate the performance of the proposed method. Regarding grids or clusters as units, Richmond's method (2002) counts the number of point pairs ($\boldsymbol{n}$) for a certain distance ($\boldsymbol{d}$) between two grids ($\boldsymbol{g}_i$ and $\boldsymbol{g}_{i'}$) or two clusters ($\boldsymbol{c}_i$ and $\boldsymbol{c}_{i'}$). Then the two-point declustering method assigns the inverse proportion of the total pair number $w_{\alpha\alpha'} = 1/n$ as the weights to those specified point pairs ($\boldsymbol{x}_\alpha \in \boldsymbol{g}_i/\boldsymbol{c}_i$ and $\boldsymbol{x}_{\alpha'} \in \boldsymbol{g}_{i'}/\boldsymbol{c}_{i'}$) to adjust a variogram. The subsampling approach (Olea 2007), which is to pick subsamples free of clusters to build a representative histogram, is used for evaluation as well. Samples are divided into two subsets: subset 1 (free of clusters) and subset 2 (only including clusters). Based on the maximum nearest neighborhood distance between the two subsets, some points are moved from subset 2 to subset 1, iteratively. The distance distribution of the expanded subset 1 should match that of the original subset 1. Then, the expanded subset 1 can be used to model a variogram for prediction purposes.

The proposed method in this chapter adopts the Box–Cox transformation to improve the variogram estimation for a heterogeneous surface when clustered preferential sampling usually is used. Instead of the traditional Box–Cox transformation (Sakia 1992), which treats a sample as coming from a homogeneous population and which aims at achieving a symmetric or Gaussian distribution of the transformed sample observations, the objective of the proposed method is to stabilize the variances in both (1) the clustered sample points and (2) those sample observations associated with the remainder of a study area.

2 Methodology

The spatial distribution of the clustered dataset used in this research is introduced in details in Sect. 2.1. Then Sect. 2.2 introduces the variance-stabilizing approach based on the revision of the standard Box–Cox transformation. This section discusses how to identify the optimal transformation exponent λ based on the characteristics of the data. After conducting kriging predictions in the transformed scale, the predicted results are transformed back to the original measurement scale and are compared with the predictions for the other three methods (Richmond 2002; Olea 2007).

2.1 Data

Clustered data from the GSLIB (Deutsch and Journel 1997) have been used with the two-point declustering method (Richmond 2002) and Olea's subsampling approach (2007). The total sample contains 140 observations (Fig. 1b) selected from a 50-by-50 regular grid image (Fig. 1a), including 86 single points (subset 1) from the disperse background of the study area and the remainder from ten clusters (subset 2), which were identified during the sample collection. Most clusters consist of five sample points having a distance of one around a central sample point, with the exception of one nine-point cluster, which is a combination of two five-point clusters with one edge point being cut off (Olea 2007). Figure 1c displays a three-dimensional (3D) map of the reference population data, with the sharp high peaks representing clusters of large values. Figure 1d displays the abnormal variogram cloud of sample point pairs in three dimensions: the spatial distance between two sample points (z_i and z_j), the average attribute value $(z_i + z_j)/2$ identifying the different baseline levels in a cluster and the surrounding background data values, and the absolute difference between two sample points $|z_i - z_j|$, which measures their dissimilarity. Three groups of point pairs may be identified in the variogram cloud: (1) the relationship between clustered-surrounding sample point pairs in green, (2) clustered-clustered point pairs in red, and (3) surrounding-surrounding point pairs in black. Due to the larger variance in the clustered-surrounding and clustered-clustered groups, the variogram cloud does not display the usually increasing attribute dissimilarity trend with increasing interpoint distance. Compared to the variogram of the full referenced population (Fig. 1e), the estimated variogram of the clustered sample (Fig. 1f) displays an irregular pattern with high dissimilarities at small distance intervals.

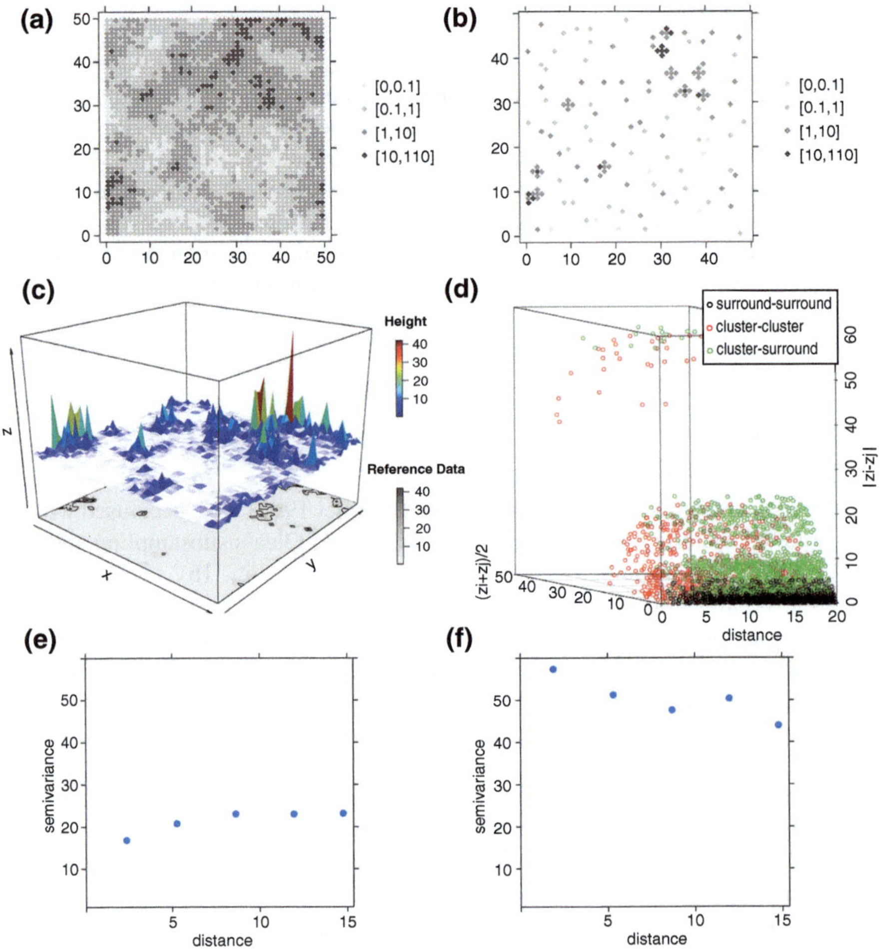

Fig. 1 Exploratory data analysis, including variogram construction with the untransformed data. **a** Geographic distribution of the reference population. **b** Geographic distribution of the sample locations with clusters. **c** 3D surface of the reference population. **d** Exploratory heterogeneity 3D plot of the sample point dissimilarities. **e** The variogram for the reference population. **f** The variogram for the sampling locations with clusters

2.2 The Box–Cox Transformation and Kriging Prediction

The observed sample values range from 0.06 to 58.32; nearly half of the observations are smaller than one, which implies two different underlying distributions are contributing to this heterogeneous surface. In Fig. 2a and b, the probability density distribution of subset 1 (86 single points) is displayed in red, and that of subset 2 (ten clusters) is represented by green. The means and standard deviations of subsets 1 and 2 are calculated before and after the Box–Cox transformation, separately. Most of sample points are from subset 1, which has a small mean and small variance, whereas the clusters are from subset 2, which has a large mean and large variance (Fig. 2a). Therefore, the dissimilarities captured at short distances cannot reflect the true differences within the entire study area. To address this problem, we can conduct a Box–Cox transformation (Eq. (1)) with the objective of making the variances of both subsets as similar as possible (Fig. 2b). Compared to the standard application of the Box–Cox transformation, the proposed approach uses a different objective function. Rather than maximize the likelihood to obtain a normally distributed sample after the transformation (Sen and Srivastava 1990), the objective function of the proposed approach is to obtain two subsets of classified sample observations that have identical variances. The joint distribution of both subsets, the identified clusters and the surrounding background observations, still may remain bimodal, even in the transformed measurement system. The joint probability density function before transformation can be represented by $\textit{Joint-density} = \underbrace{\pi_1 \cdot \mathcal{N}(\mu_1, \sigma_1^2)}_{background} + \underbrace{\pi_2 \cdot \mathcal{N}(\mu_2, \sigma_2^2)}_{cluster}$, where π_1 and π_2 respectively denote the proportions of background and clustered observations. After conducting the Box–Cox transformation, the transformed joint probability density function (denoted by a*) is changed to $\textit{Joint-density}^* = \underbrace{\pi_1 \cdot \mathcal{N}(\mu_1^*, \sigma^{*2})}_{background} + \underbrace{\pi_2 \cdot \mathcal{N}(\mu_2^*, \sigma^{*2})}_{cluster}$, with the identical variances after the transformation and the same mixture proportion for each subset before and after the transformation.

The equation for the standard Box–Cox transformation is

$$y_i^{(\lambda)} = \begin{cases} (y_i^\lambda - 1)/\lambda & \lambda \neq 0 \\ \ln y_i & \lambda = 0 \end{cases}, \tag{1}$$

where y_i denotes the sample data i on the original scale that needs to be transformed, and λ is the optimal transformation exponent. The endogenous variable in this example has a positive support, and zero refers to its natural origin. Therefore, an application of an offset parameter in a shifted Box–Cox transformation is neither necessary nor recommended (Sen and Srivastava 1990; Tiefelsdorf 2013). We regard subsets 1 and 2 to be representative for the underlying heterogeneous population. The candidate transformation exponent λ was set to range from –2 to 2, with 0.01 increments. A grid search (Thisted 1988) has been conducted iteratively for all candidate values of λ, and the variances in both subsets were evaluated at each iteration. This approach renders the optimal λ value for which both variances are approximately identical. In this case, the optimal transformation parameter becomes $\lambda = 0.19$ (see Fig. 2b), which makes the standard deviations of both transformed subsets approximately identically (i.e., $\sigma = 1.09$). After the optimal λ has been identified, we estimate the variogram model parameters in the transformed scale and conduct ordinary kriging to predict interpolated values for the entire study

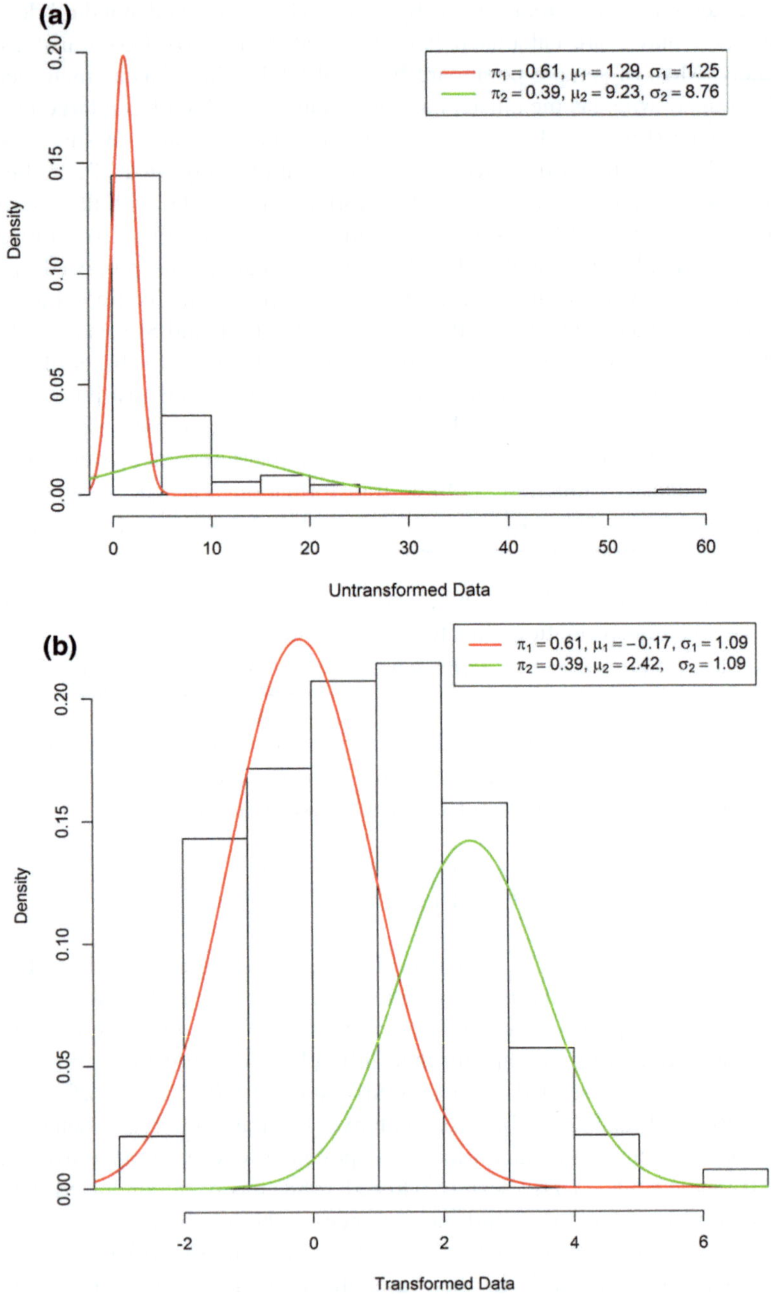

Fig. 2 Distributions of background and clustered sample observations. **a** Density curves before the transformation. **b** Density curves after the transformation

area. In addition to these predicted values in the transformed scale, prediction standard errors and 95 % prediction interval bounds ($CI_{0.025}$ and $CI_{0.975}$) also are calculated for each prediction location.

Although the model is estimated in the transformed measurement scale, and initial predictions (including the prediction variances and the prediction interval bounds) are performed in the transformed scale, the final predictions are calculated using an adjusted inverse Box–Cox transformation of these initial predictions (Yang 1999; Cho and Loh 2006; Perry and Walker 2015). The adjustment is needed because the distributional shape, on which the expectations are based, changes. This is due to the nonlinear nature of the inverse Box–Cox transformation. While the Jacobian factor (Sen and Srivastava 1990; Tiefelsdorf 2013) relates to the transformed distribution, a simpler approach is taken here because we are only interested in the expected value of the predictions rather than the entire distribution of the predictions. A second-order Taylor series approximation of the expected value is used to obtain unbiased predictions in the original measurement system (Shumway et al. 1989; Freeman and Modarres 2006; Tiefelsdorf 2013). In Eqs. 2 and 3, μ_λ is the predicted value in the transformed scale, and $\sigma^2_{(\lambda)}$ is the local prediction variance at each predicted location (Freeman and Modarres 2006; Tiefelsdorf 2013).

$$E(Y) \approx (\lambda \cdot \mu_\lambda + 1)^{1/\lambda} \cdot \left(1 + \frac{1}{2} \cdot \sigma^2_{(\lambda)} \cdot \frac{(1-\lambda)}{\left(\lambda \cdot \mu_{(\lambda)} + 1\right)^2}\right) \tag{2}$$

$$Var(Y) \approx \sigma^2_{(\lambda)} \cdot \left(\lambda \cdot \mu_{(\lambda)} + 1\right)^{\frac{2}{\lambda} - 2} \tag{3}$$

In particular, the variance expression may exert a noticeable truncation error (Freeman and Modarres 2006; Tiefelsdorf 2013) because it is based only on a first-order Taylor series expansion. Therefore, the back-transformed prediction interval (Yang 1999; Cho and Loh 2006; Perry and Walker 2015) also is reported.

Table 1 summarizes the variogram modeling parameters estimated (Richmond 2002; Olea 2007) for the three earlier methods and those of the proposed variance-stabilizing approach. Richmond (2002) and Olea (2007), by default, apply a log-transformation to their original data, which is equivalent to choosing $\lambda = 0$, before estimating their variograms. In contrast, the optimal variance-stabilizing transformation parameter for these sample observation is $\lambda = 0.19$.

Table 1 Variogram parameter estimates for the different declustering methods

	Model type	Nugget	Partial sill	Range
Cell (Richmond 2002)	Spherical	0.10	1.90	9.75
Cluster (Richmond 2002)	Spherical	0.10	1.90	10.25
Subsampling (Olea 2007)	Spherical	0.06	1.77	9.52
Variance stabilization	*Spherical*	*0.56*	*2.07*	*11.41*

3 Results

The distributional characteristics of the predicted values in the original measurement scale of the four declustering methods are different from the true characteristics of the underlying reference population data (Table 2). The means and medians of the predicted back-transformed values for all four methods are biased, but the mean and skewness of the proposed variance-stabilizing method are closer to those of the reference population.

After the back-transformation of the predicted interpolation values to its original measurement scale, the aggregated width of the prediction intervals $\sum(CI_{0.975} - CI_{0.025})$, root mean squared errors (RMSE), and sum of prediction variances are calculated as evaluation criteria. Although the predictions based on the proposed variance-stabilizing transformation have the largest RMSE, the aggregated widths of the prediction intervals and the sum of local predication variances are much lower than those of the other three methods (Table 3).

Figure 3 shows the prediction variances of the four declustering methods. Figure 3d displays results for the proposed variance-stabilizing approach, highlighting a decrease in the prediction uncertainty in nonclustered subregions. In other words, the method proposed in this chapter achieves the highest prediction certainty while showing a slight bias, as indicated by the RMSE.

Table 2 Distributional characteristics for the four declustering methods and the true reference distribution

	Mean	1st quartile	Median	3rd quartile	Standard deviation	Skewness
Reference population	***2.58***	***0.34***	***0.96***	***2.56***	***5.15***	***6.83***
Cell (Richmond 2002)	2.43	0.89	1.70	3.12	2.65	5.85
Cluster (Richmond 2002)	2.40	0.86	1.65	3.07	2.66	5.80
Subsampling (Olea 2007)	2.33	0.84	1.63	2.97	2.60	6.16
Variance stabilization	*2.42*	*1.11*	*1.80*	*2.89*	*2.45*	*6.99*

Table 3 Predicted accuracy and uncertainty for four declustering methods

	RMSE	$\sum(Q_{95\%} - Q_{5\%})$	Sum of uncertainty
Cell (Richmond 2002)	3.98	36936.81	37762.89
Cluster (Richmond 2002)	3.98	34962.81	35410.85
Subsampling (Olea 2007)	4.00	31798.75	29281.93
Variance stabilization	*4.16*	*26602.85*	*9504.08*

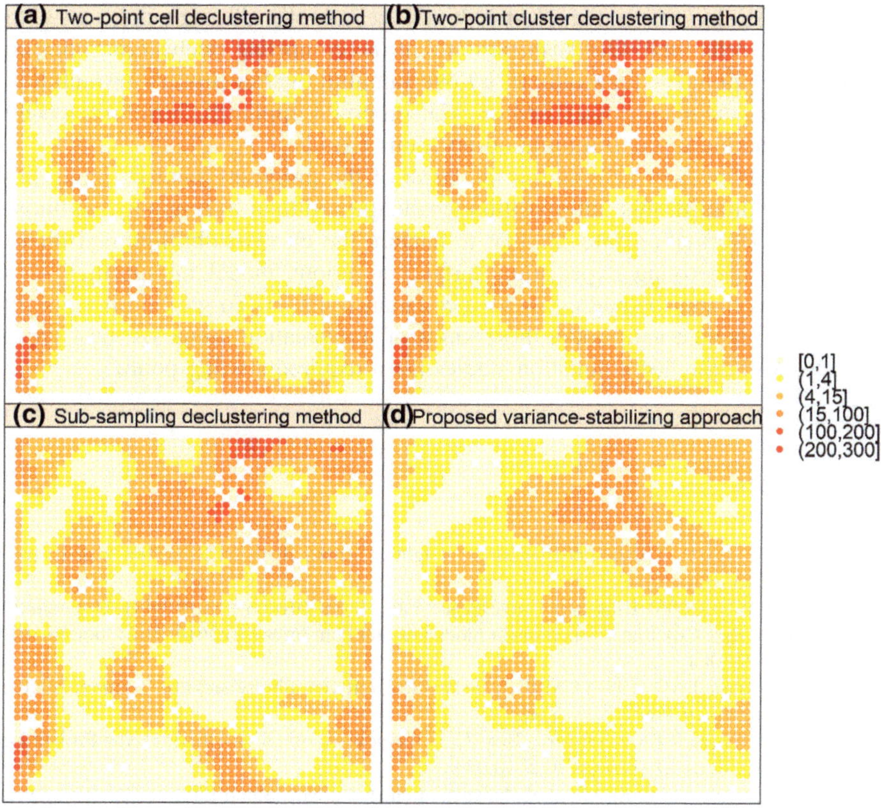

Fig. 3 Predicted uncertainty of four declustering methods

4 Conclusions

Highly heterogeneous spatial surfaces are observed not only in the natural sciences but also in other disciplines. For example, population density mainly is flat in rural regions and then sharply peaks once one enters urbanized areas; alternatively, air pollution measurement stations focus on areas with a high emission potential. Methods to sample representatively these heterogeneous surfaces and subsequently to perform accurate interpolations are highly relevant. The proposed variance-stabilizing transformation offers one approach to handle these diverse scenarios. The RMSE, as evaluated in this chapter, is computed over all grid cells, which comprise both the cluster cells with extremely large data values and the background cells with moderate to low data values. A well-known property of using squared errors is that large errors are extremely influential on the RMSE. Therefore, the goal becomes a matter of balancing. Are we paying more attention (1) to the fewer data values within the clusters, which exhibit larger deviations or (2) to a majority of the grid cells in the background category? An interpolation method may achieve a lower RMSE by fitting the clustered grid cells well, although not fitting the background cells well. Our suspicion is that the competing declustering methods follow this rationale—for example, by oversampling within clusters—which leads to a cluster-dominated lower RMSE. The proposed approach, however, weights clustered and background sample observations equally.

In conclusion, we feel that, so far, none of the proposed methods have satisfactorily addressed this interpolation challenge: without taking any additional exogenous information into account to capture the cluster-to-background dichotomy of an underlying expectation surface, a serious model misspecification is present than cannot be addressed by a covariance model. To gain further insights into how to handle and model these kinds of interpolation scenarios most appropriately, well-designed simulation studies are a required starting point.

References

Bourgault G (1997) Spatial declustering weights. Math Geol 29:277–290. doi:10.1007/BF02769633

Cho K, Loh WY (2006) Bias and convergence rate of the coverage probability of prediction intervals in Box-Cox transformed linear models. J Stat Plan Infer 136:3614–3624. doi:10.1016/j.jspi.2005.03.004

Deutsch CV, Journel AG (1997) GSLIB: geostatistical software library and user's guide, 2nd edn. Oxford University Press, New York

Freeman J, Modarres R (2006) Inverse Box–Cox: the power-normal distribution. Stat Probab Lett 76:764–772. doi:10.1016/j.spl.2005.10.036

Kovitz JL, Christakos G (2004) Spatial statistics of clustered data. Stoch Environ Res Risk Assess 18:147–166. doi:10.1007/s00477-003-0133-9

Menezes R, Garcia-Soidán P, Febrero-Bande M (2008) A kernel variogram estimator for clustered data. Scand J Stat 35:18–37. doi:10.1111/j.1467-9469.2007.00566.x

Olea RA (2007) Declustering of clustered preferential sampling for histogram and semivariogram inference. Math Geol 39:453–467. doi:10.1007/s11004-007-9108-6

Perry MB, Walker ML (2015) A prediction interval estimator for the original response when using Box-Cox transformations. J Qual Technol 47:278–297

Richmond A (2002) Two-point declustering for weighting data pairs in experimental variogram calculations. Comput Geosci 28:231–241. doi:10.1016/S0098-3004(01)00070-X

Sakia RM (1992) The Box-Cox transformation technique: a review. Statistics 41:169–178

Sen A, Srivastava M (1990) Transformation. Regression analysis: theory, methods, and applications. Springer, New York, pp 180–217

Shumway R, Azari A, Johnson P (1989) Estimating mean concentrations under transformation for environmental data with detection limits. Technometrics 31:347–356. doi:10.1080/00401706.1989.10488557

Thisted AR (1988) Nonlinear statistical methods. Elements of statistical computing: numerical computation. Chapman and Hall/CRC, New York, pp 200–202

Tiefelsdorf M (2013) A note on the reverse Box–Cox transformation. http://www.tiefelsdorf.spatialfiltering.com/Notes/TheoryReverseBoxCox.pdf. Accessed 17 Feb 2016

Yang Z (1999) Predicting a future lifetime through Box-Cox transformation. Lifetime Data Anal 5:265–279. doi:10.1023/A:1009676116509

Estimating a Variance Function of a Nonstationary Process

Eunice J. Kim and Z. Zhu

Abstract Non-constant variance functions are a common feature in geophysical processes, and estimating the variance function is important to provide accurate prediction intervals. We propose a nonparametric approach to estimating the variance function from a single continuous-space process where the mean function is smooth and the additive errors are correlated and heteroskedastic. We explore a few configurations of difference filters and recommend filter weights depending on the strength of the correlation in the error process. Symmetric-weight filters are preferred when the errors are strongly correlated, and Hall–Kay–Titterington weight filters are preferred when the errors are weakly correlated or independent. The proposed method provides efficiency in computing, especially with strongly correlated data.

Keywords Nonstationary process · Variance function · Difference · Symmetric weight

1 Introduction

We develop a method to estimate a variance function of a nonstationary continuous process. For example, consider a map of the average daily high temperature in April in the United States. Using a big brush, we can paint the South warm and the North cool. Near the Great Lakes and the Rockies, the temperature field exhibits a dip in comparison to the coasts at the same latitudes. Models reflecting geography and driven by physics should account for these large-scale trends. However, the average daily high temperatures in small patches deviate from the forecasts of a physical

E.J. Kim (✉)
Department of Mathematics and Statistics, Amherst College,
Box 2239, Amherst, MA 01002, USA
e-mail: eunicejk@alumni.upenn.edu

Z. Zhu
Department of Statistics, Iowa State University,
Snedecor Hall 1211, Ames, IA 50010, USA
e-mail: zhuz@iastate.edu

D.A. Griffith et al. (eds.), *Advances in Geocomputation*,
Advances in Geographic Information Science,
DOI 10.1007/978-3-319-22786-3_25

process model. We call these deviations of the observed from the model forecasts *errors*. They exhibit *heteroskedasticity* (the size of the errors vary by region) and positive spatial *autocorrelation* (the values are similar in close proximity). We are interested in estimating the variance function of such errors to provide accurate confidence and prediction intervals for a temperature field.

Estimating the variance by differencing data is a method first detailed in von Neumann et al. (1941). When data have a natural order of presentation and display a gradual change in the mean function, the variance can be estimated by simple differencing of neighboring points assuming that the errors are additive, independent, and identically distributed (i.i.d.). Gasser and Müller (1984), Buckley et al. (1988), Hall and Carroll (1989), Hall et al. (1990), Brown and Levine (2007), and Cai and Wang (2008) estimate a one-dimensional variance function, applying the differencing idea to independent and heteroskedastic errors. Hall et al. (1991) employ the idea in image processing to estimate the variance of i.i.d. errors. When the data are positively autocorrelated, Martin (1974) shows that the variance is underestimated due to the redundant information in the data. Here, we propose a difference-based variance function estimation in two-dimension where the errors are positively correlated and heteroskedastic. We define a local variogram, which measures spatial dispersion by differencing in localized context to assume stationarity. Zhu and Stein (2002) introduce a generalized variogram and use difference filters to estimate the fractal dimension of fractional Brownian fields. Our contribution is the introduction of a local variogram and the estimation of a non-constant variance function embedded in a nonstationary spatial process.

2 Data Model and Variance Function Estimator

In Sect. 2, we describe our data model, define a linear filter, and introduce a filter variogram and a local filter variogram. Then, we present our variance function estimator.

2.1 Data Model

Consider a continuous process on a two-dimensional plane. Our data model assumes that the observed process $Z(\boldsymbol{s})$ has a smooth mean function $\mu(\boldsymbol{s})$ and an additive non-constant error $\sigma(\boldsymbol{s})X(\boldsymbol{s})$, which comprises a smooth standard deviation function $\sigma(\boldsymbol{s})$ and a second-order stationary process $X(\boldsymbol{s})$ with mean 0, variance 1, and $cor\left(X(\boldsymbol{s}), X(\boldsymbol{s}')\right) = c\|\boldsymbol{s} - \boldsymbol{s}'\|^{\alpha}$, where $0 < \alpha < 2$ and $0 < c < 1$. In other words, the model for a nonstationary process $\{Z(\boldsymbol{s})\}$ is

$$Z(\boldsymbol{s}) = \mu(\boldsymbol{s}) + \sigma(\boldsymbol{s})X(\boldsymbol{s}), \tag{1}$$

and satisfies the regularity conditions of a stationary process when $Z(\boldsymbol{s})$ is standardized and the set of locations $\boldsymbol{s}$ lie in $\mathbb{R}^2$. To simplify, we assume that observations are on a regular lattice grid.

2.2 Notation and Definitions

A linear filter function L is defined by a set of points $\mathcal{J}$ about the center of a configuration $\boldsymbol{p}_0$ and a set of non-zero weights $\mathcal{A}$ assigned to each point in $\mathcal{J}$. In mathematical set notation,

$$\mathcal{J} = \left\{ \boldsymbol{p}_j = (p_{1j}, p_{2j}) \in \mathbb{Z}^2 : \sum_j (\boldsymbol{p}_j - \boldsymbol{p}_0) = \boldsymbol{0} \right\} \text{ and } \mathcal{A} = \{a_j : a_j \neq 0,\, p_j \in \mathcal{J}\}. \tag{2}$$

Then,

$$L(Z(\boldsymbol{s})) = \sum_{j \in \mathcal{J}} a_j Z(\boldsymbol{s} + \boldsymbol{p}_j) \tag{3}$$

represents a filter L applied to a process $\boldsymbol{Z}$ at $\boldsymbol{s}$. Before we go into details, we introduce shorthand notation to be used throughout this chapter: $Z(\boldsymbol{s}_i + \boldsymbol{p}_j) = Z_{i+j}$, $cor\left(Z(\boldsymbol{s}_i), Z(\boldsymbol{s}_j)\right) = \rho_{\|i-j\|}$, and $j \in \mathcal{J}$ instead of $\boldsymbol{p}_j \in \mathcal{J}$. Next, filter weights in $\mathcal{A}$ should satisfy the following three basic conditions:

1. $\sum_{j \in \mathcal{J}} a_j = 0$, which implies that $E(\sum_{j \in \mathcal{J}} a_j X_{i+j}) = 0$.
2. $\sum_{j \in \mathcal{J}} a_j^2 = 1$.
3. $\sum_{j \in \mathcal{J}} a_j \boldsymbol{p}_j = (0, 0)$.
4. The weights are symmetrically distributed about $\boldsymbol{p}_0$.

When the number of nodes in a filter $|\mathcal{J}|$ is greater than the number of conditions imposed on the nodes, the weights in $\mathcal{A}$ are not uniquely determined. We impose a *symmetry condition* that brings constraints both in the x- and y- directions.

Definition 1 A filter is called a symmetric-weight filter when the set of weights on each node of a filter satisfies Conditions 1–4.

Some filter configurations may satisfy Conditions 1–3 but not 4. In such cases, we recommend rotating the filter about $\boldsymbol{p}_0$ and averaging the filtered results because this adjustment leads to a lower mean squared error in variance estimation.

Assuming errors are i.i.d. and additive to a large-scale trend, Hall et al. (1991) propose a set of weights that satisfies Conditions 1 and 2, and that minimizes the mean squared error of the variance estimation. We refer to these configurations as *Hall–Kay–Titterington* (HKT) weights. To simplify the number of terms involved in variance calculation, Hall et al. (1991) recommend compact, linear, and disjoint configurations, or sparse configurations where the maximal overlap from a filter translation is a single node. We take the compact filter recommendations that can easily adjust to Conditions 1–4 and show in Fig. 2 of Sect. 3. HKT filters have a lopsided weight distribution in absolute values, and therefore the center of weights is close to the heavy node, whereas symmetric-weight filters have the center of weights at $\boldsymbol{p}_0$ by Condition 3.

Definition 2 Define an L-filter variogram at scale h as

$$\varrho_L(h) = \sigma^2 \left(1 - 2 \sum_{j \in \mathcal{J}} \sum_{\substack{k \neq j \\ k \in \mathcal{J}}} a_j a_k \rho_{h\|k-j\|} \right). \tag{4}$$

The L-filter variogram describes the dispersion in correlated data as a function of lag size h and filter L. Figure 1 portrays variograms for a stationary Gaussian process with an exponential correlation function. An L-filter variogram with symmetric weights is in red and shows dampened dispersion compared to a regular variogram in blue, while that with HKT weights in black exaggerates the dispersion for mid-range distances. The lopsided distribution of weights decreases the size of coefficients $a_j a_k$ in Eq. (4) and results in a reduced sum and hence elevates the arc of the function. The shape of a filter plays a small role in the L-filter variogram; the solid and the dotted lines represent different L's and are hardly distinguishable.

For a heteroskedastic error process, dispersion should depend on local variances. We define a local L-filter variogram for such processes where differencing is centered at a certain location and aggregated locally.

Definition 3 Define a local L-filter variogram at $\boldsymbol{s}$ in the domain of interest and lag h for a two-dimensional nonstationary process as the leading term of $E\left[L(Z(\boldsymbol{s}, h))^2\right]$.

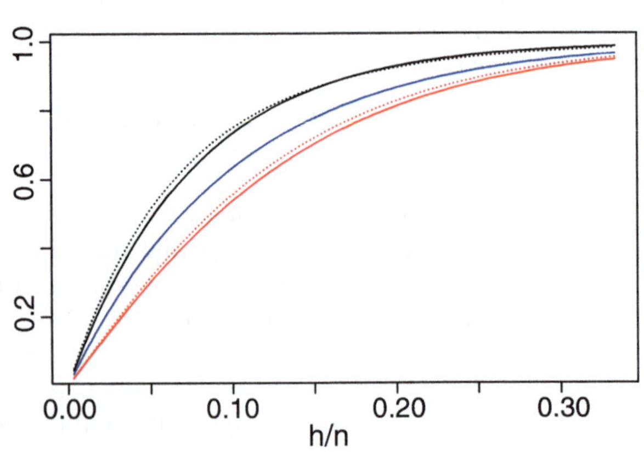

Fig. 1 *Blue* denotes a basic variogram; *red* denotes an L-filter variogram with symmetric weights; and *black* denotes one with HKT weights

$$\Gamma_{\Lambda}(s; L(Z(s,h))) = \sigma^2(s)\left(1 - \sum_{j\in\mathcal{J}}\sum_{\substack{k\neq j\\ k\in\mathcal{J}}} a_j a_k \rho_{h\|j-k\|}\right) \approx \sigma^2(s)\varrho_L(h). \tag{5}$$

By Definition 3 the local L-filter variogram is a product of a variance function and an L-filter variogram embedded in a nonstationary process. We use this feature to derive an estimator for the variance function.

2.3 A Variance Function Estimator

From Eq. (5), we derive the estimator of a variance function $\sigma^2(s)$ as the ratio of the local L-filter variogram, $\Gamma_{\Lambda}(s; L(Z(s,h)))$, and the L-filter variogram, $\varrho_L(h)$. The empirical estimation of a local L-filter variogram is

$$\hat{\Gamma}_{\Lambda}(s_0;\, L(h)) = \sum_i K_{(\lambda_x,\lambda_y)}(s_i, s_0) L(Z(s_i,\, h))^2. \tag{6}$$

The steps involved in estimating a local L-filter variogram are in Eq. (6). First, apply a linear filter L to gridded data and square the resulting differenced data; then, apply a two-dimensional kernel $K_{(\lambda_x,\lambda_y)}(\cdot)$ formed as a Kronecker product of Gasser and Müller (1984) kernels with bandwidths λ_x and λ_y respectively in the x- and y-directions. The bandwidth controls the spatial range of averaging and greatly affects the quality of estimation. We suggest taking the cross-sections of the data in x- and y-directions and doubly perform a bandwidth selection. To remove or reduce the edge effect Gasser and Müller (1984) provide asymmetrically shaped kernels. With the appropriate choice of bandwidths, we obtain the functional estimate of the local L-filter variogram. The purpose of a difference filter in estimating a variance function is to reduce autocorrelation in the process, which helps with bandwidth selection for smoothing.

The estimator of the variance function at location s_0 is

$$\hat{\sigma}_{\Lambda}(s_0) = \frac{\hat{\Gamma}_{\Lambda}(s_0; L(h))}{\varrho_L(h; \hat{\theta})}. \tag{7}$$

Estimating the L-filter variogram $\varrho_L(h;\theta)$ is equivalent to estimating the embedded correlation function $\rho(\cdot)$ and its correlation parameter θ of the error process. Kim and Zhu (2016) presents a methodology to estimate a variance function in one-dimension, and shows that finding the exact correlation function is not necessary because the estimation is done only at one fixed lag-size h, which is often the smallest grid size. Without knowing the limit of the correlation as the distance between two points $\|s_j - s_k\| \to 0$, determining the exact correlation structure is difficult. We assume a simple parametric correlation function and estimate the correlation parameter $\hat{\theta}$ using maximum likelihood.

Given that the mean function is smooth and has a smaller degree of differentiability than the variance function, a difference filter may reduce the bias in the variance function estimation with a high-order kernel by filtering out a low-order and slowly-varying mean function, as studied in Brown and Levine (2007).

3 Exploring Filter Options and an Application

We describe five filter configurations with both symmetric and HKT weights in Sect. 3.1 and explore the differing conditions of estimation via simulation in Sects. 3.2 and 3.3. Then, in Sect. 3.4, we apply our method to yearly precipitation data of the U.S. Midwest, and estimate the standard deviation function of locations from a single map.

3.1 The Filter Configuration and Weights

The number of possible filter configurations increases exponentially as the dimension of the domain increases. When choosing a filter configuration, we should consider the number of nodes, the number of possible overlapping nodes as the filter moves in all directions, and the main direction of influence imposed by the filter. We limit the span of a filter in a 3-by-3 grid because increasing the span increases the number of nodes in a filter, and hence increases the number of terms in a local L-filter variogram (see Eq. 5). Incorporating differences at multiple intervals should induce bias in the estimation of correlation parameters and a variance function.

Figure 2 displays symmetric weights in the top panel, and HKT weights in the bottom panel for the same five filter configurations. Three filter configurations are symmetric about four axes that naturally stem from the layout of the grid: Square2 (four-node), Square3 (eight-node), and cross-shaped (five-node) filters. Two configurations are directionally arranged: Y-shaped (four-node) and three-point line filters. The symmetric weights, in absolute value, are distributed evenly about the center. A central weight, if present, equals the sum of the remaining weights having the opposite sign. In contrast, HKT weights are lop-sided, with a large weight in absolute value (represented in dark blue) appearing at one end of each filter. HKT weights are designed to minimize the mean squared error of the variance when the errors are i.i.d.

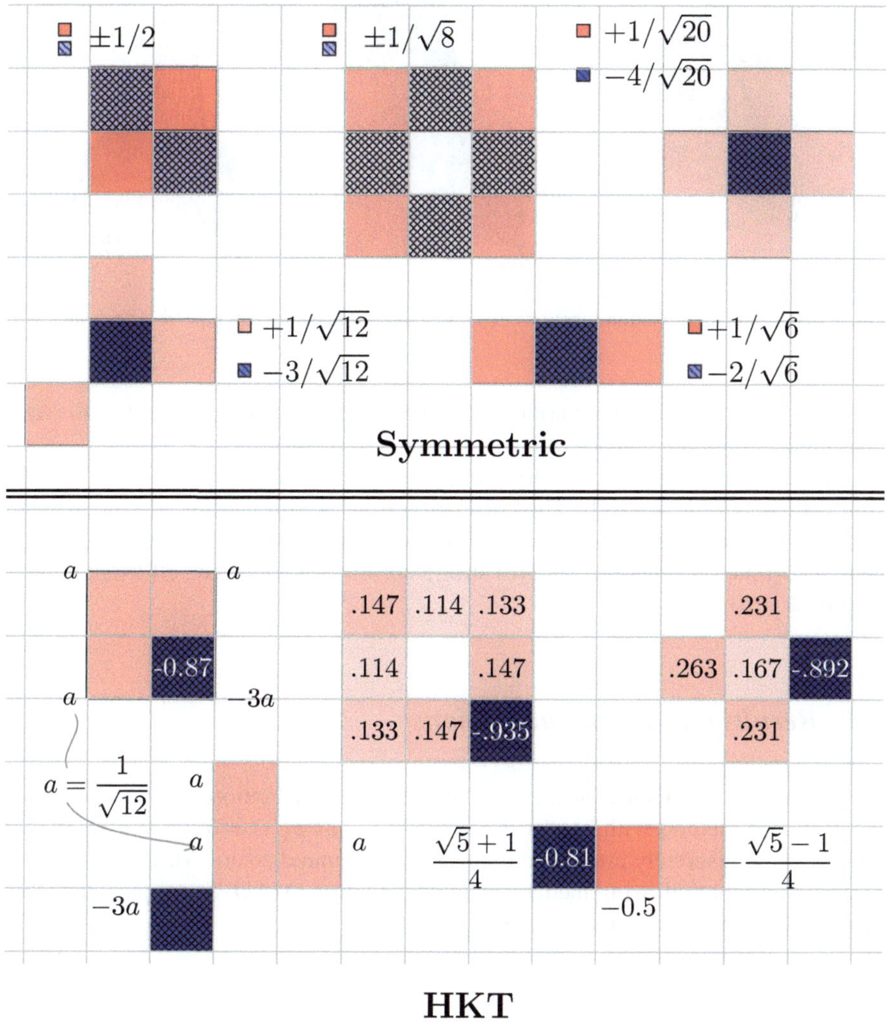

Fig. 2 Filter configurations with symmetric- and HKT-weight filters. The intensity of the colors corresponds to the absolute values of the weights, and the *red* and *blue* hues represent the opposite, but not fixed, signs of the weights

3.2 Simulation Set-up

We compare the performance of symmetric-weight to HKT-weight filters in the estimation of a variance function when the errors have zero mean, non-constant variance, and autocorrelation. We simulated nonstationary processes by multiplying a standard deviation function in Fig. 3 to a mean-zero stationary process. These $\sigma(s)$'s

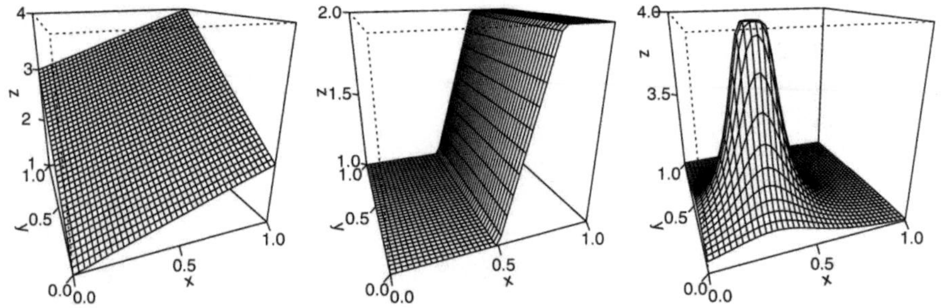

Fig. 3 The standard deviation functions $\sigma(\cdot, \cdot)$ in a three-dimensional perspective drawing

represent linearly increasing variability (left), a level shift (middle), and a tall crown marking a complement space of a Bundt pan form (right).

Data are generated at two resolutions, a grid of $N = 40$-by-40 and 100-by-100 over a unit square ($[0, 1] \times [0, 1]$). The correlation is set at three levels: (a) independent errors, (b) weak correlation where the effective range is 0.03 ($\theta = 0.01$ for the exponential correlation function), and (c) strong correlation where the effective range is 0.30 ($\theta = 0.10$).

3.3 Results and Recommendations

We can define a deviance of the estimated standard deviation function at $\boldsymbol{s}$ using the proposed method with bandwidth $\Lambda = (\lambda_x, \lambda_y)$ as $\hat{e}_\Lambda(\boldsymbol{s}) = \hat{\sigma}_\Lambda(\boldsymbol{s}) - \sigma(\boldsymbol{s})$. We use the following discretely integrated relative mean squared error (rMSE) as an overall measure, and the following median absolute deviation (MAD) as a typical deviation in a full functional estimation summary:

$$rMSE\,(\sigma(\cdot), \Lambda) = \frac{1}{N} \sum_i \left(\frac{\hat{\sigma}_\Lambda(\boldsymbol{s}_i) - \sigma(\boldsymbol{s}_i)}{\sigma(\boldsymbol{s}_i)} \right)^2$$

$$MAD\,(\sigma(\cdot), \Lambda) = \frac{1}{N} \text{median}_i \left| \hat{e}_\Lambda(\boldsymbol{s}_i) - \text{median}_j \hat{e}_\Lambda(\boldsymbol{s}_j) \right| .$$

At each location $\boldsymbol{s}_i$, $\hat{e}_\Lambda(\boldsymbol{s}_i) = 0$ means the estimation is on target, and a negative and a positive quantity means under- and over-estimation, respectively.

Figure 4 depicts the results of the estimation where two sets of weights for each of the five filters were applied to 100 nonstationary processes generated at three levels of autocorrelation. The top row of plots used coarse data, and the bottom used a 2.5-by-2.5-fold finer resolution than the top. The results of HKT-weight filters are in light blue, and those of symmetric weights are in white. The order of filters from left to right is a three-point line, cross, Y-shaped, Square2, and Square3. The three-point line and Y-shaped filters had been rotationally directed at every 90° (four rotations)

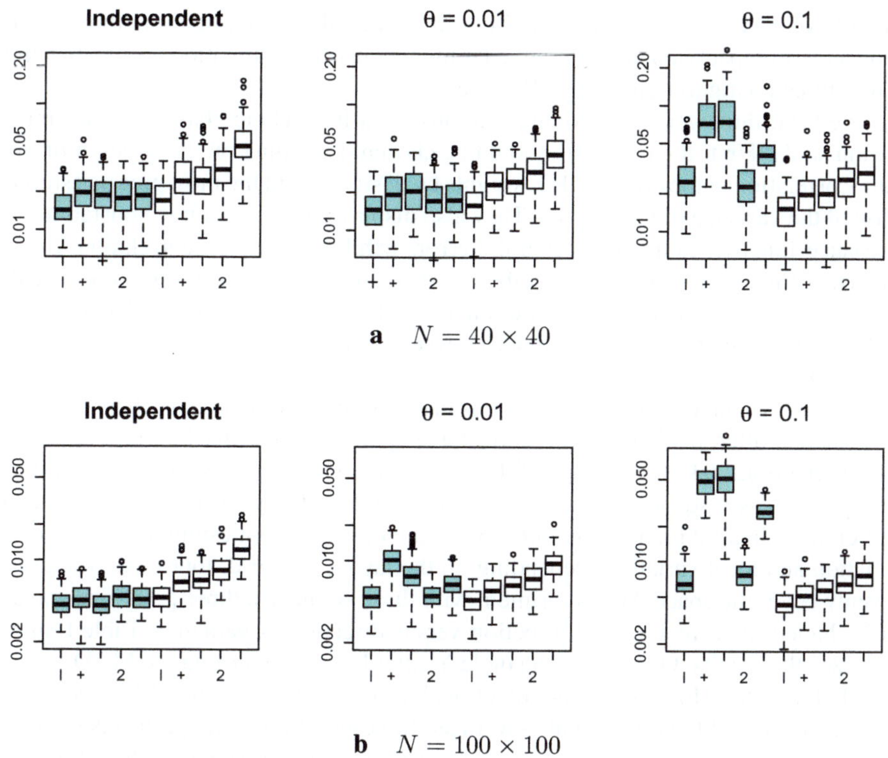

Fig. 4 Relative MSE of five filters with the HKT-weight in *light blue* and the symmetric-weight in *white*. The filter configurations are (directionally averaged) three-point line, +, (directionally averaged) Y, Square2, and Square3. The *top row* used data on a coarse grid of N = 40-by-40, and the *bottom* used data on a fine grid of N = 100-by-100

and averaged, and the results are obtained incorporating nine nodes with the central node being counted most heavily.

In Fig. 4, the left plots display the estimation results of independent errors, and the HKT weights yield a smaller rMSE than the symmetric weights, except for the directionally-averaged three-point line filter. The middle plots are the case of weak autocorrelation, and the HKT weights tend to render a smaller rMSE than the symmetric weights when the data are nearly independent and on a coarse grid of 40-by-40. On a finer grid of 100-by-100, the estimation results with symmetric- and HKT-weight filters seesaw: the cross- and Y-shaped filters produce a smaller rMSE with symmetric weights than with HKT weights; the rotated three-point line-filters are similar; and for Square2 and Square3, HKT weights are overall more efficient. The right plots contain the estimation results of a nonstationary error process with strong autocorrelation, and the estimation results are reversed from the cases of independent error processes. Symmetric weights show a more consistent and smaller rMSE than HKT weights. Also, as pointed out in Hall et al. (1991), both the three-point line and a compact Square2 filter give far more consistent estimation results than the other three configurations using HKT weights.

We have three general recommendations for a filter choice based on the simulation study. First, filter weights should be chosen depending on the strength of the autocorrelation in the errors. HKT weights by design output the most precise estimation of the variance of i.i.d. errors. They also display high precision in nonstationary variance function estimation when the errors exhibit very weak correlation. Symmetric weights are preferred when the errors are moderately-to-strongly correlated. The strength of the correlation should not only reflect the variation in data but also refer to the scale of a minimal distance between points vis-à-vis to the overall domain of a map.

Second, when a filter configuration is oriented in one direction, such as a line or a Y-shaped filter, performing directional averaging is best. After filtering a process in all directions, take the average of the squared filtered processes and use the resulting process as a squared error process. Assuming isotropy of the errors, the data differenced in all directions should provide a more consistent estimation of the correlation and local variance. Depending on the layout of a grid, four to six rotations evenly cover the space. We experiment with these settings: a three-point line versus a Y-shaped filter, and a fixed direction versus directional-averaging. Table 1 summarizes the results. On average, a single Y filter has a smaller relative MAD than a single line filter. However, when directional averaging is done, the MAD decreases significantly, and the directionally-averaged three-point line filter performs the best among all filter configurations tested. In turn, this finding suggests that choosing

Table 1 rMAD comparisons between line versus Y-shaped filters, and with and without directional rotation and averaging

MAD (%)		Fixed direction						Dir. avg.
Shape	θ	Min.	Q1	Median	Q3	Max.	Mean (Stdev)	Mean (Stdev)
	0	5.10	7.10	8.20	9.20	12.20	8.20 (1.40)	6.52 (1.02)
	0.01	5.00	7.10	7.80	8.60	11.80	7.80 (1.20)	6.13 (0.87)
	0.1	5.70	7.20	7.90	8.60	12.10	8.00 (1.20)	6.11 (1.10)
	0	4.80	6.10	7.00	7.80	9.30	6.90 (1.10)	6.69 (1.08)
	0.01	3.80	6.10	6.60	7.20	9.50	6.70 (1.00)	6.26 (0.95)
	0.1	4.50	5.80	6.60	7.30	10.10	6.60 (1.20)	6.17 (1.06)

an appropriate weight distribution can improve the statistical efficiency of variance estimation.

Third, a compact configuration helps to reduce bias and variance in estimation. Overall, Square3 filter has the largest rMSE in every combination of scenarios we examined, and the line filter consistently shows the smallest rMSE. We also investigated the effect of lag size h on estimation but for brevity did not include a summary table. In short, when the errors are independent, a lag size of either $h = 1$ or 2 results in a reasonable estimation. In contrast, when errors are strongly correlated, the smallest scale of $h = 1$ results in the tightest MAD and rMSE. In other words, a filtered process is better de-correlated when an applied filter uses the smallest possible lag size.

3.4 An Empirical Example

We apply the difference-based variance estimation method to the yearly precipitation data of the United States. The data have been collected at nearly 2,100 weather stations, which do not form a regular grid. We allow filters to lean and bend slightly to convolve with the point locations across the domain. In Fig. 5, the left plot shows linearly interpolated rainfall across the South and Midwest. Because the method assumes smooth mean and variance functions, it is sensible to limit the area of interest to this region where the topography stays relatively flat and brings few abrupt changes to a precipitation map. A greater amount of local variability exists along the east-west axis in the south than in the north, and variability is greater in a wider geographical area in the east than in the west. Using the proposed method, we obtain

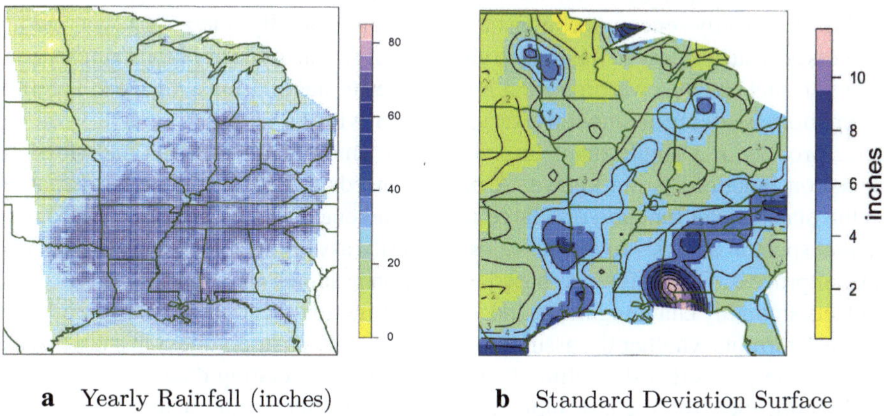

Fig. 5 **a** Annual precipitation of the U.S. Midwest and South in 1990 linearly interpolated from 2,082 observations, **b** an estimated standard deviation function of the yearly precipitation across the domain

the estimate of a standard deviation function shown in Fig. 5b. The top half of this area has the standard deviation hovering between 0.5 and 6 in., whereas the standard deviation in the south is greater and ranges mainly between 3 and 11 in. The estimation result reflects the variability we observe in the actual yearly precipitation record. The bandwidth was set to 2.6°, as the latitude ranges between 27° N and 45° N and the longitude between 100.45° W and 80° W. Because precipitation records exist for multiple years, we may test the consistency of the local variability estimation assuming the outcomes for each year are independent.

4 Conclusions

The presented difference-based variance function estimation procedure provides a simple and fast nonparametric method of estimation. We explored five filter configurations combined with two weights via simulation. We recommend using a compact, symmetric shape or a line configuration. Directional averaging should provide more precise estimation than a single filter, especially when a filter is directionally arranged to reflect information in all directions. Also, when data contain weak to no autocorrelation in the errors, HKT weights estimate a variance function more precisely. When data contain moderate to strong spatial autocorrelation, symmetric weights are preferred.

We applied our method to an annual precipitation map of the U.S. Midwest and the South to estimate the local variability of yearly precipitation. The observations are not distributed on a regular grid but are dispersed uniformly densely. So, we applied a line filter with slight angle bends. Our future work includes studying the conditions of data distribution to guarantee the consistency of variance function estimation, especially when data are collected from irregularly dispersed sampling locations.

We frequently encounter nonstationarity in spatially-referenced continuous processes, which refers to having varying levels, scales, or correlations across locations. In this chapter, we propose a nonparametric method for a variance function estimation assuming a smoothly varying mean with an additive, scaled, isotropic error process in a single map. In practice, the smoothness of mean and standard deviation functions is not known. A modeler needs to decide how much variation in the map a mean function or a large-scale trend encapsulates. Fuentes (2005) uses spatial spectral analysis to test for nonstationarity assuming a constant variance in a spatial process. Our difference-based estimation assumes an isotropic error process and lets smoothly changing mean and variance functions account for nonstationarity.

To determine whether the assumption of continuity or smoothness in a large-scale trend is met, we suggest plotting data or linearly interpolating data. If the decision is difficult, then we suggest applying the method directly and analyzing the estimated variance surface. If the original data map has a shift of levels in a large-scale trend, our method overestimates the variability in the data around the shift. If the map contains a shift in levels of variability, our method estimates the surface with a diffused shift. Because our method employs smoothing for estimation, when a nonstationary process has a parallel shift in $\sigma(\cdot)$ (refer to the middle plot of Fig. 3), the estimated surface becomes a bridge between two levels. If an unexpected overestimation occurs, we suspect a level shift in the large-scale process exists and suggest splitting up the domain of analysis. If a sharp connection exists between two or more levels, then we suspect that the underlying process has a fully smooth variance function. These decisions are best made based on the context of data, and, ideally, with covariate information.

References

Brown LD, Levine M (2007) Variance estimation in nonparametric regression via the difference sequence method. Ann Stat 35(5):2219–2232. doi:10.1214/009053607000000145. http://arxiv.org/pdf/0712.0898.pdf

Buckley MJ, Eagleson GK, Silverman BW (1988) The estimation of residual variance in nonparametric regression. Biometrika 75(2):189–199

Cai TT, Wang L (2008) Adaptive variance function estimation in heteroscedastic nonparametric regression. Ann Stat 36:2025–2054

Fuentes Monserrat (2005) A formal test for nonstationarity of spatial stochastic processes. J Multivar Anal 96:30–66

Gasser Theo, Müller Hans-Georg (1984) Estimating regression functions and their derivatives by the kernel method. Scand J Stat 11(2):171–185

Hall P, Carroll RJ (1989) Variance function estimation in regression: the effect of estimating the mean. J R Stat Soc Ser B 51(1):3–14

Hall P, Kay JW, Titterington DM (1990) Asymptotically optimal difference-based estimation of variance in nonparametric regression. Biometrika 77:521–528. doi:10.1093/biomet/77.3.521

Hall P, Kay JW, Titterington DM (1991) On estimation of noise variance in two-dimensional signal processing. Adv Appl Probab 23:476–495

Kim EJ, Zhu Z (2016) One-dimensional nonstationary process variance function estimation. https://arxiv.org/pdf/1605.06579v1.pdf

Martin Ronald L (1974) On spatial dependence, bias and the use of first spatial differences in regression analysis. Area 6:185–194

von Neumann J, Kent RH, Bellinson HR, Hart BI (1941) The mean square successive difference. Ann Math Stat 12(2):153–162

Zhu Zhengyuan, Stein Michael L (2002) Parameter estimation for fractional brownian surfaces. Stat Sinica 12:863–883

The Statistical Distribution of Coefficients for Constructing Eigenvector Spatial Filters

Parmanand Sinha, Monghyeon Lee, Yongwan Chun
and Daniel A. Griffith

Abstract This chapter presents an exploratory simulation experiment to investigate the distribution of coefficients that are used to construct eigenvector spatial filters. The experiment involves five hexagonal tessellations and three levels of autocorrelation. The results of this experiment show that coefficients for eigenvectors selected to construct a spatial filer approximately follow a gamma distribution. The shape and scale parameters of the gamma distribution fitted to coefficient frequency distributions are further investigated.

Keywords Spatial filtering · Spatial autocorrelation · Eigenvector · Eigenvector spatial filtering (ESF)

1 Introduction

The fundamental idea of eigenvector spatial filtering (ESF; Griffith 2000, 2003) exploits the decomposition of a spatial variable into the following three components: trend, a spatially structured random component (i.e., a spatial stochastic signal), and random noise. The spatially structured component is modeled with a linear combination of synthetic proxy variables, which are extracted as eigenvectors from a spatial weights matrix that ties geographic objects together in space; it is

M. Lee · Y. Chun · D.A. Griffith
The University of Texas at Dallas, 800 W. Campbell Rd, Richardson, TX, USA
e-mail: monghyeon.lee@utdallas.edu

Y. Chun
e-mail: ywchun@utdallas.edu

D.A. Griffith
e-mail: dagriffith@utdallas.edu

P. Sinha (✉)
The University of Tennessee, 304 Burchfiel Geography Building
1000 Phillip Fulmer Way, Knoxville, TN 37996-0925, USA
e-mail: pnsinha@utk.edu

D.A. Griffith et al. (eds.), *Advances in Geocomputation*,
Advances in Geographic Information Science,
DOI 10.1007/978-3-319-22786-3_26

added as a control variable to a model specification to account for spatial autocorrelation (SA). This control variable identifies and isolates the stochastic spatial dependencies among the georeferenced observations, thus allowing model building to proceed as if the observations are independent. Identifying the distribution of coefficients used to construct this linear combination may support simulation experiments employing the generation of spatially autocorrelated random variables using ESF.

2 Eigenvector Spatial Filtering

The Moran coefficient (MC)-based ESF methodology utilizes the properties of eigenvectors and their corresponding eigenvalues of the transformed spatial weights matrix $(\mathbf{I} - \mathbf{11}^T/n)\mathbf{C}(\mathbf{I} - \mathbf{11}^T/n)$, where $\mathbf{I}$ is an n-by-n identity matrix, $\mathbf{1}$ is an n-by-1 vector of ones, $\mathbf{C}$ is an n-by-n spatial weights matrix, and superscript T denotes the matrix transpose operator. Studies, including Tiefelsdorf and Boots (1995) and Griffith (1996), show that its n mutually orthogonal and uncorrelated (Griffith 2000) eigenvectors, $\mathbf{E} = \{\mathbf{E}_1, \mathbf{E}_2, \ldots, \mathbf{E}_n\}$, and n corresponding eigenvalues, $\boldsymbol{\lambda} = \{\lambda_1, \lambda_2, \ldots, \lambda_n\}$, relate to SA. Important properties of these vectors include: (1) they furnish distinct map pattern descriptions visualizing latent SA in georeferenced variables, and (2) the eigenvalues index the level of SA of a map pattern that is generated when the corresponding eigenvector is mapped on the given tessellation. That is, the MC of the map pattern produced by eigenvector $\mathbf{E}_j$ is $MC_j = \lambda_j . n/\mathbf{1}^T\mathbf{C1}$.

The ESF linear regression model specification can be written as

$$\mathbf{Y} = \mathbf{X}\boldsymbol{\beta} + \mathbf{E}_k\boldsymbol{\beta}_E + \boldsymbol{\varepsilon}, \quad (1)$$

where $\mathbf{E}_k$ is an n-by-K matrix containing K eigenvectors, $\boldsymbol{\beta}_E$ is the corresponding vector of regression parameters, $\mathbf{E}_k\boldsymbol{\beta}_E$ is the eigenvector spatial filter, and $\boldsymbol{\varepsilon} \sim N(\mathbf{0}, \mathbf{I}\sigma_\varepsilon^2)$ is an n-by-1 error vector whose elements are independently and identically distributed (i.i.d.) normal random variates. Because the linear combination of the eigenvectors, $\mathbf{E}_k\boldsymbol{\beta}_E$, accounts for SA, the ESF linear regression specification does not suffer from spatially autocorrelated residuals. The K eigenvectors, $\mathbf{E}_k$, can be identified with a stepwise regression technique (Griffith 2003).

3 Methodology

Random numbers using a simultaneous autoregressive (SAR) data-generating mechanism can be simulated with

$$\mathbf{Y} = (\mathbf{I} - \rho\mathbf{W})^{-1}\boldsymbol{\varepsilon}, \tag{2}$$

where matrix $\mathbf{W}$ is the row standardized geographic connectivity matrix, and ρ is a parameter the nature and degree of SA. Theoretically, $(\mathbf{I} - \rho\mathbf{W})^{-1}$ can be approximated by a linear combination of its eigenvectors, $\mathbf{E}_k\boldsymbol{\beta}_E$. (Griffith 2003). This property leads the quest for finding the distribution of ESF coefficients $\boldsymbol{\beta}_E$ while controlling the SA. The corresponding coefficients can be used to generate spatially autocorrelated random variables using a combination of eigenvectors as follows:

$$\mathbf{Y} = \mathbf{E}_k\boldsymbol{\beta}_E + \boldsymbol{\varepsilon}. \tag{3}$$

The candidate set of eigenvectors, from which $\mathbf{E}_k$ is selected, varies based on the magnitude of the SA parameter ρ (Griffith and Chun 2009). The number of selected eigenvectors and coefficients $\boldsymbol{\beta}_E$ depend on the result of a stepwise regression of the SAR-generated random numbers with the candidate set of eigenvectors.

4 A Simulation Experiment

A simulation experiment was conducted using 10-by-10, 25-by-25, 50-by-50, 75-by-75, and 100-by-100 hexagon tessellations and three levels of SA, with 10,000 replications. The following are the steps involved:

1. Generate 10,000 spatially autocorrelated random variables, $\mathbf{Y}$, for each tessellation using an SAR data-generating mechanism with $\rho = 0.1$, 0.5, and 0.9.
2. Calculate the candidate set of eigenvectors for each tessellation and each level of SA using the threshold criterion (Chun et al. 2016)

$$\frac{n_+}{1 + e^{[2.1480 - 6.1808[(z_{MC} + 0.6)^{0.1742}]/n_+^{0.1298} + 3.3534/(z_{MC} + 0.6)^{0.1742}]}}, \tag{4}$$

   where, z_{MC} is the MC z-score of $\mathbf{Y}$, n_+ is the number of eigenvectors with positive eigenvalues, and e is an exponential function.
3. Select the significant eigenvectors, using the $\mathbf{Y}$ as dependent variables in stepwise regressions with the candidate set of eigenvectors, and selection based upon an Akaike information criterion (AIC) maximizing goodness-of-fit.
4. Calculate the probability of selection of each significant eigenvector in 10,000 simulation replications, retaining those with a minimal empirical probability of 0.01 for further analysis. Eigenvectors selected fewer than 100 times out of 10,000 simulations are assumed not to be significant.
5. For the significant eigenvectors coefficients, calculate the shape and scale parameters of a gamma distribution.

5 Results

Table 1 shows the candidate set and selected eigenvectors using a stepwise selection procedure for different tessellations and different levels of SA. As the level of SA (ρ) increases, the candidate set of eigenvectors increases in size, and more eigenvectors are selected to account for a higher degree of SA.

Figure 1 displays the distributions of coefficients, $\boldsymbol{\beta}_E$, for eigenvectors randomly selected from their respective candidate sets. Each of these histograms has 10,000 coefficients for a given eigenvector. Because the histograms show that the coefficients closley conform to a gamma distribution, they have been overlaid with a corresponding fitted gamma distribution. The fitted shape and scale parameter values are recorded for coefficients of individual eigenvectors.

Figure 2 portrays the distributions of fitted gamma distribution shape and scale parameters for 25-by-25 tessellations across the index number of the eigenvectors. Yellow denotes the trend line. The shape and scale parameters for $\rho = 0.1$ are fitted with a straight line that shows a near-constant trend: this trend is reminiscent of a uniform distribution. The parameters for $\rho = 0.5$ or 0.9 are fitted with a nonlinear trend line. The parameter values tightly scatter about the nonlinear trend lines for $\rho = 0.5$ and $\rho = 0.9$, whereas they more widely scatter about the trend line for $\rho = 0.1$. In general, the shape parameter value tends to increase, and the scale parameter tends to decrease, as the index number of eigenvectors increases. The shape parameter increment for higher levels of SA tracks an exponential curve, whereas the scale parameter decrement tracks a hyperbolic curve.

Table 1 The number of selected eigenvectors for different tessellations

Tessellation size	Expected rho	Number of eigenvectors	
		In the candidate set	Selected
10-by-10	0.1	9	2
	0.5	25	10
	0.9	33	21
25-by-25	0.1	44	10
	0.5	145	61
	0.9	207	125
50-by-50	0.1	44	10
	0.5	145	61
	0.9	207	125
75-by-75	0.1	373	92
	0.5	1,214	529
	0.9	1,769	1,065
100-by-100	0.1	670	159
	0.5	2,133	941
	0.9	3,103	1,873

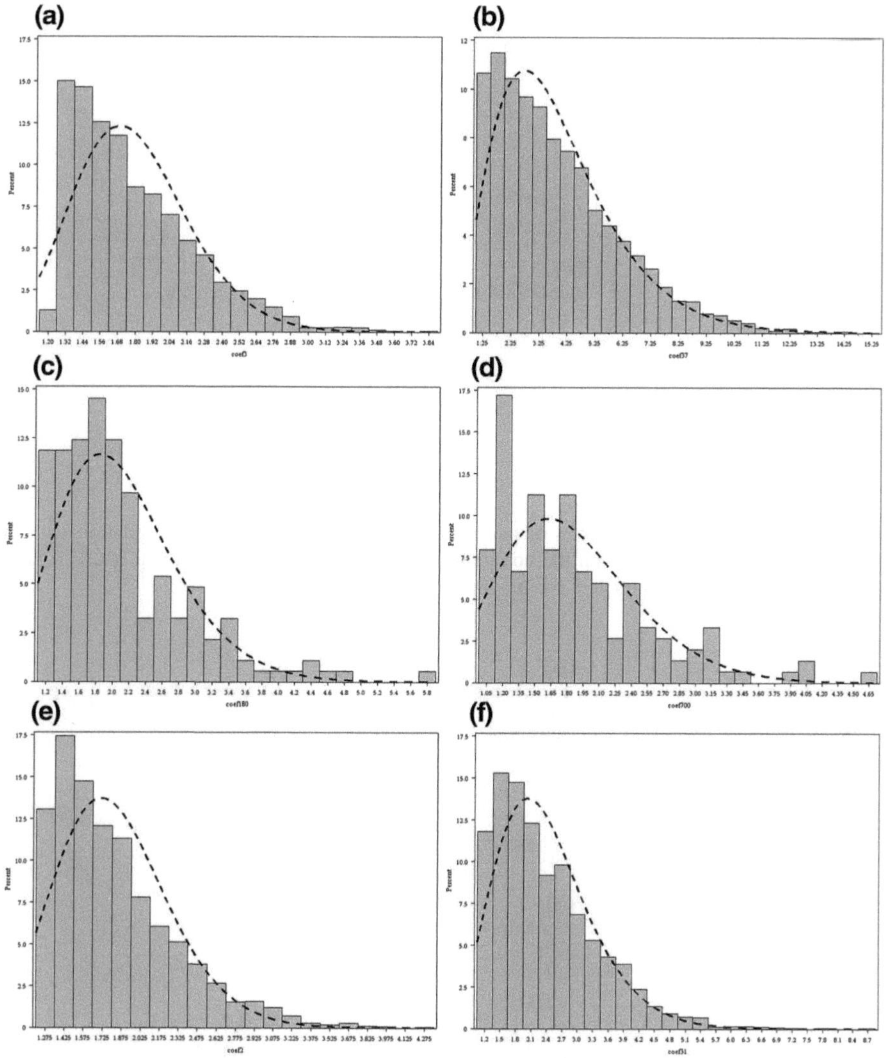

Fig. 1 Estimated coefficient distributions for selected eigenvectors: **a** 3rd of a 10-by-10 tessellation for $\rho = 0.1$; **b** 37th of a 25-by-25 tessellation for $\rho = 0.9$; **c** 180th of a 50-by-50 tessellation for $\rho = 0.5$; **d** 700th of a 50-by-50 tessellation for $\rho = 0.9$; **e** 2nd of a 75-by-75 tessellation for $\rho = 0.1$; and **f** 31st of a 100-by-100 tessellation for $\rho = 0.5$

Figures 3 and 4 show the distributions of fitted gamma distribution shape and scale parameters for 50-by-50 and 75-by-75 tessellations, respectively. Their overall patterns are similar to those of the 25-by-25 tessellation. The shape parameter values tend to increase, while the scale parameter values tend to decrease. The parameters tend to deviate from the trend lines more for $\rho = 0.1$ than for the

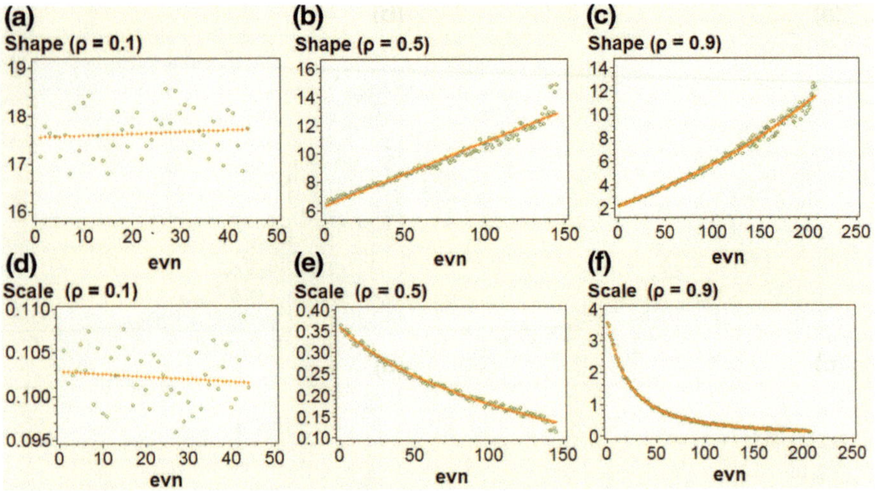

Fig. 2 Shape and scale parameters of gamma distributions for a 25-by-25 tessellation with the index number of eigenvectors on the horizontal axis: **a–c** shape parameters for $\rho = 0.1, 0.5, 0.9$ respectively; **d–f** scale parameters for $\rho = 0.1, 0.5, 0.9$ respectively

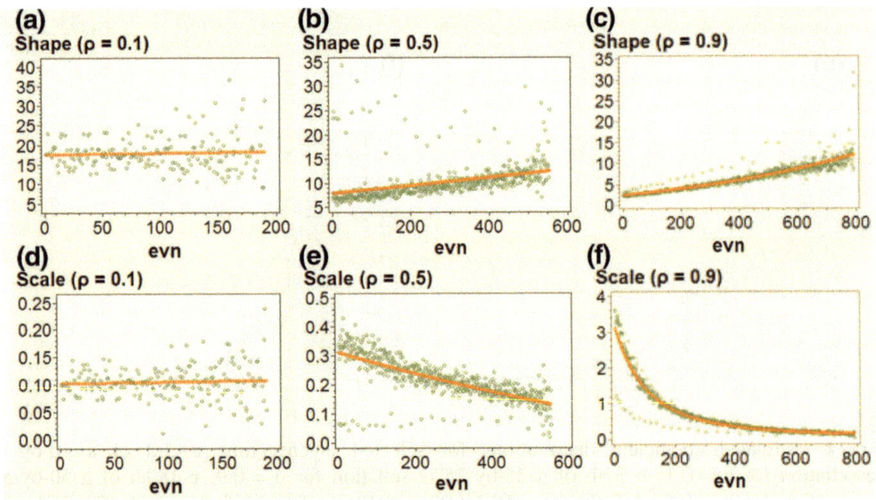

Fig. 3 Shape and scale parameters of gamma distributions for a 50-by-50 tessellation with the index number of eigenvectors on the horizontal axis: **a–c** shape parameters for $\rho = 0.1, 0.5, 0.9$ respectively; **d–f** scale parameters for $\rho = 0.1, 0.5, 0.9$ respectively

other ρ levels. However, a systematic deviation exists for the 50-by-50 tessellation results, especially for the $\rho = 0.5$ and $\rho = 0.9$ cases. This systematic deviation merits further investigation.

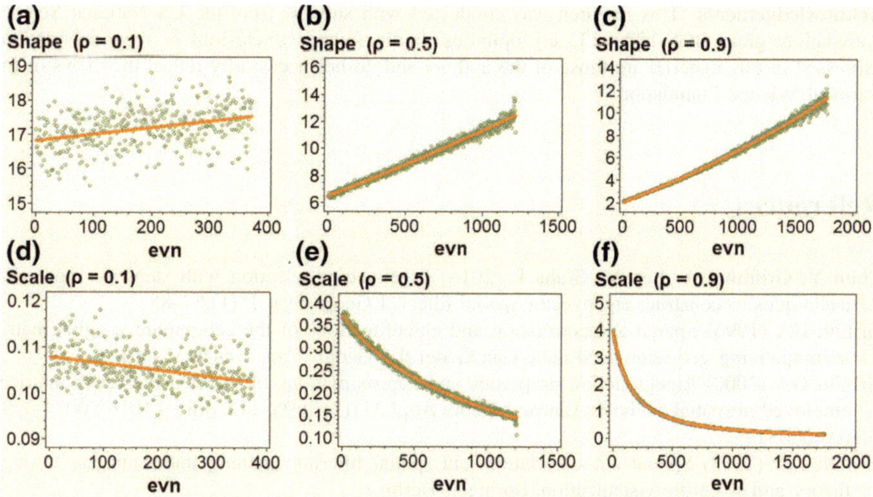

Fig. 4 Shape and scale parameters of gamma distributions for a 75-by-75 tessellation with the index number of eigenvectors on the horizontal axis: **a–c** shape parameters for ρ = 0.1, 0.5, 0.9 respectively; **d–f** scale parameters for ρ = 0.1, 0.5, 0.9 respectively

6 Implications

Visualization of results from a simulation experiment suggests that the frequency distribution of the elements of vector $\boldsymbol{\beta}_E$ is similar to a gamma distribution with its scale and shape parameters varying with n and the level of SA. The selection of a candidate set of eigenvectors also depends on the level of SA. For higher levels of SA, eigenvectors with the largest eigenvalues were selected more often; for lower levels of SA, the probability of selecting eigenvectors appears to be nearly uniform. Correctly identifying the statistical distribution of coefficients will support generating spatially autocorrelated random variables using ESF while controlling for the level of SA. Theoretically, in the generation of spatially autocorrelated random numbers, an SAR mechanism uses all of the eigenvectors, whereas, with an ESF, the stepwise selection is done with eigenvectors reflecting the nature of SA. The statistical distribution of the coefficients would be useful for designing Monte Carlo simulation studies using ESF because this could allow for more experimental control in a simulation experiment.

Future research includes replicating the experiment for rectangular and irregular tessellations. Also, larger size tessellations beyond 100-by-100 need to be explored. Furthermore, generation of spatially autocorrelated random numbers based on the statistical distribution of coefficients will be included in future research.

Acknowledgements This research was conducted with support from the US National Science Foundation, grant BCS-1229223; any opinions, findings, and conclusions or recommendations expressed in this material are those of the authors and do not necessarily reflect the views of the National Science Foundation.

References

Chun Y, Griffith DA, Lee M, Sinha P (2016) Eigenvector selection with stepwise regression techniques to construct eigenvector spatial filters. J Geogr Syst 18(1):67–85

Griffith DA (1996) Spatial autocorrelation and eigenfunctions of the geographic weights matrix accompanying geo-referenced data. Can Geogr./Le Géogr. Can. 40(4):351–367

Griffith DA (2000) Eigenfunction properties and approximations of selected incidence matrices employed in spatial analyses. Linear Algebra Appl 321(1–3):95–112. doi:10.1016/S0024-3795(00)00031-8

Griffith DA (2003) Spatial autocorrelation and spatial filtering: gaining understanding through theory and scientific visualization. Springer, Berlin

Griffith DA, Chun Y (2009) Eigenvector selection with stepwise regression techniques to construct spatial tilters. In: 105th annual association of American geographers meeting, Las Vegas, NV, March 25

Tiefelsdorf M, Boots B (1995) The exact distribution of Moran's I. Environ Plan A 27(6):985–999

Spatial Data Analysis Uncertainties Introduced by Selected Sources of Error

Monghyeon Lee, Yongwan Chun and Daniel A. Griffith

Abstract Spatial data analysis uncertainty has been examined with various sources of error through simulation experiments. The general sources of the uncertainty are sampling error, measurement error, specification error, and location error. Location error is a unique error in spatial data analysis and occurs when an observed location deviates from its true location. We simulate spatial data analyses with different levels of location and measurement error, and compare the simulation results. Geographically aggregated pediatric blood lead level point data for Syracuse, New York, are utilized for the simulation together with a simultaneous autoregressive model. The results show that even with different levels of error, regression coefficients are quite robust. However, coefficient standard errors become larger with higher levels of location error and smaller administration units, such as census blocks.

Keywords Spatial data analysis · Spatial data uncertainty · Location error · Measurement error

1 Introduction

Various sources of error lead to spatial data analysis uncertainty. Like aspatial data analysis, sampling error (i.e., deviations of sample statistics from their corresponding population parameter values) is one major source of uncertainty. The scoring of attributes also contains measurement error (i.e., differences between pairs

M. Lee (✉) · Y. Chun · D.A. Griffith
School of Economic, Political and Policy Sciences, The University of Texas at Dallas, Dallas, USA
e-mail: monghyeon.lee@utdallas.edu

Y. Chun
e-mail: ywchun@utdallas.edu

D.A. Griffith
e-mail: dagriffith@utdallas.edu

D.A. Griffith et al. (eds.), *Advances in Geocomputation*,
Advances in Geographic Information Science,
DOI 10.1007/978-3-319-22786-3_27

of true and measured values). Another major source is specification error, which is the difference between reality and a model's representation of it. Furthermore, due to the locational aspect of spatial data, location error is also a source of uncertainty. Spatial data are georeferenced data that consist of aspatial as well as spatial information. Aspatial information refers to the nonlocational characteristics of features, whereas spatial information describes relative and/or absolute positioning of these features. Here, location error may affect a spatial data analysis because it might introduce deviations from true locations. These four sources of error interact and affect the quality of a spatial data analysis. In addition, features or geographical units can be merged or aggregated, perhaps due to confidentiality, data management, and/or representational concerns. Geographic aggregation can exacerbate and propagate error in spatial data from all four of these sources.

Any output from a model in a spatial data analysis is likely to deviate from its reality, because errors or uncertainties in inputs can propagate to their outputs. However, knowing how large the uncertainty is for a model is important, as is the degree to which uncertainty affects output. Uncertainties, at best, can render slightly incorrect modeling results and, at worst, can be completely fatal to an analysis of georeferenced data and undermine any outcome of a spatial data analysis (Fisher 1999). Another critical issue in spatial data analysis is that uncertainties behave differently at different scales (e.g. county, state, and country) and resolutions of geographical units (i.e., different administrative regions, such as census block, census block group, and census tract).

This chapter summarizes our investigation through simulation experiments about how these errors impact upon spatial data analyses. The experiments deal with location error and measurement error for geographically aggregated variables.

2 Literature Review

The general categories of spatial data analysis uncertainty can be specified by two dimensions: the locational definition of a spatial entity and the attribute of a spatial entity (Robinson et al. 1985). The combination of these two dimensions, attribute and locational exactness/inexactness, yields four categories of possible uncertainty: no uncertainty only, an uncertain location only, an uncertain attribute, and both an uncertain location and an attribute (Robinson et al. 1985). Several common reasons for uncertainty in spatial data are measurement, assignment, class and spatial generalization, miscoding, temporal changes, and processing (Fisher 1999). The combinations of these errors in spatial data analysis propagate to the output of an analysis, causing that output to be unreliable for correct conclusions (Heuvelink 1998). However, few studies exist about how uncertainty propagates through an analysis to its output (Crosetto et al. 2000).

Location error in a spatial regression context has been discussed in Griffith et al. (2007). They utilize pediatric blood lead level (BLL) data for Syracuse, New York, and confirm the presence of noticeable positional error propagation in spatial regression analysis results, although statistical estimates are not dramatically different from their original estimates in the spatial regression results. Model misspecification also may propagate to analysis results. If we assume the "true" model is nested within a model specification, model estimations converge to the true model with a large sample size; however, if a specification does not contain the "true" model, then estimates tend to be over- or underestimated (Barry and Elith 2006). Another major source of uncertainty is measurement error. Reeves et al. (1998) argue that data are subject to measurement error and/or errors of imputation, and these errors can change estimates and results.

3 Data and Simulation Experiments

The data that we use in our simulation experiments are pediatric BLL measurements collected with lead poisoning screening tests (capillary, via finger prick, or venous, via a blood draw) for children in Syracuse, New York, during 1992–1996. This database, which originally was obtained from the Onondaga County Health Department, also was utilized by Griffith et al. (2007). This database contains 16,691 BLL test results for children who resided in the city of Syracuse during the six-year period, as well as the residential addresses for these children (Fig. 1). The (x, y) coordinates of the residential addresses were generated through a rigorous geocoding process and have undergone extensive cleaning. These geographic points with their individual data can be aggregated into census blocks, census block groups, and census tracts for ecological regression analysis purposes. These data serve as the points that are perturbed in the simulation experiments.

These experiments were conducted with 1,000 replications of each error type. The response variable is the arithmetic mean BLL. The simultaneous autoregressive (SAR) model furnishes a description for the mean BLLs. The independent variables vary according to the level of census unit aggregation. Table 1 lists these variables.

3.1 Location Error Simulation Experiment Design

With the development of geographic information system (GIS) technology, geocoding is utilized to match address-labeled spatial data with census or other geographic area data (Cromley and McLafferty 2002). During this process, the locational and ecological accuracy becomes a critical concern in spatial data analyses. For example, Cayo and Talbot (2003) show that some positional errors caused by selected geocoding methods are unacceptably large.

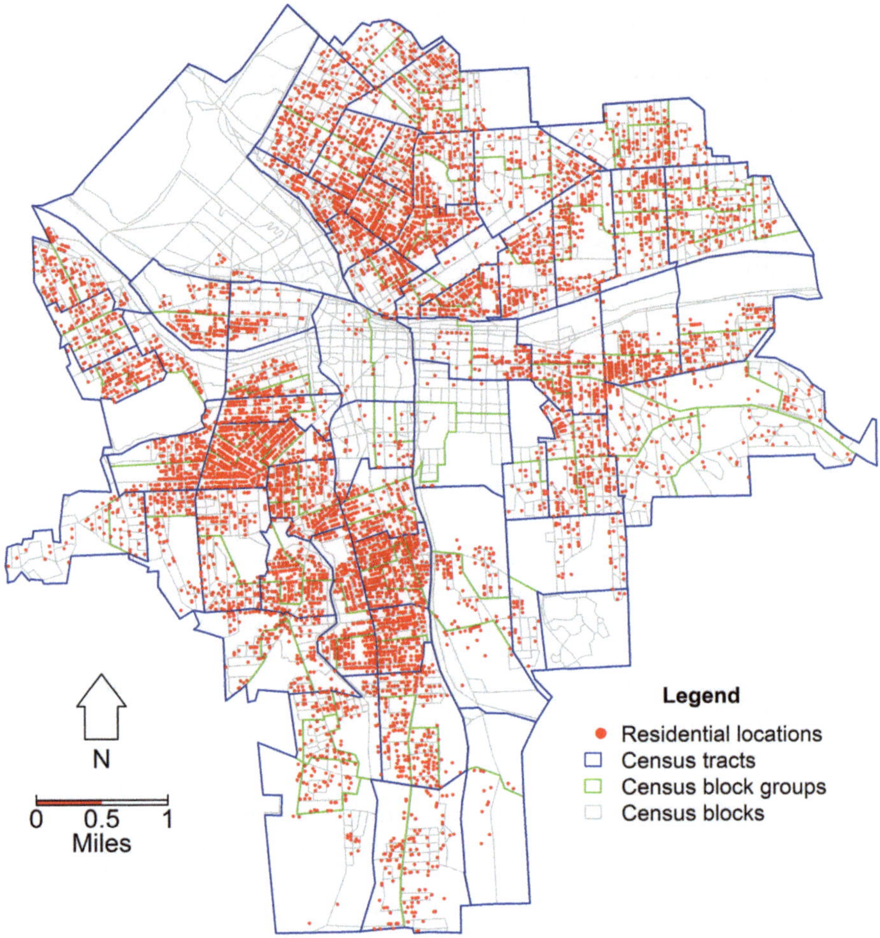

Fig. 1 The distribution of pediatric BLL locations across the city of Syracuse, NY, 1992–1996

Our simulation experiment design adds location error as follows:

(1) Randomly select 10 % of BLL points
(2) Assign 10 m of location error with a random direction to the selected points, constraining them to remain within the City of Syracuse limits
(3) Estimate a SAR model with the dependent variable, the log-transformed average BLL in the region based on the error added points' locations and socio-economic variables of the areal units as covariates
(4) Repeat steps (2)–(3) for different location error levels: i.e., 25, 50, 75 and 100 m
(5) When step (4) is completed, return to step (1) and sample 20, 30, 40 and 50 % of the points, repeating steps (2) to (4) each time

Table 1 Independent variables by different census administration levels

Administration unit	Independent variable
Census block	Average house value Percentage in cohort under 5 years of age Percentage of black population Percentage of Hispanic population Population density Zero indicator (0 with no BLL observation and 1 otherwise)
Census block group	Average house value Percentage black Population density East-west coordinate logarithm of the number of cases
Census track	Average house value Percentage in cohort under 18 years of age

The distances for location error were systematically chosen with a 25 m interval up to 100 m, in addition to 10 m. Studies (e.g., Danskin et al. 2009; Wing 2011) show that the accuracy level of global positioning system (GPS) receivers is usually less than 10 m, although recreation-grade GPS receivers can have larger location errors (up to approximately 40 m). However, location errors by automated geocoding errors can be much larger. Zimmerman and Li (2010) report that, in their study of 9,298 residential addresses in Carroll County, Iowa, the mean of geocoding location errors is 127 m, and its median is 57 m. These steps were executed for the three different administration units of census block, census block group, and census track.

3.2 Measurement Error Simulation Experiment Design

Measurement error within georeferenced data leads to spatial data uncertainty (Griffith et al. 2009). We added measurement error to the response variable, BLL, following guidelines from the Centers for Disease Control and Prevention (CDC). Federal regulations allow laboratories that perform blood lead testing to operate with a total allowable error of either ±4 μg/dL or ±10 %.[1] Like the location error process, the measurement error simulation experiment was conducted for five different sample corruption sizes: 10, 20, 30, 40 and 50 %. The errors follow an approximately bell-shaped beta distribution with a range from −5 to 5. After adding measurement error, we recalculated mean BLLs for the response variable and repeated spatial regression analyses.

[1]http://www.cdc.gov/mmwr/preview/mmwrhtml/rr5608a1.htm.

4 Results

This section reports the results of location and measurement error simulations. Regression coefficients and their standard errors for error-embedded dependent variables are compared with the corresponding coefficients and standard errors for the original locations and/or measurements. Among 25 error combinations for location errors and 10 error combinations of measure errors, this section focuses on only a small number of error combinations, especially the least and most error cases, because the results of the other cases lie between those of the extreme cases.

4.1 *Location Error*

Figure 2 portrays the census block-level regression coefficients of independent variables (red vertical lines) and their 95 % confidence intervals (blue vertical lines). The histogram bars represent the coefficients of 1,000 simulation replicates with distance errors of 10 m and 100 m only. According to the results, all of the simulated coefficients are inside of their confidence intervals and are concentrated around their original coefficients when a small amount of error is introduced (Fig. 2a). When 100 m displacement error is added to 50 % of the observations, a majority of coefficients are inside of their confidence intervals. However, some of them, such as the zero indicator variable, are outside of their confidence intervals (Fig. 2b). The results of the other location error levels are not presented; their corruptions fall between the results of these two extreme location error level cases.

Figure 3 portrays the coefficient standard errors. The red lines are the standard error values of the original data without any location error added, and the blue lines are their 95 % confidence intervals. Histogram bars represent the coefficients from 1,000 simulation replicates. With the 10 m location error, all of the standard errors fall within their confidence intervals (Fig. 3a). However, many of the standard errors are outside of their confidence intervals with 100 m location error (Fig. 3b). Most of the coefficient standard errors have been inflated by relatively large location error.

Results from the coarser geographic resolutions—census block groups and census tracts—have smaller uncertainty than census blocks, because the larger regions contain more observations and have fewer observation points that cross the region boundaries when adding distance error. These results are not included in this chapter.

4.2 *Measurement Error*

Results from this simulation experiment are similar to those from the location error experiment. These results show that measurement errors do not cause the SAR

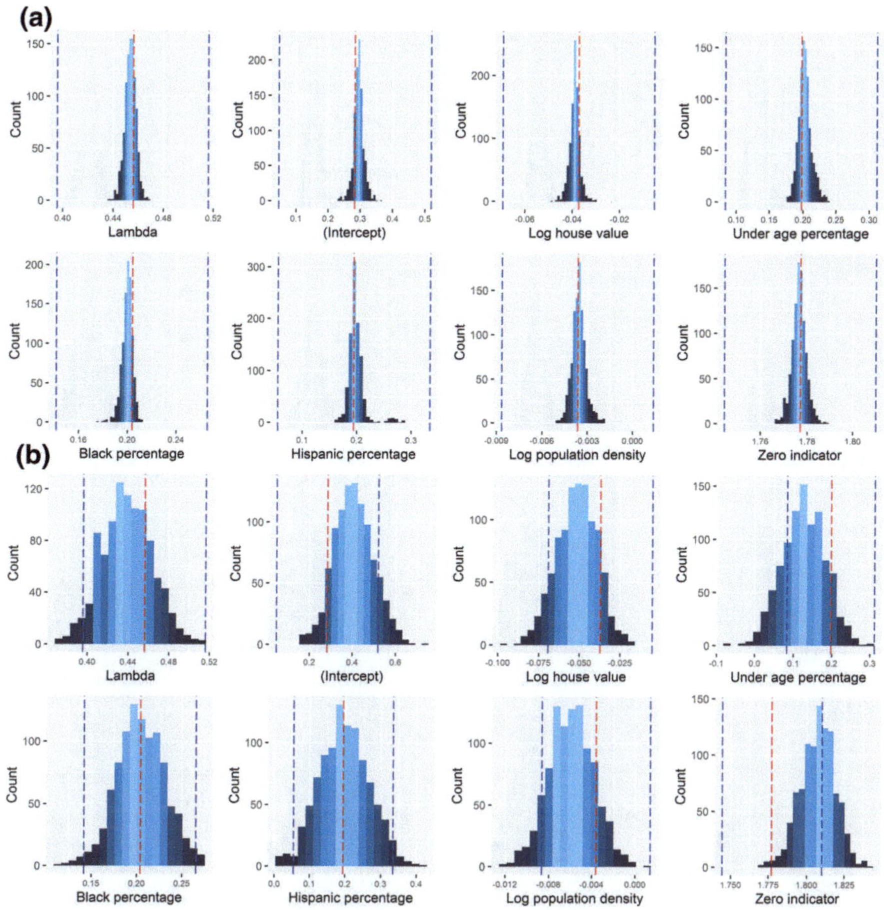

Fig. 2 SAR coefficient estimates for census blocks. **a** 10 m distance error added to 10 % of the observations, **b** 100 m distance error added to 50 % of the observations

results to deviate significantly from the original SAR results. In all cases, every parameter from 1,000 simulation replicates is located inside its 95 % confidence interval, even when 50 % of them have added measurement error. These values are more dispersed than those with only 10 % error (Fig. 4). The results of the other error levels (i.e., measurement errors in 20, 30 and 40 % of the samples) are not presented here; their results are between the results of these two extreme error levels.

Figure 5 portrays the coefficient standard errors from the measurement error simulation experiment. Unlike measurement error coefficient results (Fig. 4), some standard errors are outside of their 95 % confidence intervals when a large amount of measurement error is present (Fig. 5b). These results indicate that measurement error also inflates the standard error of these coefficients.

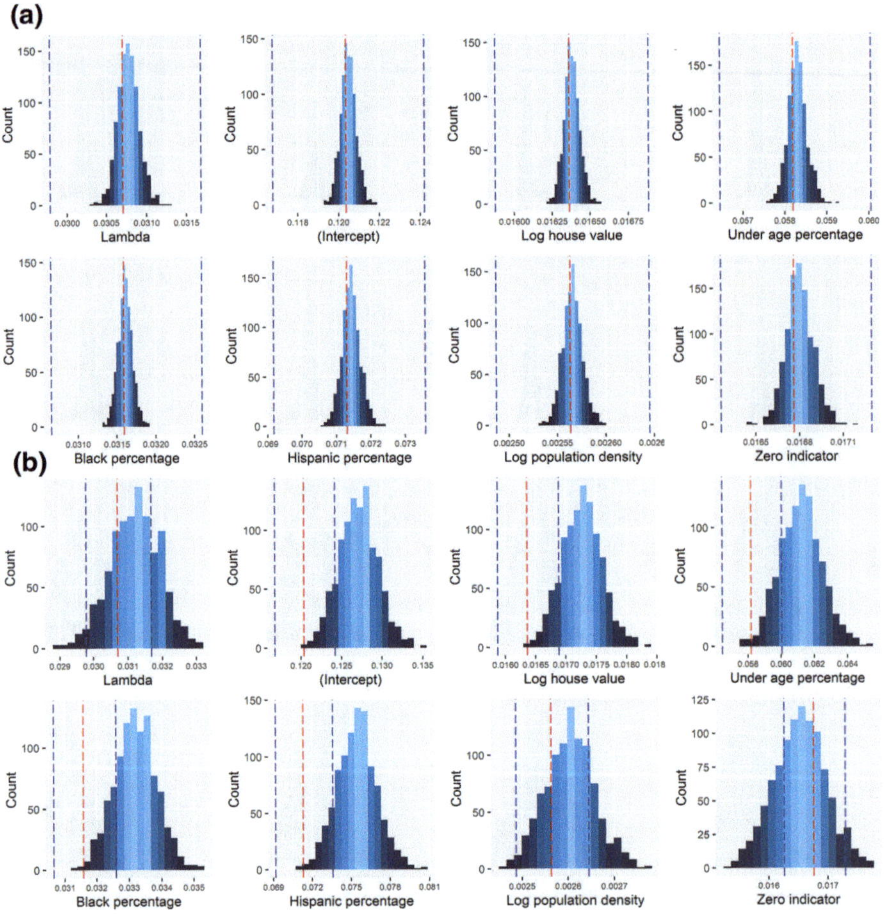

Fig. 3 SAR coefficient estimate standard errors for census blocks. **a** 10 m distance error added to 10 % of the observations, **b** 100 m distance error added to 50 % of the observations

5 Findings and Future Research

Ecological spatial regression analyses of mean BLLs appear to be robust in the presence of relatively severe but realistic levels of locational and measurement error. A majority of coefficients from the simulation experiments with different levels of location and measurement error are inside of their original counterpart coefficients' 95 % confidence intervals, but coefficient standard errors increasingly are inflated with higher levels of location error. Furthermore, although we did not show the results for other census units, such as census block groups and census tracts, the smallest unit, the census block, because of its relatively fine geographic

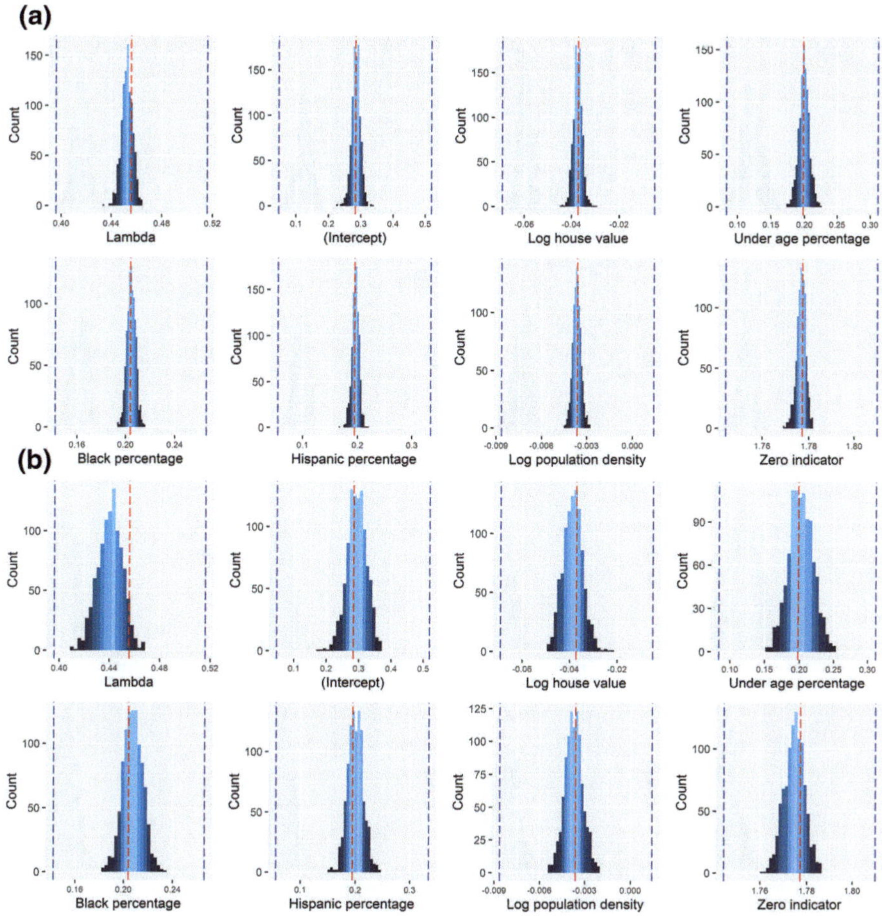

Fig. 4 SAR coefficient estimates for census blocks. **a** Measurement error added to 10 % of the observations, **b** Measurement error added to 50 % of the observations

resolution, has more significant coefficient and standard error deviations from its original results than coarser units have.

The simulation experiments can be further extended by considering combinations of multiple error sources and other error sources. The simulations whose results are summarized here investigate the impacts of location error and measurement error separately, but these errors are not separable in most empirical data analyses. Hence, the impact of combined errors needs to be investigated in future research. Furthermore, other sources of error have an impact on spatial data analysis, including model specification errors (Griffith et al. 2015), which also need to be examined in future research.

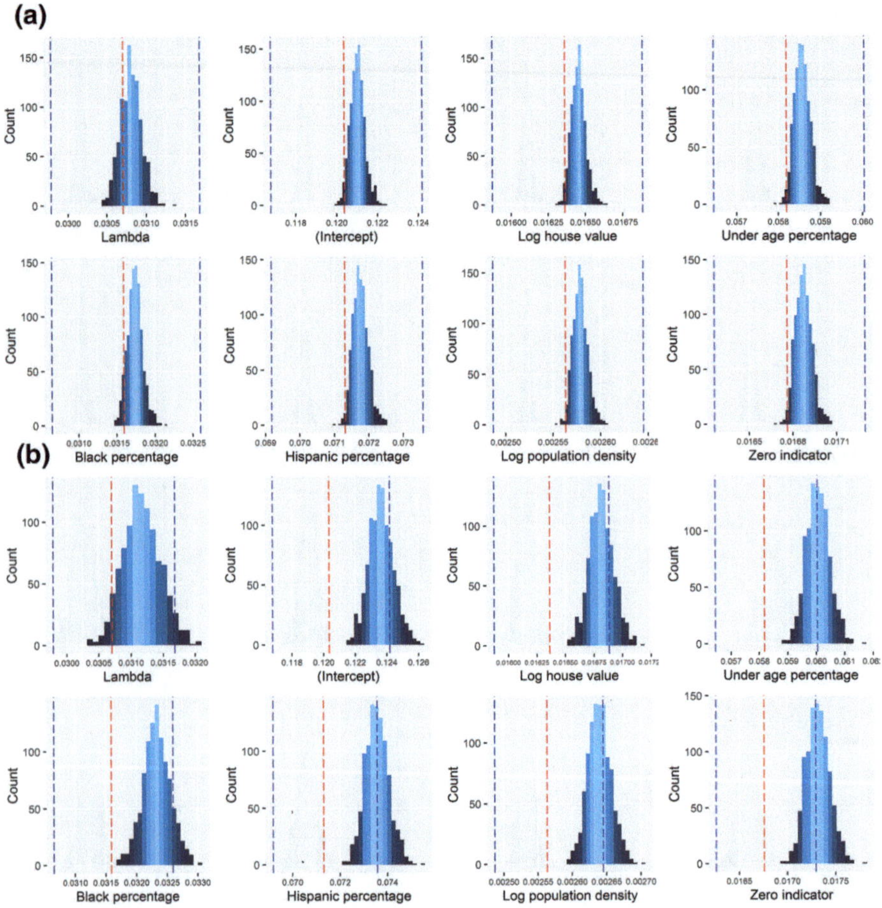

Fig. 5 SAR coefficient estimate standard errors for census blocks. **a** Measurement error added to 10 % of the observations, **b** Measurement error added to 50 % of the observations

Acknowledgements We thank Dr. Parmanand Sinha for comments concerning this research during the earlier phase of the project. This research was supported by the National Institutes of Health, grant 1R01HD076020-01A1; any opinions, findings, and conclusions or recommendations expressed in this material are those of the authors and do not necessarily reflect the views of the National Institutes of Health.

References

Barry S, Elith J (2006) Error and uncertainty in habitat models. J Appl Ecol 43:413–423

Cayo MR, Talbot TO (2003) Positional error in automated geocoding of residential addresses. Int J Health Geogr 2(1). doi:10.1186/1476-072X-2-10

Cromley EK, McLafferty SL (2002) GIS and public health. The Guilford Press, New York

Crosetto M, Tarantola S, Saltelli A (2000) Sensitivity and uncertainty analysis in spatial modelling based on GIS. Agric Ecosyst Environ 81:71–79

Danskin SD, Bettinger P, Jordan TR, Cieszewski C (2009) A comparison of GPS performance in a southern hardwood forest: exploring low-cost solutions for forestry applications. Southern J Appl For 33(1):9–16

Fisher PF (1999) Models of uncertainty in spatial data. Geogr Inf Syst 1:191–205

Griffith DA, Johnson DL, Hunt A (2009) The geographic distribution of metals in urban soils: the case of Syracuse, NY. GeoJournal 74(4):275–291

Griffith DA, Millones M, Vincent M, Johnson DL, Hunt A (2007) Impacts of positional error on spatial regression analysis: a case study of address locations in Syracuse, New York. Trans GIS 11(5):655–679

Griffith DA, Wong D, Chun Y (2015) Uncertainty related research issues in spatial analysis. In: Shi J, Wu B, Stein A (eds) Uncertainty modelling and quality control for spatial data. Taylor & Francis Group/CRC Press, London, pp 3–11

Heuvelink GBM (1998) Error propagation in environmental modelling with GIS. Taylor & Francis, London

Reeves GK, Cox DR, Darby SC, Whitley E (1998) Some aspects of measurement error in explanatory variables for continuous and binary regression models. Stat Med 17:2157–2177

Robinson VB, Avenue P, Frank AU (1985) About different kinds of uncertainty in spatial information systems. In: Auto-Carto 7, American society for phtogrammetry and remote sensing and American congress on surveying and mapping, pp 440–449

Wing MG (2011) Consumer-grade GPS receiver measurement accuracy in varying forest conditions. Res J For 5(2):78–88

Zimmerman DL, Li J (2010) The effects of local street network characteristics on the positional accuracy of automated geocoding for geographic health studies. Int J Health Geogr 9. doi:10.1186/1476-072X-9-10

Spatiotemporal Epidemic Modeling with libSpatialSEIR: Specification, Fitting, Selection, and Prediction

Grant D. Brown and Jacob J. Oleson

Abstract In an increasingly connected world, epidemic modeling techniques are an important tool for understanding epidemic phenomenon and forecasting future pathogen spread. While an extensive literature concerning the application of these techniques to a variety of pathogen life cycles and population contexts has arisen, general-purpose epidemic modeling software has not been forthcoming, especially for computationally challenging stochastic epidemic models. We introduce a general-purpose spatial epidemic modeling framework applicable to many pathogens and describe an open source package for the R statistical computing environment designed to perform such analyses.

Keywords Epidemics • Compartmental models • MCMC • Statistical computing

1 Introduction

The ability to effectively model the spread of infectious diseases over space and time is an important tool in an increasingly connected world. Epidemic modeling allows analysts to estimate the size of ongoing outbreaks, quantify pathogen reproductive behavior, evaluate public health interventions, and predict the degree of future spread. Despite these attractive features, epidemic models can be difficult to specify and computationally impractical to fit.

G.D. Brown (✉) · J.J. Oleson
University of Iowa, 145 N. Riverside Drive, Iowa City, IA, USA
e-mail: grant-brown@uiowa.edu

J.J. Oleson
e-mail: jacob-oleson@uiowa.edu

D.A. Griffith et al. (eds.), *Advances in Geocomputation*,
Advances in Geographic Information Science,
DOI 10.1007/978-3-319-22786-3_28

We propose a general class of epidemic models that allows a straightforward specification of a wide range of spatial structures and have developed the open source libSpatialSEIR epidemic modeling software to implement it. Set in the stochastic compartmental modeling framework, these techniques track the transition of populations through the susceptible, exposed, infectious, and removed (SEIR) disease states, and can address a wide array of spatial and nonspatial hypotheses (Porter and Oleson 2013; Lekone and Finkenstädt 2006; Chowell et al. 2004). Our software aims to expand the class of models that may be fit using modest hardware and the number of researchers with the ability to use such models. It includes both a simplified, high-level application programming interface (API) for standard analyses and a set of tools for specifying the model components individually to allow for maximum flexibility.

In this work, we introduce the conceptual framework behind these models, discuss several important computational tools, and illustrate the model fitting, selection, and prediction process in the context of the ongoing Ebola epidemic in West Africa.

2 Stochastic Compartmental Models

Compartmental techniques have a long history in the epidemic modeling literature and were originally introduced in the context of deterministic systems of differential equations (Kermack and McKendrick 1927; Hethcote 2000). Stochastic formulations of these techniques introduce probabilistic transitions between disease states, allowing for full estimation of the uncertainty of important model parameters. Although discrete time stochastic epidemic models sometimes are thought of as approximations to differential equation techniques, stochastic methods reflect the fact that epidemics are highly variable processes (Jacquez and O'Neill 1991). Our work concerns the stochastic spatial SEIR model class, which incorporates measurements of disease processes that occur over discrete time and space. Spatial heterogeneity has been incorporated into stochastic compartmental models in numerous ways, often focusing on the development of homogeneous clusters or equivalent, network-based, concepts (Cauchemez et al. 2004; Chis Ster et al. 2009; Cook et al. 2007; Deardon et al. 2010; Hooten et al. 2011; Jewell et al. 2009; van Boven et al. 2010; Verdasca et al. 2005).

This discrete formulation often corresponds directly to the spatiotemporal scale on which epidemic data are available: periodic counts of disease incidence in administrative regions. The temporal process employed by these models is given in Eqs. 1 and 2, for time points $\{t_i : i = 1, \ldots, T\}$ and spatial locations $\{l_j : j = 1, \ldots, T\}$. In this notation, capital letters denote the disease compartment (susceptible, exposed, infectious, or removed), and subscripts provide the location and time indices of these counts. Terms associated with an asterisk denote transition counts corresponding to individuals transitioning into the compartment of the same label. For example, E^*_{ij} gives the number of newly exposed individuals in location l_j and time t_i. Transition probabilities are denoted with Greek letters using superscripts, identifying the compartments losing and gaining individuals.

$$\begin{aligned} S_{ij} &= S_{i-1,j} + S^*_{ij} - E^*_{ij} \\ E_{ij} &= E_{i-1,j} + E^*_{ij} - I^*_{ij} \\ I_{ij} &= I_{i-1,j} + I^*_{ij} - R^*_{ij} \\ R_{ij} &= R_{i-1,j} + R^*_{ij} - I^*_{ij} \end{aligned} \tag{1}$$

$$\begin{aligned} S^*_{ij} &\sim binom\left(R_{ij}, \pi_j^{(RS)}\right) \\ E^*_{ij} &\sim binom\left(S_{ij}, \pi^{(SE)}\right) \\ I^*_{ij} &\sim binom\left(E_{ij}, \pi^{(EI)}\right) \\ R^*_{ij} &\sim binom\left(R_{ij}, \pi^{(IR)}\right) \end{aligned} \tag{2}$$

We introduce spatial and temporal heterogeneity into the form of the prior distribution for the exposure probabilities, $\left\{\pi_{ij}^{(SE)}\right\}$, by including a linear predictor term to capture shared and location-specific explanatory variables, and a set of distance matrices, $\{D_z: z = 1, \ldots, Z\}$, to model contact between spatial locations. Each distance matrix is associated with a single spatial parameter, ρ_z, which determines the contribution of its corresponding distance metric to the epidemic mixing process. With the introduction of several minor distributional assumptions, these components can be used to calculate the exposure probability given by Eq. 3 (Brown et al. 2015). Here, θ_{ij} denotes the value of the linear predictor corresponding to the time point t_i and spatial location l_j, and N_j is the total population size at location l_j. An example of practical use of this linear predictor term is given in Sect. 4.

A simple exponential form is employed for the E to I and I to R transition probabilities, as illustrated in Eq. 4. In this parameterization, $1/\gamma_{(EI)}$ and $1/\gamma_{(IR)}$ correspond to the average latent and infectious periods, respectively.

$$\pi_{ij}^{(SE)} = 1 - \exp\left(\left\{-\eta_{i.} - \sum_{z=1}^{Z} \rho_z (D_Z \eta_{i.})\right\}_j\right) \tag{3}$$

$$\eta_{i.} = \left(\frac{I_{i1}}{N_1} e^{\theta_{i1}}, \ldots, \frac{I_{in}}{N_n} e^{\theta_{in}}\right)$$

$$\pi^{(EI)} = 1 - \exp\left(-\gamma_{(EI)}\right); \pi^{(IR)} = 1 - \exp\left(-\gamma_{(IR)}\right) \tag{4}$$

3 Software

The libSpatialSEIR modeling software, freely available online (Brown 2014), comprises a C++ library and R-package interface, and aims to provide an easy-to-use and flexible implementation of the previously described spatial models. The software was designed to allow researchers to begin modeling epidemics quickly, while allowing enough flexibility for them to explore numerous hypotheses. Example 1 illustrates how simple the R code required to perform such an analysis can be. The code presented here reads incidence data for the 1995 Ebola outbreak in the Democratic Republic of the Congo and fits a model to evaluate the effectiveness of the intervention efforts. A complete analysis of these data using libSpatialSEIR may be found in Brown et al. (2015).

Example 1 High-level API

```
library(spatialSEIR)
cases = read.csv("http://bit.ly/1wdm3Xr")

results = fit.qSEIR(Count ~ daysSinceIntervention,
                    p_ei=1-exp(-1/5),
                      p_ir=1-exp(-1/7),
                      data = cases,
                      N = 5.36e6,
                      transition_ess=1000,
                      seed=12345,
                    n.cores=3,
                      return.cluster=TRUE)
```

4 Analysis

To explore the range of decisions informing epidemic data analysis in the spatial SEIR setting, we demonstrate a complete analysis of the 2014–2015 Ebola epidemic in West Africa. Data originally were collected from World Health Organization situation reports and were preprocessed and smoothed for clarity. In Example 2, we begin by loading the requisite R packages and reading in the processed data. As we have found no complete and detailed account of all public health intervention activities, we employ a set of basis splines to capture changes in population behavior over the course of the epidemic. Using the model selection techniques discussed in Brown et al. (2015), we selected a three-degrees-of-freedom basis, created with the *ns* function from the *splines* package. Next, this temporal basis is combined with a separate intercept for each nation to form the intensity process design matrix, Z. Finally, we define a distance matrix for each national border. This spatial structure is defined by a set of indicator functions, equal to one when the spatial locations with corresponding row and column indices share a border, and zero otherwise. While such a spatial structure is discrete, one instead could include informative weights motivated by, for example, a gravity model. Our devclopment may be seen explicitly in Example 2.

Specification of hierarchical models is always complex. In libSpatialSEIR, rather than require a user to provide all data and configuration options to a single complex function, we define an array of model objects that correspond to each level of the model. These include: a data model to relate the observations to the underlying epidemic parameters; process models to capture exposure, any reinfection activity, latent and infectious durations, and spatial heterogeneity; starting values for unknown population counts; and, instructions for selection and configuration of the Markov chain Monte Carlo (MCMC) samplers. These components are created in Example 3 and combined into a functional model by the *buildSEIRModel* function. Additional information about the construction of each model component is available in the package documentation, but this example illustrates the basic use of the required functions.

Example 2 Ebola model preparation

```
library(spatialSEIR)
library(splines)

# Read in Data
processedData = read.csv("http://bit.ly/1DYcLQx")

# Build temporal basis
basis = ns(cumsum(processedData$offset), df = 3)
Z = cbind(diag(3)[rep(1:3, each = nrow(processedData)),],
     basis[rep(1:nrow(basis), 3),])

# Declare neighborhood matrices.
DM1 = matrix(c(0,1,0,
        1,0,0,
        0,0,0), nrow=3, byrow=TRUE)
DM2 = matrix(c(0,0,1,
        0,0,0,
        1,0,0), nrow=3, byrow=TRUE)
DM3 = matrix(c(0,0,0,
        0,0,1,
        0,1,0), nrow=3, byrow=TRUE)

# Define the data set
cases = cbind(processedData$Guinea, processedData$Liberia,
       processedData$SierraLeone)

# Declare the population matrix and initial infectious counts
N = matrix(c(1.005e7, 4.1285e6, 6.1902e6), nrow = nrow(cases),
      ncol = 3, byrow=TRUE)
I0 = c(86,0,0)
```

Although the high-level *qSEIR* and *qSpatialSEIR* functions automatically run several MCMC chains until convergence, the models constructed as in Example 3 require a user to configure and request samples. Example 4 illustrates this process, demonstrating how an end user interacts with the model created by *buildSEIRModel.* This code also produces the example infectious count summary shown in Fig. 1.

Example R code to generate predictions using MCMC samples is available in the supplemental companion package to Brown et al. (2015). Predictions made using the latest available data in January 2015 show a decrease in infection size in all three nations, with an epidemic extinction time in late April. Updated models at the end of March validate the observed decrease in Sierra Leone and Guinea, in particular, but indicate that different dynamics have taken over as the epidemic has

Example 3 Low-level API

```
DataModel = buildDataModel(cases, type = "overdispersion", phi = 1)
ExposureModel = buildExposureModel(Z, nTpt=nrow(processedData),
              nLoc=3, offset=processedData$offset)
ReinfectionModel = buildReinfectionModel("SEIR")
SamplingControl = buildSamplingCont rol(iterationStride=500)
InitContainer = buildInitialValueContainer(data=cases, N=N,
                    S0=N[1,]-3*I0,
                    E0 = I0,
                    I0 = I0)
DistanceModel = buildDistanceModel(list(DM1, DM2, DM3),
                                 priorAlpha = 1,
                 priorBeta = 10)
TransitionPriors = buildTransitionPriorsFromProbabilities(
                    p_ei = 1-exp(-1/5), p_ir = 1-exp(-1/7),
                    p_ei_ess = 1000,  p_ir_ess = 1000)
SEIRModel = buildSEIRModel("samples.csv", DataModel, Exp osureModel,
                       ReinfectionModel, DistanceModel,
                       TransitionPriors, InitContainer,
                       SamplingControl)
```

Example 4 Sampling and inference

```
# Keep track of compartment counts, not just basis parameters
sapply(0:2, function(i){SEIRModel$setTrace(i)})

# Configure samplers; these are reasonable values for many analyses.
SEIRModel$compartmentSamplingMode=17
SEIRModel$useDecorrelation=10
SEIRModel$performHybridStep=11
SEIRModel$setRandomSeed(999774)

# Run the model
SEIRModel$simulate(100000)
# Read in the MCMC samples
mcmc.samples = read.csv("samples.csv")

# Clean up C++ objects
rm(SEIRModel, DataModel, ExposureModel, ReinfectionModel,
  DistanceModel, TransitionPriors, InitContainer,
  SamplingControl)

# Example summary measure, shown in figure 1
hist(mcmc.samples$I_2_46[100:nrow(mcmc.samples)],
   main = paste("Estimated Remaining Infectious Individuals",
                     "\n Sierra Leone - 3/29/2015"),
   freq=FALSE, breaks = 20, xlab = "Currently Infectious")
```

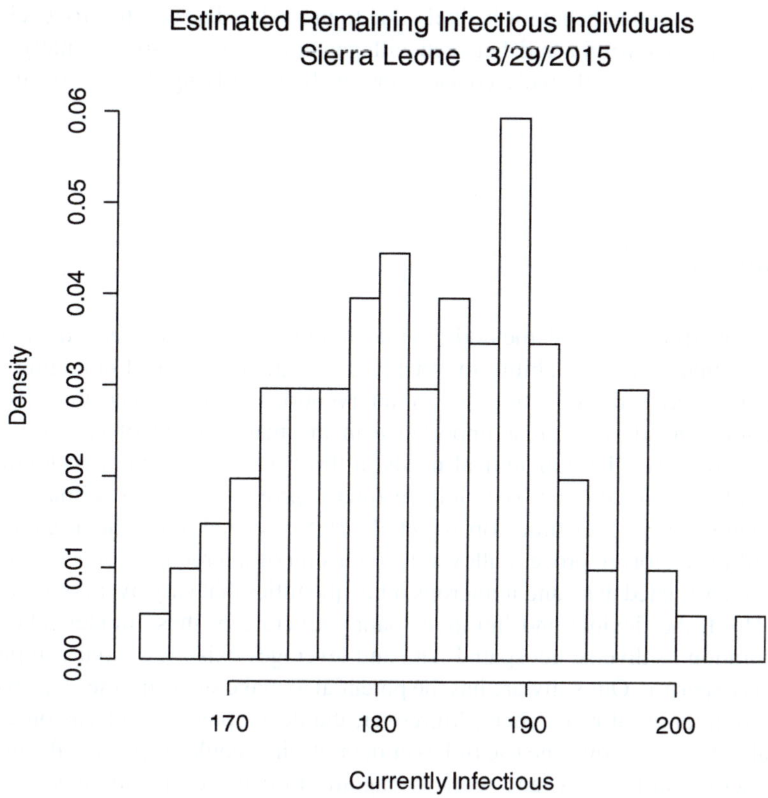

Fig. 1 Example output: estimated remaining infectious individuals, Sierra Leone, 3/29/2015

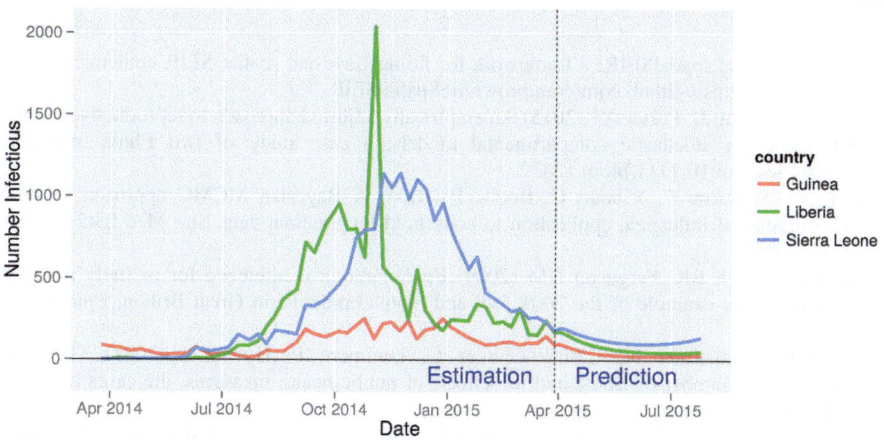

Fig. 2 Current predictions

shrunk (Fig. 2). Such behavior is likely due to unmodeled spatial heterogeneity; the epidemic spreads within and between villages and counties, so nationally aggregated data provide a relatively coarse view of the underlying disease dynamics.

5 Impact

These powerful analytical methods, while intuitive, have long lacked high-level computational tools. The ability to quickly assess the behavior of emerging pathogens, characterize the effectiveness of interventions, and evaluate the drivers of geographic spread is of great importance in the management of epidemics. Our software addresses these analytical needs in three ways. First, the development of empirically driven reproductive number estimates provides a tool to estimate changes in epidemic spread over time (Brown et al. 2015). Second, the general parameterization of the exposure process allows the inclusion of intervention effects, climactic and weather-related data, and numerous other quantities that vary over both space and time. Third, the flexible and intuitive spatial structure of these models allows the incorporation of diverse geospatial data and leverages existing spatial computation and data expertise. Our software has the potential to make such analyses feasible for a wider community of researchers. Moreover, the development emphasis on computational efficiency allows the use of larger datasets than could be practically analyzed in this setting in the past, and work is ongoing to better utilize the heterogeneous computing architectures increasingly available even on consumer devices.

References

Brown GD (2014) spatialSEIR: a framework for fitting Bayesian spatial SEIR epidemic models. R package. https://github.com/grantbrown/libSpatialSEIR

Brown GD, Oleson JJ, Porter AT (2015) An empirically adjusted approach to reproductive number estimation for stochastic compartmental models: a case study of two Ebola outbreaks. Biometrics. doi:10.1111/biom.12432

Cauchemez S, Carrat F, Viboud C, Boëlle P (2004) A Bayesian MCMC approach to study transmission of influenza: application to household longitudinal data. Stat Med 23(22):3469–3487

Chis Ster I, Singh BK, Ferguson NM (2009) Epidemiological inference for partially observed epidemics: the example of the 2001 foot and mouth epidemic in Great Britain. Epidemics 1 (1):21–34

Chowell G, Hengartner N, Castillo-Chavez C, Fenimore P, Hyman J (2004) The basic reproductive number of Ebola and the effects of public health measures: the cases of Congo and Uganda. J Theor Biol 229(1):119–126

Cook AR, Otten W, Marion G, Gibson GJ, Gilligan CA (2007) Estimation of multiple transmission rates for epidemics in heterogeneous populations. PNAS 104(51):20392–20397

Deardon R, Brooks SP, Grenfell BT, Keeling MJ, Savill NJ et al (2010) Inference for individual-level models of infectious diseases in large populations. Stat Sinica 20(1):239–261
Hethcote HW (2000) The mathematics of infectious diseases. SIAM Rev 42(2):599–653
Hooten MB, Anderson J, Waller LA (2011) Assessing North American influenza dynamics with a statistical SIRS model. Spat Spatiotemporal Epidemiol 1(2):77–185
Jacquez JA, O'Neill P (1991) Reproduction numbers and thresholds in stochastic epidemic models. 1. Homogeneous populations. Math Biosci 107(2):161–186
Jewell CP, Keeling MJ, Roberts GO (2009) Predicting undetected infections during the 2007 foot and mouth disease outbreak. J R Soc Interface 6(41):1145–1151
Kermack W, McKendrick A (1927) A contribution to the mathematical theory of epidemics. Proc R Soc A 115:700–721
Lekone PE, Finkenstädt BF (2006) Statistical inference in a stochastic epidemic SEIR model with control intervention: Ebola as a case study. Biometrics 62(4):1170–1177
Porter AT, Oleson JJ (2013) A path-specific SEIR model for use with general latent and infectious time distributions. Biometrics 69(1):101–108
Van Boven M, Donker T, van der Lubben M, van Gageldonk-Lafber RB, te Beest DE, Koopmans M, Meijer A, Timen A, Swaan C, Dalhuijsen A, Hahné S, van den Hoek A, Teunis P, van der Sande MAB, Wallinga J (2010) Transmission of novel influenza A (H1N1) in households with post-exposure antiviral prophylaxis. PLoS ONE 5(7):e11442
Verdasca J, da Gama T, Nunes A, Bernardino NR, Pacheco JM, Gomos MC (2005) Recurrent epidemics in small world networks. J Theor Biol 233(4):553–561

Geostatistical Models for the Spatial Distribution of Uranium in the Continental United States

Sara Stoudt

Abstract Although the United States Geological Survey (USGS) samples geochemical properties across the country, a complete understanding of the distribution of uranium remains elusive. Such an understanding would be useful to many government agencies because uranium can be both harmful to the environment and used to produce nuclear energy. I compare the performance of several nonparametric models for describing the geographic distribution of uranium deposits across the continental United States including the K nearest neighbors method, local regression models, generalized additive models, and Gaussian process models (kriging). I optimize model parameters using cross-validation with a training set and choose the final, most accurate model by comparison of predictions with a test set. I recommend using a kriging model, implemented with lattice krig, and utilizing an optional logarithmic transformation for uranium interpolation. Evidence for successfully avoiding overfitting through this cross-validation process is seen in the applicability of the optimal parameters for the prediction of substances other than uranium.

Keywords Uranium prediction · Kriging · Local regression · Generalized additive models

1 Introduction

Predicting the amount and concentration of uranium was first motivated by the need to minimize mining exploration expenditures (Drew 1977). Government interest soon followed, placing a heavier emphasis on integrating different datasets and analysis methods (Tang et al. 1986). Kane et al. (1982) perform an extensive optimization of the parameters for inverse distance weighting interpolation of geochemical properties. However, this method is only appropriate for regular

S. Stoudt (✉)
University of California, 331 Evans Hall, Berkeley, CA 94704, USA
e-mail: sstoudt@berkeley.edu

D.A. Griffith et al. (eds.), *Advances in Geocomputation*,
Advances in Geographic Information Science,
DOI 10.1007/978-3-319-22786-3_29

sampling patterns and breaks down when true samples are not uniform across the region of interest. Wu et al. (2011) compare different kriging methods for skewed data, but they replaced extreme values of the data with the median value instead of incorporating these values into their analysis.

This chapter includes an extensive comparison of kriging methods and other methods less widely used in traditional geostatistics yet still reasonable for modeling uranium. The interpolation of uranium has been done to meet various motivations but often on a small scale. For example, Garza et al. (2014) focus on uranium prediction in a portion of New Mexico. I do this on a large scale, using samples from the continental United States (US) [provided by the USGS (2004)] that are not uniformly distributed across the region, and without removing any "troublesome" large values. I find that although each method produces an interpolation that is visually distinct, the performance of each on the test set, as measured by the root mean squared error (RMSE), only negligibly differs.

I want the ability to predict the amount of uranium in parts per million (ppm) at any point within the continental United States with minimal error and uncertainty. Modeling uranium deposits faces several challenges. First, the samples are not uniformly distributed across the United States (Fig. 1), which introduces uncertainty to any model of sparsely sampled areas. Second, standard kriging assumptions for drawing inference are not tenable, because the distribution of uranium is neither symmetric nor normal, and furthermore cannot be easily transformed into a quasi-normal distribution. Figure 1 shows that the distribution that the data follow most closely is a beta distribution after transformation to the interval [0, 1]. Third, the large sample size of more than 40,000 uranium measurements can make traditional implementations of kriging slow on a personal computer.

In this chapter, I make the following three contributions to the existing knowledge about uranium in the lower 48 states:

1. Development of a scheme for determining the optimal set of parameters for predicting uranium (in ppm) that does not compromise the results' generalizability. This includes the use of training and test sets, 15-fold cross-validation, and a parameter sweep operation that checks all possible combinations of a subset of possible parameters yet avoids overfitting. The avoidance of overfitting is demonstrated by the optimal parameters chosen per model being similar if not exactly the same as those chosen for the optimal prediction of other deposit substances.
2. A comparison and contrasting of the K nearest neighbors (KNN) method, local regression (LR), generalized additive models (GAMs), and Gaussian process (kriging) models, as well as modifications to them that mitigate untenable assumptions.
3. Interactive results made available through the Google Earth interface.

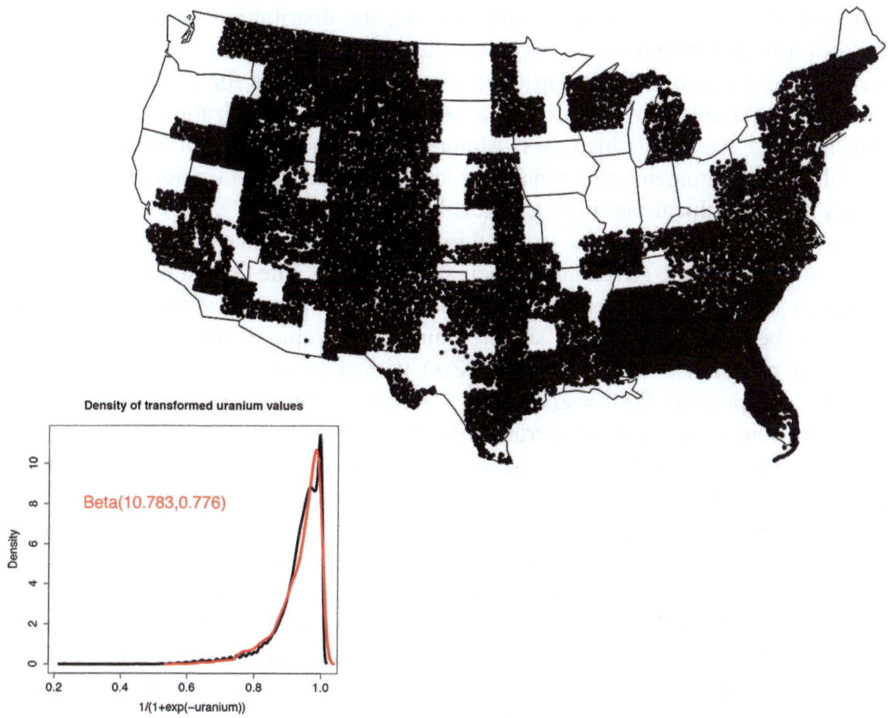

Fig. 1 Challenges: nonuniform sampling, nonnormal distribution

2 Methods

The KNN and LR methods have a neighborhood parameter. KNN assigns the value at an unknown point to be the average of some number of closest known sample points (Hastie et al. 2009). An extension of the KNN method is LR. In this method, the value of uranium at a particular location s is predicted by fitting a low-degree polynomial to a subset of the known data. Here the latitude and longitude coordinates are the predictor variables. The polynomial is fit using weighted least squares, giving more weight to points near (in Euclidean distance) to s and less weight to points farther away (Loader 1999). For both methods, I find the optimal parameters to be those that correspond to the smallest neighborhood, giving us the least smooth interpolation locally.

A GAM specifies the linear predictor as a sum of smooth basis functions of predictor variables. GAMs relax the assumption of normality in a response variable and allow for the response variable to follow distributions other than the normal distribution (Wood 2006). This feature is particularly useful because the distribution of uranium is nonnormal. For the exploration of GAMs, I map the uranium values to the interval [0,1] so that I can use a beta distribution and the logit link function. When I perform this transformation, the distribution of uranium closely follows a beta distribution. In each specification of the GAMs, I find the optimal bases to be unpenalized. Having unpenalized bases allows for local "wiggliness"; there is no parameter that enforces a penalty on the optimization criteria as the dimension of each basis increases. Although these models do not have an explicit neighborhood parameter, not penalizing local curvature gives us similar results: the ability to model local variation is valuable.

Gaussian processes allow us to think of the coordinates as more than just covariates and to take advantage of their spatial nature. The Gaussian process model (kriging) is defined by a spatial covariance matrix that takes into account relationships between neighboring sample points. The value of uranium at a location s can be modeled as a random field $Z(s)$ with a trend component, $m(s)$, and a residual component, $R(s) = Z(s) - m(s)$. Kriging estimates the residual at s as a weighted sum of residuals at surrounding data points. Kriging weights are derived from a covariance function (Cressie 1993). Similar to local regression, for which a parameter that determines what proportion of the data is to be considered "neighbors," in kriging, s_a represents points in the known data that lie within a neighborhood used to estimate the value at an unknown point s. This neighborhood can be thought of as the points within a distance of s, where spatial correlation is still noticeable (Bohling 2005). For the kriging methods, optimal large variance weights and a small range value translate to a rough spatial field and the least smooth covariance structure. Again, I find that the optimal parameters allow for local variation while preserving a smooth interpolation overall.

Traditional kriging fundamentally is a matrix inversion problem; to find the optimal weights, the covariance matrix is inverted. With a large dataset, this can get computationally expensive because inverting an n-by-n matrix is $O(n^3)$ in the number of operations, where n is the number of data points in the sample (Ikramov 2007). Therefore, implementation choice is crucial in order to have a reasonable execution time.

I use lattice krig (LK) for all of my kriging analyses It implements a multiresolution Gaussian process model that takes advantage of sparse matrices to speed up the computations (Nychka 2014). This implementation makes analysis of more than 40,000 data points feasible on a personal computer. Recall that I use 15-fold cross-validation together with a parameter sweep operation to determine the optimal parameters; each implementation must be reasonably fast so that the parameter choice process is not too computationally intensive.

3 Results

The preceding discussion emphasizes that the optimal parameters for each method correspond to those that allow for the most flexibility locally while maintaining a visually smooth interpolation globally. I want my interpolation to be flexible

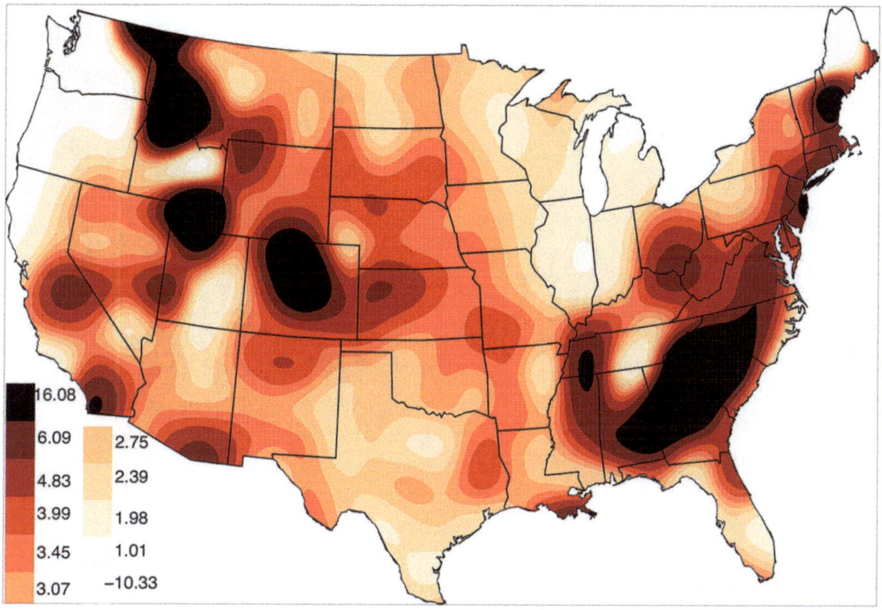

Fig. 2 Interpolated uranium using an optimal kriging method

because the true distribution of uranium is likely not completely smooth. However, I am careful to avoid overfitting when fitting to local details.

I find that kriging is the most effective method for predicting uranium (see Fig. 2). This is perhaps unsurprising, because kriging is the best linear unbiased predictor (BLUP). Kriging yields a RMSE for the test set of just under six ppm. This is underwhelming, as most of the uranium samples are less than five ppm; but, a large portion of this error comes from a few influential points. This method overestimates the value of uranium more often than it underestimates, but I can balance the residuals by using a transformation at the expense of a slightly larger RMSE. I also see large residuals in areas where I would expect more uranium from a geological point of view (see Fig. 3). If this geological intuition could be formalized, co-kriging on this information could be helpful. Table 1 contains the full results from other methods. The optimal parameters for uranium extend to the accurate prediction of aluminum, chromium, gallium, lithium, and magnesium.

Uncertainty in my predictions increases as the sparsity of the data sample locations increases. Unsurprisingly, and as a good commonsense check, low predicted values of uranium occur in the Midwest because I do not have any data in that region. Reassuring is that I only predict the largest values of uranium in areas where I have much information about uranium.

The best method assumes that the data follow a Gaussian distribution, so I want to find adaptations of the standard methods that do not require a normality assumption. I tried various transformations to make the uranium data conform more

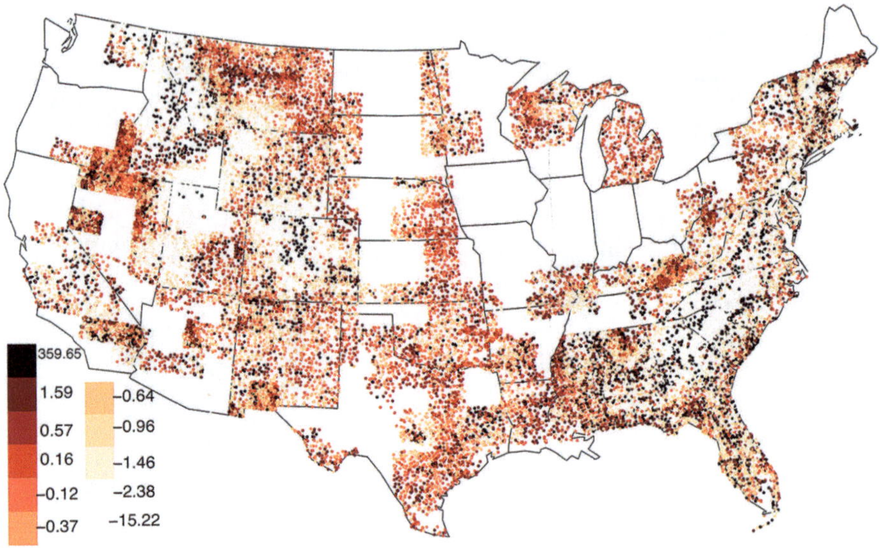

Fig. 3 Residuals for the training set using an optimal kriging method

Table 1 Results for the test set

Method	Test RMSE (ppm)	Max (ppm)	Qualitative assessment
LK	5.967	16.08	Most detailed yet smooth interpolation
LR	6.033	29.13	Can be marginally better than LK by using $\alpha < 0.2$ but then gives up some generalizability
KNN	6.182	18.50	In sparse regions relies heavily on samples that may not be representative
GAM	6.268	34.47	Smallest standard errors

closely to a normal distribution. I also explored methods that loosen the normality assumptions.

Through a logarithmic data transformation, I increase my RMSE yet remove the pattern of overestimating more than underestimating (see Fig. 4). Transformations remove patterns in the residuals but fail to lower the overall RMSE. I also can transform the data into its z-scores to remove any dependence on units. I then back-transform to assess performance. Similarly, n-score kriging ranks the sample data and assigns each a value based on the expectation of the order statistic of the same rank in a standard normal random variable, forcing the uranium data to follow a normal curve exactly.

Indicator kriging removes the dependency on normality but introduces an assumption that I am predicting a bounded random variable (Cressie 1993). I krigged to predict the probabilities that the amount of uranium lay in ten quantiles

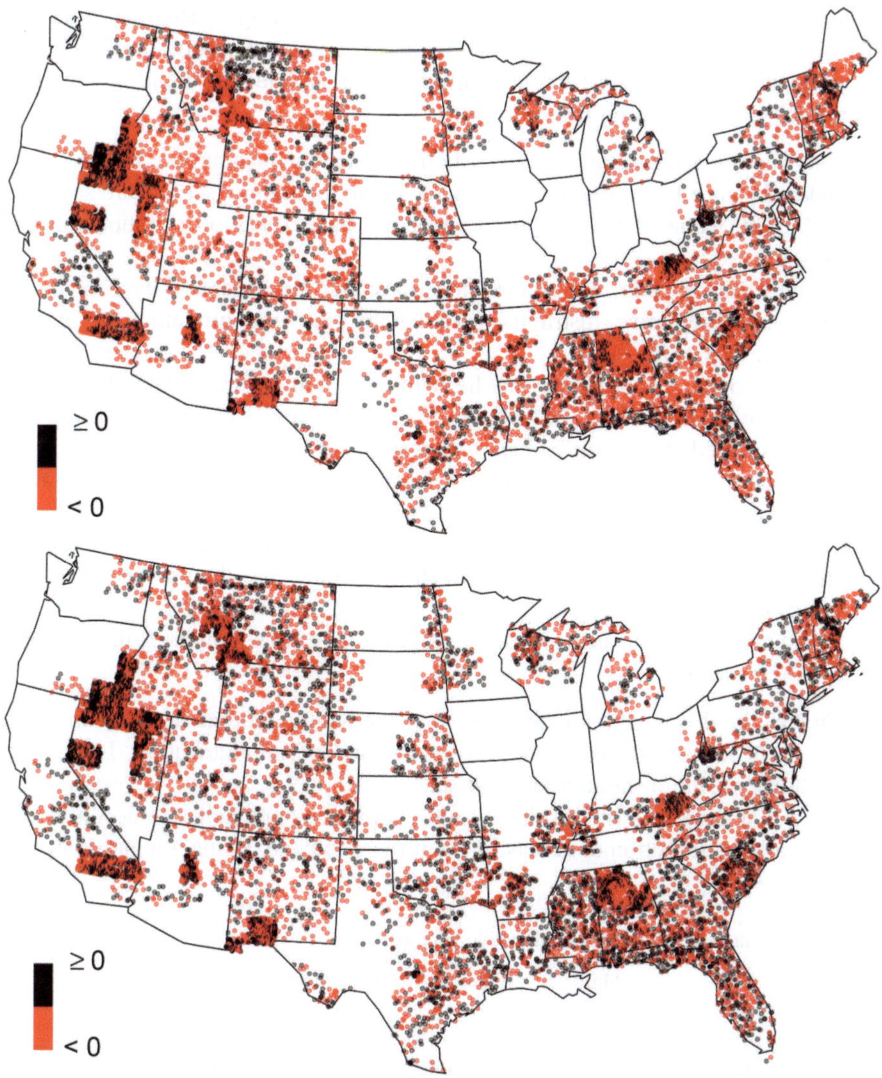

Fig. 4 Sign of residuals (observed–predicted) for the testing set using an optimal kriging method (*top*) coupled with a logarithmic data transform (*bottom*)

of the uranium distribution. I then combined these predicted values to determine an expected value of uranium for each location. The expected value interpolation looks reasonable and produces similar spatial patterns as other methods. However, the RMSE is extremely high, and the residuals reach large values throughout the continental United States. These large overestimates are most likely due to the many adjustments needed to make the estimates adhere to probability rules.

A copula can generalize the Gaussian random field in kriging (Kazianka and Pilz 2010). Creating a data-driven empirical copula is computationally intensive, but it is reasonable for a subset of the uranium data at a state level. I build an empirical copula for Colorado using the following steps:

1. Find the empirical cumulative distribution function for uranium **Y**: $F(Y)$.
2. Pick a distance **h** and find locations $\mathbf{s_i}$ and $\mathbf{s_j}$ in the training set that are separated by **h**. The value of **h** and the width of the band for approximation must be chosen.
3. Create a set of pairs representing locations that are separated by **h** using the empirical cumulative distribution values for the uranium amounts in each location: $(F(y_i), F(y_j))$.
4. This set contains coordinates that lie within the unit square. When plotted, they form a bivariate density, or copula, of interest. Now I can use this empirical copula to predict uranium values for locations in the test set.
5. For a test location $\mathbf{s_t}$ I find the nearest neighbors in the training set. The number of neighbors is a choice that can be optimized. For each neighbor $\mathbf{n_k}$, I draw a random value from the copula conditioned on the empirical cumulative density value of uranium at $\mathbf{n_k}$, $F(y_{n,k})$. I can choose to increase the number of random values drawn and aggregate them in some way.
6. I must choose how to summarize the values from each neighbor. A first step is to use their mean.

Choices in this method include **h**, the bandwidth around **h**, the number of neighbors **N** to use in prediction, and the method of summarization. I optimized these choices using a training and test set.

Table 2 summarizes the results for my alternative approaches to accommodate the normality assumption or at least make it more tenable. Using a logarithmic data

Table 2 Results for the test set—alternative approaches

Method	Test RMSE (ppm)	Max (ppm)	Qualitative assessment
LK	5.967	16.08	Most detailed yet smooth interpolation
LR	6.033	29.13	Can be marginally better than krig by using $\alpha < 0.2$ but then gives up some generalizability
Log (LK)	6.035	>1000	More balanced residuals but predicts some unrealistically large values
Log (LR)	6.118	237	More balanced residuals but predicts some unrealistically large values
Z-score krig	6.240	16.18	Normality adjustment does not help
Local copula	6.280	NA	Computationally intensive for full interpolation
N-score krig	6.930	365.70	Large predictions do not correspond to true extreme values

transformation performs the best of the alternative approaches. Performing kriging with z-scores and a local copula yields similar results, while performing kriging with the n-scores yields the largest RMSE. Kriging still outperforms the other approaches, but the differences in RMSE are not more than about one ppm. Nevertheless, these results provide further insight into where I could further improve these methods. I was most worried about a lack of normality, but, in reality, my main problems do not seem to come from this.

4 Conclusions

Through 15-fold cross-validation with a training set, I was able to optimize the parameters in the KNN, LR, GAMs, and kriging methods implemented with LK. These optimal values for the parameters are similar to those for predicting other deposit substances, providing evidence that overfitting was successfully avoided. The optimal parameters for predicting the amount of other deposit substances that have similar distributions to uranium, such as aluminum, chromium, gallium, lithium, and magnesium, are also those that correspond to local flexibility. In trying to improve the RMSE, I addressed the untenable normality assumption through various transformations and more general methods that do not require normality. These methods did not improve the RMSE for the test set. I found kriging to be the most accurate method according to the RMSE criteria. The RMSE values for the test set were very similar across all methods, but the prediction surfaces were visually distinct. Upon closer investigation of the final RMSE, I determined that about 60 % of the RMSE comes from ten points in the test set, showing

In Fig. 5, the bottom layers of each rectangle show the influence (the fraction of total MSE) of extreme values on my results, while the map shows the geographic location of these influential points. The topmost layer on each rectangle represents the combined effect of the rest of the points; about 60 % of the MSE comes from only ten locations. All of the methods fail to predict the same extremely large values

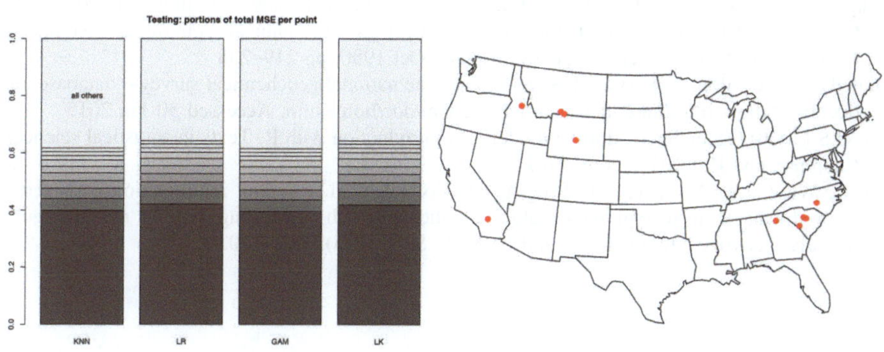

Fig. 5 Top ten most influential points

of uranium. For context, the location with the largest influence used to be a commercial uranium mine. Future work will address these few, but influential, samples. One way to do this is to use geological context and uranium mine history to classify areas likely to contain large amounts of uranium. Google Earth can be leveraged to obtain relevant information for classification. Each of the interpolations can be exported to a Google Earth layer and compared with the local landscape. Then, I can incorporate this information into the model as directional dependence and spatial heterogeneity in the variance.

Acknowledgements Thank you to Ben Baumer, Nick Horton, and Antonio Possolo for advice and guidance on this project. Thank you to NSF Travel Support for funding my participation in the 13th International Conference of GeoComputation (i.e., Geocomputation 2015). Thank you to the editors for providing constructive feedback on this work.

References

Bohling G (2005) Kriging. http://people.ku.edu/~gbohling/cpe940/Kriging.pdf. Accessed 30 Jan 2015

Cressie N (1993) Statistics for spatial data. Wiley series in probability and mathematical statistics. Wiley, New York

Drew M (1977) US uranium deposits: a geostatistical model. Resour Policy J 3(1):60–70

Garza O, Cabrera M, Sanin L, Cortes M, Meyer, E (2014) Spatial analysis techniques applied to uranium prospecting in Chihuahua State, Mexico. In: AIP conference proceedings, Chiapas, Mexico, April 2014, vol 1607. AIP Publishing LLC, Chiapas, Mexico, pp 116–122

Hastie T, Tibshirani R, Friedman J (2009) The elements of statistical learning. Springer, New York

Ikramov K (2007) Inversion of a matrix. http://www.encyclopediaofmath.org/index.php/Inversion_of_a_matrix. Accessed 30 Jan 2015

Kane V, Begovich C, Butz T, Myers D (1982) Interpretation of regional geochemistry using optimal interpolation parameters. Comput Geosci J 8(2):117–135

Kazianka H, Pilz J (2010) Geostatistical modeling using non-Gaussian copulas. In: Accuracy symposium, Leicester, UK, July 2010, vol N, pp 49–52

Loader C (1999) Local regression and likelihood. Statistics and computing. Springer, New York

Nychka D (2014) R Package lattice krig. http://cran.r-project.org/web/packages/LatticeKrig/LatticeKrig.pdf. Accessed 20 April 2015

Tang S, Xue Y, Meng J (1986) Application of the geostatistical analyses to uranium geology. In: Geological data integration techniques: proceedings of technical committee meeting of International Atomic Energy Agency, Vienna, Oct 1986, pp 219–238

United States Geological Survey (USGS) (2004) The national geochemical survey—database and documentation. http://mrdata.usgs.gov/geochem/doc/home.htm. Accessed 30 Jan 2015

Wood S (2006) Generalized additive models: an introduction with R. Texts in statistical sciences. Chapman and Hall/CRC, New York

Wu C, Wu J, Luo Y, Zhang H, Teng Y, DeGloria S (2011) Spatial interpolation of severely skewed data with several peak values by the approach integrating kriging and triangular irregular network interpolation. Environ Earth Sci J 63(5):1093–1103

Modeling Land Use Change Using an Eigenvector Spatial Filtering Model Specification for Discrete Responses

Parmanand Sinha

Abstract Spatial discrete choice models best suit unordered categorical response variables like land use change, but little literature is available about their application to large datasets. This chapter focuses on addressing the computational issues of a spatial discrete choice model and large datasets. An eigenvector spatial filter specification with coarser resolution has been used that accounts for spatial autocorrelation between neighboring land pixels. Variables of the built environment and socioeconomic and demographic characteristics are used as covariates in the spatial multinomial logistic regression.

Keywords Spatial filtering • Land use change • Spatial discrete choice model • Spatial logit

1 Introduction

Population growth is the basic driving force for land use change. According to the United States (US) Census 2010, every state grew in population, with growth continuing to be concentrated both within and adjacent to metropolitan areas. This increase in population near metropolitan areas results in land use change that has social, economic, and environmental impacts. Based on the current population projections, the same trend of substantial growth in metropolitan region is suggested over the next 50 years. This research is focused on incorporating spatial dynamics in urban land use change models for large datasets that could be used for forecasting with higher accuracy.

To model land use change, an analysis of variables causing this change is required. This analysis requires considering spatial effects, which becomes challenging in a discrete choice framework. Ignoring spatial effects leads to inefficient

P. Sinha (✉)
The University of Tennessee, 304 Burchfiel Geography Building 1000 Phillip Fulmer Way, Knoxville, TN 37996-0925, USA
e-mail: pnsinha@utk.edu

D.A. Griffith et al. (eds.), *Advances in Geocomputation*,
Advances in Geographic Information Science,
DOI 10.1007/978-3-319-22786-3_30

and sometime biased estimators. Many recent studies have been done in land use change, mostly involving binary choice. The binary models are more widely used due to their computational simplicity relative to multinomial models. This simplicity comes at the expense of assuming the independence of irrelevant alternatives in every case. With recent advances in computational algorithms, it is now possible to implement multinomial models accounting for more than one unordered alternative.

Sidharthan and Bhat (2012) apply the maximum approximate composite marginal likelihood (MACML) approach for a multinomial probit (MNP) model that involves spatiotemporal land use data. This methodology worked for a small dataset but can be cumbersome to apply for a large dataset. Another recent study by Carrion-Flores et al. (2009) superimposes a spatial lag structure on a multinomial logit (MNL) model using a linearized version of Pinkse and Slade's (1998) generalized method of moments (GMM) approach. However, it is based on a two-step instrumental variable estimation technique after linearizing around zero interdependence and so works well only in the case of large sample size estimation and weak spatial dependence. Chakir and Parent (2009) use Bayesian Markov chain Monte Carlo (MCMC) methodology to estimate an MNP model of land use change, which requires extensive simulation, is time-consuming, and is not straightforward to implement. Griffith (2004) estimates parameters of a spatially lagged autologistic model using eigenvector spatial filtering (ESF; Getis and Griffith 2002; Griffith 2003), which is a comparatively newer technique that does not involve inversion of matrices and is much simpler to apply (Wang et al. 2013). This chapter focuses on finding a way to reduce the computational burden of ESF for large multinomial data.

2 Multinomial Autologistic Regression for Land Suitability Analysis

Land suitability modeling exploits suitability analysis as development potential by applying multinomial logistic regression. Theoretically, if we establish a relationship between explanatory variables and the choice of land use development (e.g., residential versus commercial) from past data, we can use this relation to project the development of land parcels with similar characteristics.

Suppose a finite set of land pixels, $|\mathbf{N}| = n$ exists, and each pixel can be converted into one of the land uses in the set $\mathbf{M}$, $|\mathbf{M}| = m$. The logit equation can be written as

Ln (odds of land use change from vacant to j for location i at time t_2) = f (factors of the Built Environment, and Socioeconomic Characteristics),

where Ln denotes the natural logarithm, and f denotes a function. Included covariates measure, to an extent, some of the potential spatial dependence. How well they account for it can be seen from the statistical significance of the estimate of the spatial lag parameter in the MNL model:

$$U_{ij} = \rho \sum_{k=1}^{n} w_{ik} U_{ik} + \mathbf{X}_i \boldsymbol{\beta}, \tag{1}$$

where U_{ij} is a latent dependent variable representing the underlying utility from choosing a given alternative j, $\mathbf{X}_i$ is a vector of explanatory variables, $\boldsymbol{\beta}$ is a vector of regression coefficients, w_{ik} are elements of the spatial weight matrix w, and parameter ρ measures the nature and degree of spatial dependence. The spatial weight matrix is typically row standardized such that $\sum_{k=1}^{n} w_{ik} = 1$ for $k \neq i$, and $w_{ii} = 0$. For $P(Y_{ij} = 1)$, the probability of land parcel i having land use j:

$$P(Y_{ij} = 1) = \frac{\exp(U_{ij})}{\sum_{j=1}^{m} \exp(U_{ij})} \tag{2}$$

$$= \frac{\exp\left(\rho \sum_{k=1}^{n} w_{ik} Y_{ik} + \mathbf{X}_i \boldsymbol{\beta}\right)}{\sum_{j=1}^{m} \exp\left(\rho \sum_{k=1}^{n} w_{ik} Y_{ik} + \mathbf{X}_i \boldsymbol{\beta}\right)} \tag{3}$$

3 Estimation Method

As per the ESF model specification, a utility function can be decomposed into $\mathbf{U} = \mathbf{X}\beta + \mathbf{E}_K \boldsymbol{\beta}_E$, where X are explanatory variables, $\boldsymbol{\beta}$ is a vector of regression coefficients, $\mathbf{E}_k$ is an n-by-K matrix containing K eigenvectors, and $\boldsymbol{\beta}_E$ is the corresponding vector of regression parameters. The eigenvector spatial filtering specification for the multinomial auto-logistic model is

$$\mathbf{U}_{ij} = \mathbf{X}_i \boldsymbol{\beta} + \mathbf{E}_{Ki} \boldsymbol{\beta}_E \tag{4}$$

$$\mathrm{P}(\mathbf{Y}_{\mathrm{ij}} = 1) = \frac{\exp(\mathbf{U}_{\mathrm{ij}})}{\sum_{\mathrm{j}=1}^{m} (\exp \mathbf{U}_{\mathrm{ij}})} \tag{5}$$

$$= \frac{\exp(\mathbf{X}_i \boldsymbol{\beta} + \mathbf{E}_{Ki} \boldsymbol{\beta}_E)}{\sum (\mathbf{X}_i \boldsymbol{\beta} + \mathbf{E}_{Ki} \boldsymbol{\beta}_E)} \tag{6}$$

For a binomial dataset, ESF requires shorter and more straightforward computation. In the case of a multinomial dataset, the computation time significantly increases. The K eigenvectors are selected from the candidate set using stepwise multinomial logistic regression maximizing model fit at each step, which requires

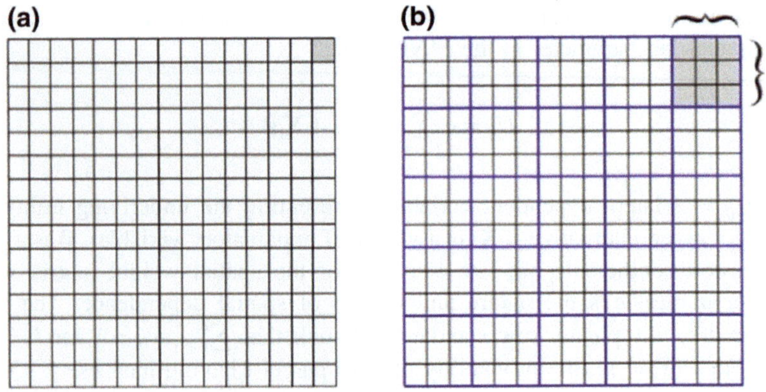

Fig. 1 **a** The finer spatial resolution for the response variable; **b** the coarser spatial resolution for the eigenvectors

lots of computation time as n increases. One possible way to reduce this computation time is to reduce the number of eigenvectors in the candidate set. This can be done by making the spatial resolution for the eigenvectors coarser while keeping the same spatial resolution for the response variable, as shown in Fig. 1a, in which nine cells of eigenvectors have been aggregated to one larger cell for the spatial effect, as shown in Fig. 1b.

4 Study Area and Data

The study area selected for this research is Collin County, Texas. With a total area of 886 mi^2 and a population of 782,341 (2010 census), this is a suburban county located in the northern part of the Dallas–Fort Worth Metroplex. For this chapter, the study area is divided into approximately 101,874 square cells, each with a dimension of 150-by-150 m. The centroid of each cell is assigned the land use of the parcel located in that grid cell. If a grid cell has more than one parcel, the cell is assigned that land use in which its centroid falls. The next coarser pixel resolution of eigenvectors, 450 m-by-450 m, has been used. At this coarser resolution, 2,419 sets of eigenvectors have been selected in the candidate set, each of which has a Moran's value greater than or equal to 0.25.

Due to safety and environmental concerns, several natural environmental factors,[1] such as the degree of slope, floodplain, and sensitive habitats prohibit development. The water body layer and 100-year floodplain layer mapped by the Federal Emergency Management Agency (FEMA) have been masked from the study area. After masking floodplain, highways, and utilities land use, 89,793 pixels

[1]These data have been collected from the North Central Texas Council of Governments (NCTCOG) regional data center. http://rdc.nctcog.org.

Table 1 Collin County land use distribution in 2000 and 2010

Land use	2000	2010	Change	% change
Single family (SF)	1,4745	32,660	17,915	121.50
Multifamily (MF)	1,141	3,407	2,266	198.60
Commercial (C)	2,080	4,828	2,748	132.12
Industrial (I)	604	1,415	811	134.27
Open spaces (OS)	1,094	3,113	2,019	184.55
Vacant (V)	70,129	44,370	25,759	–33.76

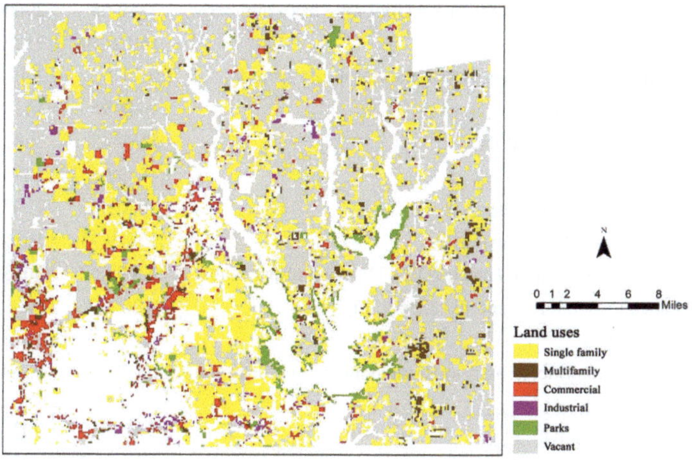

Fig. 2 Collin County land use change between 2000 and 2010

have been distributed in six land use categories. Table 1 and Fig. 2 show the land use change between 2000 and 2010.

For the built environmental factors, the selected variables are based on accessibility and/or proximity of land parcels to the location of specific activities. These factors typically include (but are not limited to) employment centers, airports, highways, major road intersections, schools, city centers, shopping centers, and public transit facilities. Socioeconomic variables include population, economic conditions, housing characteristics, vacancy rate, and employment characteristics. These data are available at the block group level from the US Census Bureau.[2]

Selection of the suitability factors depends upon the characteristics of the planning area and the categories of development or land use considered. Every region has unique, specific natural environmental as well as built environmental features. Table 2 displays the descriptive statistics of explanatory variables used in the study summarized in this chapter for the year 2000.

[2]Collected from http://geographicresearch.com/simplymap/.

Table 2 Explanatory variables for the year 2000

Variable	Mean	Standard. deviation
Built environmental factors (all distances are in feet)		
Proximity to shopping center weighted by gross leasing area (sqft/mile)	109476.40	442908.68
Proximity to employment center weighted by number of employees (#/mile)	171.77	840.29
Distance to school (0.6 mile or less, categorical)	0.13	0.33
Distance to highways	23962.61	18031.28
Distance to arterial roads (more than one-quarter mile from highways)	3348.68	2960.70
Distance to city center	15020.30	7129.86
Proximity to city center weighted by population of city (#/mile)	0.65	2.40
Distance from single-family land use	1648.95	1426.82
Distance from multifamily land use	5189.42	3681.67
Distance from commercial land use	7839.61	5898.43
Distance from industrial land use	10803.86	7202.47
Distance from parks and open spaces	15879.71	13268.73
Distance to Dallas Area Rapid Transit (DART)/public transit hubs	92099.62	36530.50
Ratio of distance to floodplain and distance to highway	0.26	0.84
Socioeconomic factors		
Easy Analytic Software, Inc. (EASI) total crime index on block group level (2008)	33.27	23.98
Median rent on block group level	514.13	203.76
Median value of owner households on block group level (2000 census)	131203.32	68595.65
Median year built on block group level (2000 census)	1985.63	20.77
Median household income on block group level (2000 census)	63408.00	18439.00
% employment in travel time less than 15 min on block group level (2000 census)	14.81	6.46
% employment in retail trade on block group level (2000 census)	11.97	2.81
Population on block level (2000 census)	67.73	137.33
% occupied housing on block level (2000 census)	78.81	33.63
Natural environmental factors		
Digital elevation model	192.16	21.64
Distance from floodplain	1470.34	1290.97
Distance to water bodies	11544.20	8442.61

5 Results

The results of the MNL model with and without spatial effects are presented below. The non-spatial MNL model uses only the explanatory variables. The spatial MNL model uses 661 set of coarser resolution eigenvectors in addition to the exploratory variables. These eigenvectors are selected from the candidate set of eigenvectors using stepwise selection.

5.1 The Nonspatial MNL Model

Figure 3 shows the predicted value of 2000–2010 land use change using the nonspatial MNL model.

Comparing predicted to actual changes in land use, the nonspatial MNL model is unable to predict changes in the suburban areas. All the multifamily, industrial, and open space land uses are mostly misclassified into single-family, commercial, and vacant land uses. Also, the predictions are clustered more toward the southeast part of the county that are part of more populous cities such as Richardson, Plano, Frisco, Allen, and Mckinney.

Multifamily and industrial land uses are predicted poorly compared with commercial and open space land use. The single-family land use category is slightly underpredicted, whereas the number of vacant pixels is overpredicted. Overall, with a kappa of 0.36, the accuracy of prediction is poor.

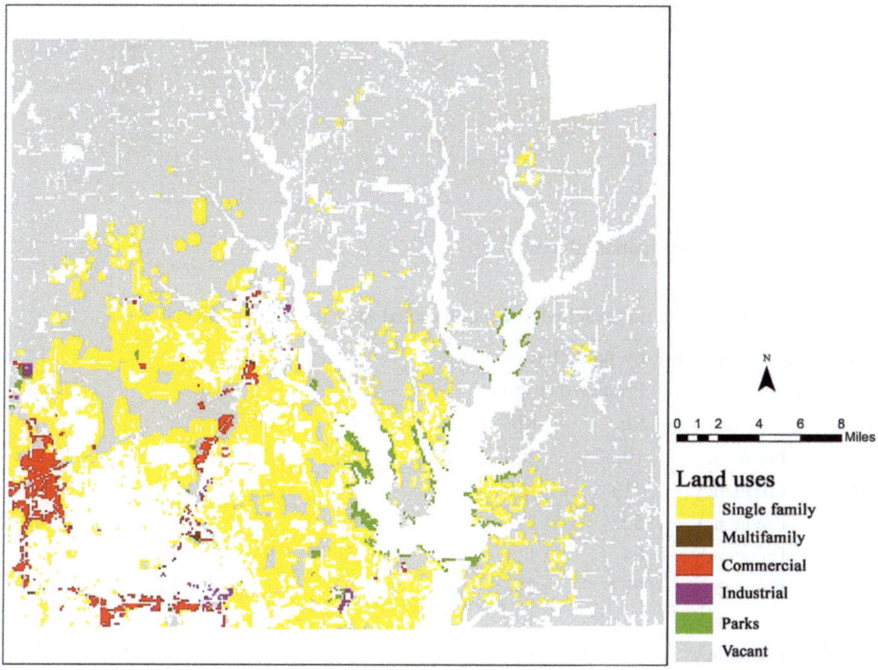

Fig. 3 Prediction without spatial effects

5.2 The Spatial MNL Model

The ESF model is used to account for spatial autocorrelation in the data as part of the MNL regression equation specification. The next coarser pixel resolution of eigenvectors grouped with 26 explanatory variables, has been used, and the result is shown in Fig. 4 and Table 3.

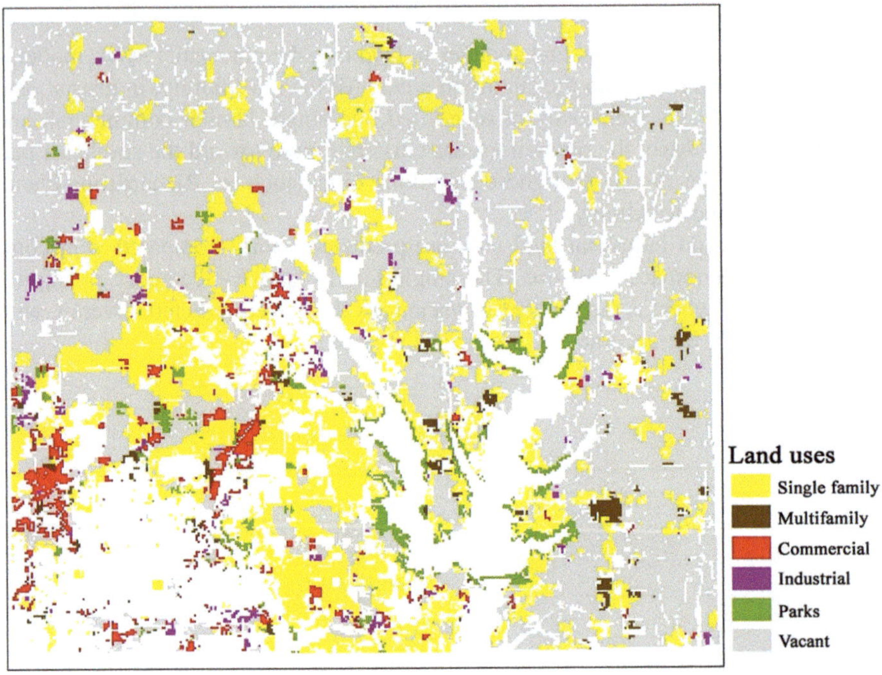

Fig. 4 Predictions with spatial effects

Table 3 A cross-tabulation of actual versus predicted land use spatial effects

Table of actual response by predicted response								
Response		Predicted response						Total
		SF	MF	C	I	OS	V	
SF	Frequency	10,375	203	221	60	236	6,820	17,915
MF	Frequency	413	569	93	10	58	1,123	2,266
C	Frequency	521	29	1,378	56	37	727	2,748
I	Frequency	64	1	33	511	9	193	811
OS	Frequency	273	23	25	8	1,506	184	2,019
V	Frequency	3,098	169	324	110	136	40,533	44,370
Total	Frequency	14,744	994	2,074	755	1,982	49,580	70,129
	Percentage	21.02	1.42	2.96	1.08	2.83	70.7	100

The sum of columns of Table 3 show the predicted land use pixels for each land use category. The diagonal elements in Table 3 show the correctly classified land uses. Prediction for multifamily land use has only 25.11 % accuracy and mostly it is misclassified into single-family and vacant land uses. This finding could be attributable to the absence of an exploratory variable that explains multifamily land use change. Single-family land use has been predicted with 58 % accuracy, commercial land use with 50.14 % accuracy, industrial land use with 60.3 % accuracy, and open spaces with 74.6 % accuracy. Taking spatial autocorrelation into account has increased the kappa statistic to 0.62.

6 Conclusion

For the ESF-based MNL model, the computational time for stepwise selection increases in polynomial time with an increase in the size of the candidate set of eigenvectors. In the case of a large spatial dataset, the candidate set of eigenvectors can be decreased by increasing the coarseness of the spatial filter resolution, which reduces computation time. Although the model with an ESF is able to capture more changes than a model without a spatial filter, it fails to capture very small changes in remote areas. This failure is probably due to the coarseness of the ESF. If the same spatial resolution as for response variable also is used for the spatial filter, it should furnish better predictions.

The kappa statistics imply that the spatial MNL gives better predictions than a nonspatial MNL. The covariate coefficient estimates differ between the non-spatial and spatial specifications, but show the same nature of the effects in both cases. After a certain number of eigenvectors enter into a model, increasing the number of eigenvectors by reducing the criteria of selection may not improve accuracy proportionately.

The ESF specification helps us to understand the marginal effect of exploratory variables more precisely. Compared to other methods, such as those based on auto-models, the ESF specification is easier to apply in a spatial discrete choice model for large datasets, and involves comparatively straightforward computation.

References

Carrion-Flores CE, Flores-Lagunes A, Guci L (2009) Land use change: a spatial multinomial choice analysis. 2009 annual meeting of agricultural and applied economics association, July 26–28, Milwaukee, Wisconsin. http://ideas.repec.org/p/ags/aaea09/49403.html

Chakir R, Parent O (2009) Determinants of land use changes: a spatial multinomial probit approach. Papers Reg Sci 88(2):327–344

Getis A, Griffith DA (2002) Comparative spatial filtering in regression analysis. Geogr Anal 34:130–140

Griffith DA (2003) Spatial autocorrelation and spatial filtering: gaining understanding through theory and scientific visualization. Springer Science & Business Media, Berlin

Griffith DA (2004) A spatial filtering specification for the autologistic model. Environ Plan A 36 (10):1791–1811

Pinkse J, Slade ME (1998) Contracting in space: an application of spatial statistics to discrete-choice models. J Econ 85(1):125–154

Sidharthan R, Bhat CR (2012) Incorporating spatial dynamics and temporal dependency in land use change models. Geogr Anal 44(4):321–49

Wang Y, Kockelman KM, Wang XC (2013) Understanding spatial filtering for analysis of land use-transport data. J Transp Geogr 31(July):123–131

Part IV
Computational Challenges and Advances in Geocomputation: High-Performance Computation and Dynamic Simulation

Conference papers in this section address computational aspects in geocomputation. A group of papers utilizes high-performance computing and/or parallel computing to handle large-sized geospatial data. Sorokine et al. present an approach for the shortest-path problem that can be used in a massive parallel environment, based on the out-of-RAM Dijkstra shortest-path algorithm. Shi discusses recent progress in parallelizing the affinity propagation algorithm on a graphics processing unit (GPU) for spatial cluster analysis, which can provide a scalable solution to process big geospatial data. Nugent et al. propose a Web-based GIS platform that can help cities to plan for risks to urban infrastructure and populations arising from climate change. They develop a distributed, high-performance geoprocessing engine to integrate a multitude of disparate, high-resolution data for analysis in a dynamic Web environment. White and Davis present a work flow for automated image registration utilizing a high-performance computing architecture.

Park et al. propose a framework for high-resolution geographic simulation models in which dynamics and behaviors of billions of humans are captured. They demonstrate the proposed method with an application of an infection model that utilizes a high-performance computing agent-based framework. Gong et al. propose an approach for scalable agent-based modeling (in which explicit spatial aspects are embedded) on a parallel computing environment. Najian and Dean present a dynamic simulation approach to investigate pedestrian travel patterns of individuals with varying levels of cognitive impairment, such as people with advanced dementia, aiming to identify travel patterns that are indicative of dementia.

This part concludes with a paper by Liao et al. that discusses pattern-based design approaches focusing on a formalized framework, the process and mechanics of pattern formation, and pattern-based planning and design methodologies. They also present a computational analysis and design method using the spatial graph grammars formalism.

From Everywhere to Everywhere (FETE): Adaptation of a Pedestrian Movement Network Model to a Hybrid Parallel Environment

Alexandre Sorokine, Devin White and Andrew Hardin

Abstract Shortest-path algorithms are hard to parallelize because they require a large number of global operations to estimate the costs of alternative routes. However, some geographic problems, such as locating archaeological sites and tracking the spread of infectious diseases, demand the ability to find a large number of the shortest paths on very large graphs or grids. Here, we present an approach based on the out-of-RAM Dijkstra shortest-path algorithm that can be employed in hybrid massively parallel or cloud environments. In this approach, we partition the graph, precompute all paths inside each partition, and then assemble the routes from precomputed paths. We demonstrate the utility of this approach by estimating travel frequency in pedestrian networks.

Keywords Single source shortest path · Population movement · Hybrid parallelization

1 Introduction

High-performance parallel computers allow geoscientists to tackle hard, computationally intensive problems and to process large datasets. In many cases, geospatial data lend themselves to parallelization through partitioning of a dataset across its geographic space. This technique of decomposing a computational process in the data domain has been used in a large number of projects, including the United States Geological Survey pRasterBlaster and PaRGO (Finn et al. 2012; Qin et al.

A. Sorokine (✉) · D. White · A. Hardin
Oak Ridge National Laboratory, One Bethel Valley Road,
P.O. Box 2008, MS-6017, Oak Ridge, TN 37831-6017, USA
e-mail: SorokinA@ornl.gov

D. White
e-mail: whiteda1@ornl.gov

A. Hardin
e-mail: hardinaj@ornl.gov

D.A. Griffith et al. (eds.), *Advances in Geocomputation*,
Advances in Geographic Information Science,
DOI 10.1007/978-3-319-22786-3_31

2014). However, some geoprocessing problems, like the shortest-path problem, are notoriously hard to parallelize due to their reliance on a large number of global operations.

The original shortest-path algorithms were developed in the 1950s (Dijkstra 1959; Warshall 1962). Since then, shortest-path algorithms have received significant attention in geography and computer science research. The computational requirements of finding the shortest path in a network are well understood and primarily driven by the network's size. In many practical cases, a network's size is relatively small, although in some cases computational requirements go beyond the capabilities of an average desktop or server computer. Examples include evacuation planning, computing alternative routes, and modeling movement without destination that involves finding a very large number of shortest paths. In such cases, computation is limited either by available random access memory (RAM), or, for case of the out-of-RAM algorithms, by an unacceptably long time to complete. These limitations can be overcome by utilizing high-performance parallel systems.

We present a strategy for porting an application of the shortest-path problem, described in White and Barber (2012), to a parallel computing environment. The purpose of the original algorithm was to predict where prehistoric population movements were likely to be channeled, thus providing insight into ancient trade/exchange networks as well as where archaeological sites may be located. This algorithm also can be applied to modern problems. Figure 1 shows a computed pedestrian network and travel frequency overlaid on shaded relief and state boundaries for the region of the 2015 Ebola outbreak in West Africa. Assessments of likelihood of pedestrian travel may help in the planning of the disease spread prevention efforts.

The algorithm works by aggregating a large set of shortest paths between many origins and destinations (from everywhere to everywhere, FETE). The set of origins and destinations can be generated as a regularly spaced grid or loaded from an external file. FETE finds Dijkstra's shortest path from each origin to every destination in a way similar to the Floyd–Warshall algorithm (Warshall 1962); that is, the priority queue is filled once for each origin and a complete set of destinations. As a result, a full set of paths from a single origin to multiple destinations is computed in a single step. Finally, all the generated paths are accumulated on a raster that represents travel probability surface.

The implementation of FETE in White and Barber (2012) can utilize shared-memory parallelism by calculating all paths from a single origin in parallel threads. The major limitation of this approach is the size of the digital elevation model (DEM) that can be loaded into RAM. Overcoming this limitation is the main intent of this study. By improving FETE's ability to process larger, high-resolution DEMs, we can improve the quality of the travel network prediction.

As stated before, we are unable to apply the technique of decomposing the problem into spatial partitions because each partition frequently would need to update a global data structure. Several approaches for decomposing the Dijkstra

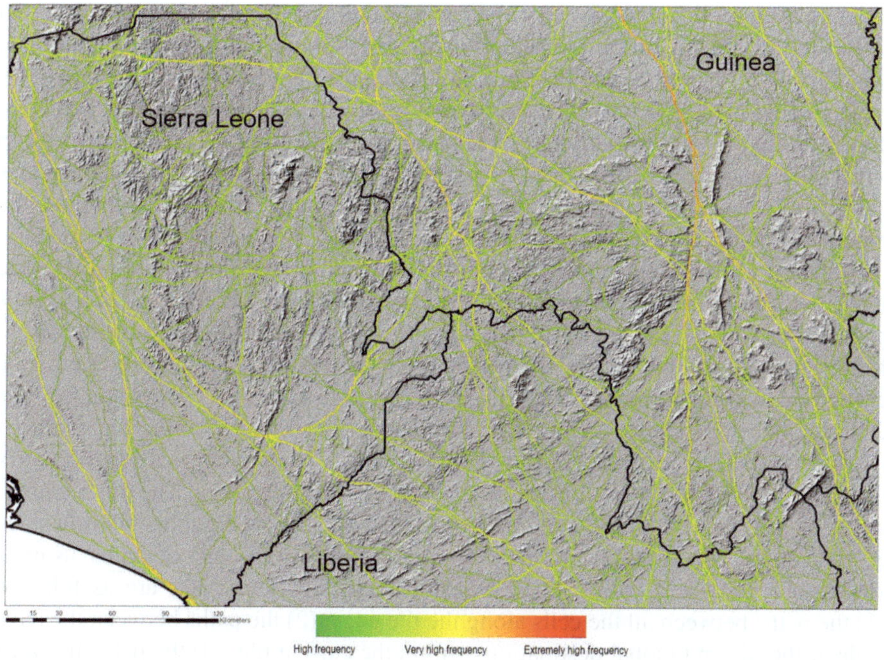

Fig. 1 A computed pedestrian network in the area of 2015 Ebola outbreak

shortest-path problem have been described in the literature (Matloff 2006; Madduri et al. 2007; Hazel et al. 2008). One approach works by splitting the priority queue into several subqueues and then distributing subqueues among the computing nodes (Matloff 2006). Another approach for out-of-RAM shortest-path computation works by decomposing an input graph into subgraphs, precomputing all paths between the entry and exit nodes inside each subgraph, and then computing the final route on the substitute graph built from the links between entry and exit nodes of the subgraphs (Madduri et al. 2007; Hazel et al. 2008).

For the purposes of this study, we have chosen the latter approach. The main drawback of this approach is the large overhead introduced by precomputing all paths in a subgraph. Because FETE works by aggregating a very large number of paths, however, the overall overhead of the path precomputation is less than the overhead resulting from the data exchange among the nodes with subqueues.

2 Proposed Solution

Our algorithm leverages a hybrid parallelism that combines shared memory multithreaded computation using OpenMP within each node, and distributed memory computation with communication facilitated by message-passing interface (MPI).

We process each subgraph on a separate node in parallel. Compared to an out-of-RAM version, our implementation involves several optimizations that utilize parallelism on varying architectures. We support the following four memory management models from which users can choose, depending upon their hardware configuration: large data structures (1) are kept in memory, (2) mapped to swap space, (3) offloaded to disk files, or (4) exchanged among the computing nodes using MPI one-sided communication. Furthermore, users can choose to limit the number of input/output threads to prevent their shared file system from overloading.

Figure 2 portrays our implementation of the FETE algorithm. First, each node loads two tiles: one from a DEM and one from a land use raster. The tile layout is provided as a model parameter, along with the physical properties of a traveler (e.g., height, weight) and the type of hiking function. As part of the initialization process, each node loads or generates a list of global origin-destination (OD) pairs. At stage 2, the connected graphs for intratile paths are created. Each connected graph is represented by a pair of matrices. The first matrix, connectivity, is a binary matrix that indicates existence of a link between a cell and its neighbors. The other matrix, costs, records the cost of traveling between each cell and its neighbors. During stage 3, each node uses multiple threads to calculate the four sets of paths using a single-source shortest path (SSSP) Dijkstra algorithm. These sets are as follows: (1) the paths between all the cells along the tile edge, (2) the paths between the edge cells of the tile and global destinations within the current tile, (3) the paths between the global origins within the current tile and destinations at the tile's edges, and (4) the paths between the global origins and destinations that both fall within the current tile. The system caches these four sets of paths and their costs.

At stage 4, the tables containing path costs are exchanged among the cluster nodes, after which each computing node constructs a substitute graph. Next, the SSSP Dijkstra algorithm is run in multiple threads on the substitute graph for each global origin that falls within the current tile. During stage 5, the computing nodes exchange the entrance and exit pairs of the paths in the substitute graph, after which they are used to reconstruct actual paths inside the tile (stage 6). Finally, the count of the paths crossing each cell is increased in the accumulated paths grid, and the grid is assembled into a single raster.

3 Results

We have successfully tested the proposed solution on a cluster and scaled the solution up to 36 computational nodes and DEMs having up to about 3 million cells. Figure 1 portays one of the resulting networks, and Fig. 3 presents a performance evaluation of the implementation, showing the relative speedup of the computation (the number of pixels processed per unit of time) versus the number of processing nodes. The black diagonal line represents a speedup of one for reference.

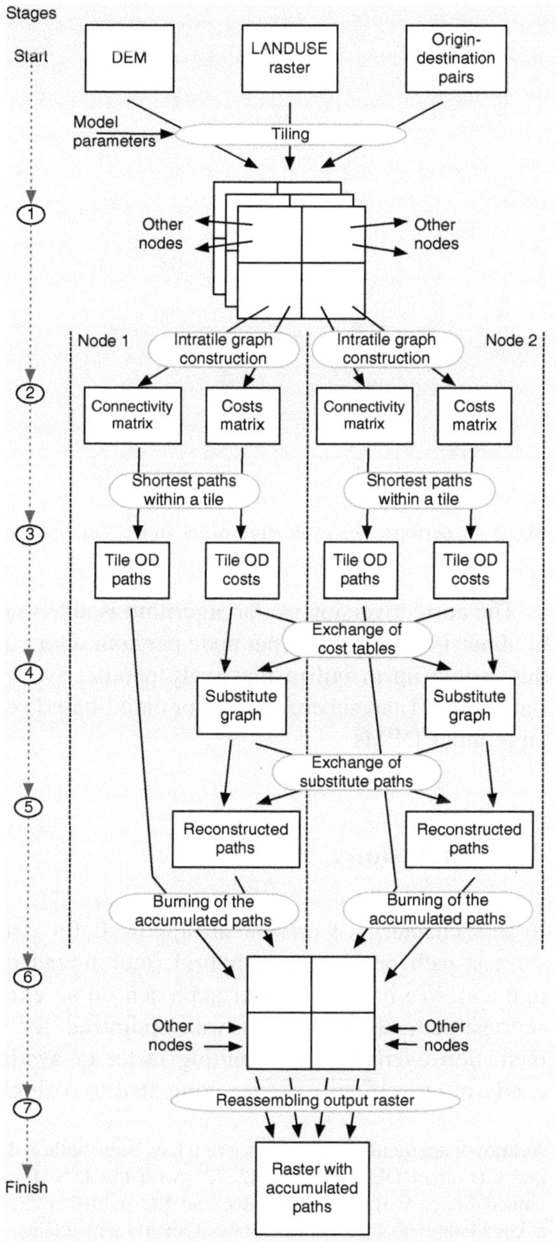

Fig. 2 The FETE algorithm with hybrid parallelism. The algorithm partitions DEM (stage 1), precomputes all paths inside each tile (stages 2–3), builds the substitute graph (stages 4–5), assembles the routes from precomputed paths (stage 5–6), and saves the results in an output raster (stage 7)

Each computation is shown as a point with color representing the size of an input DEM in number of pixels. The blue smoothed line denotes averaged computation time with a confidence interval in gray.

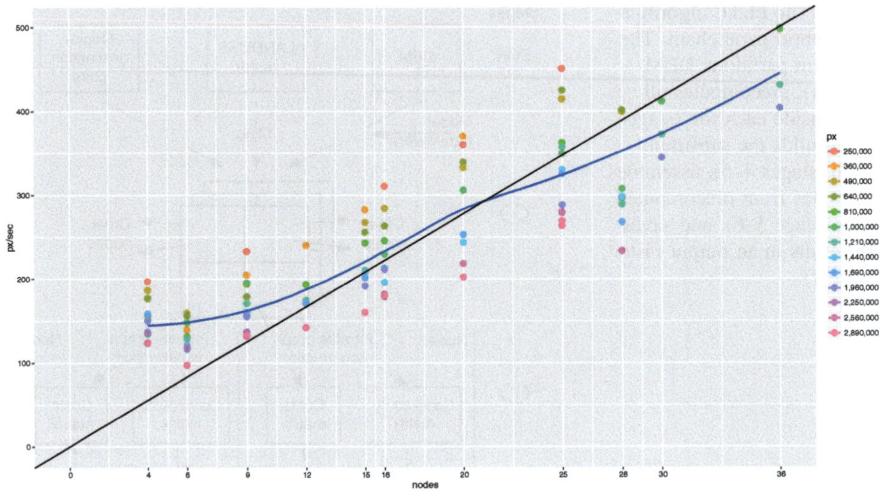

Fig. 3 A performance evaluation of the algorithm: speedup versus number of nodes

The current version of our algorithm is able to process an input DEM at the rate of about 1 million pixels per node per hour. Our current research is aimed at scaling the system up to utilize massively parallel systems like the Oak Ridge National Laboratory Titan supercomputer or cloud-based computing resources for massively large input DEMs.

4 Conclusions

In this chapter, we present an approach for parallelizing an application of the shortest-path problem in a hybrid (multithreaded and distributed memory) environment. We have based our approach on an existing out-of-RAM version of the shortest-path algorithm that was optimized for parallel computing. Our implementation overcomes the limiting factor of available memory and shows significantly improved performance compared to a single processor implementation.

Acknowledgements This manuscript has been authored by employees of UT-Battelle, LLC, under contract DE-AC05-00OR22725 with the U.S. Department of Energy. Accordingly, the United States Government retains, and the publisher, by accepting the article for publication, acknowledges that the United States Government retains, a non-exclusive, paid-up, irrevocable, world-wide license to publish or reproduce the published form of this manuscript, or allow others to do so, for United States Government purposes.

References

Dijkstra EW (1959) A note on two problems in connexion with graphs. Numer Math 1(1):269–271

Finn MP, Liu Y, Mattli DM, Guan Q, Yamamoto KH, Shook E, Behzad B (2012) pRasterBlaster: high-performance small-scale raster map projection transformation using the Extreme Science and Engineering Discovery Environment. In: Paper presented at the XXII International Society for Photogrammetry & Remote Sensing Congress, Melbourne, Australia, Aug 25–Sept 1, 2012. http://cegis.usgs.gov/projection/pdf/pRasterBlasterAbstract_ISPRS2012_V05.pdf. Accessed 3 Jan 2016

Hazel T, Toma L, Vahrenhold K, Wickremesinghe R (2008) Terracost: computing least-cost-path surfaces for massive grid terrains. J Exp Algorithmics 12:1–9

Madduri K, Bader DA, Berry JW, Crobak JR (2007) An experimental study of a parallel shortest path algorithm for solving large-scale graph instances. In: Applegate D, Brodal GS (eds) Proceedings of the ninth workshop on algorithm engineering and experiments, pp 23–35. Society for Industrial and Applied Mathematics, Philadelphia, PA. doi:10.1137/1.9781611972870.3

Matloff N (2006) Introduction to parallel processing. http://heather.cs.ucdavis.edu/~matloff/145/ParScript/MPI.pdf. Accessed 3 Jan 2016

Qin CZ, Zhan LJ, Zhu AX, Zhou CH (2014) A strategy for raster-based geocomputation under different parallel computing platforms. Int J Geogr Inf Sci 34:1–18. doi:10.1080/13658816.2014.911300

Warshall S (1962) A theorem on Boolean matrices. J ACM (JACM) 9(1):11–12

White DA, Barber SB (2012) Geospatial modeling of pedestrian transportation networks: a case study from Precolumbian Oaxaca. Mex J Archaeol Sci 39(8):2684–2696. doi:10.1016/j.jas.2012.04.017

Parallelizing Affinity Propagation Using Graphics Processing Units for Spatial Cluster Analysis over Big Geospatial Data

Xuan Shi

Abstract Introduced in 2007, affinity propagation (AP) is a relatively new machine learning algorithm for unsupervised classification that has seldom been applied in geospatial applications. One bottleneck is that AP could hardly handle large data, and a serial computer program would take a long time to complete an AP calculation. New multicore and manycore computer architectures, combined with application accelerators, show promise for achieving scalable geocomputation by exploiting task and data levels of parallelism. This chapter introduces our recent progress in parallelizing the AP algorithm on a graphics processing unit (GPU) for spatial cluster analysis, the potential of the proposed solution to process big geospatial data, and its broader impact for the GIScience community.

Keywords Spatial clustering • Affinity propagation • Parallel computing • GPU

1 Introduction

The identification of spatial clusters has been a high research priority in geoinformatics. In 2006, the National Research Council (NRC) of the National Academies released two significant publications entitled *Beyond Mapping: Meeting National Needs through Enhanced Geographic Information Science* (NRC 2006a) and *Priorities for GEOINT Research at the National Geospatial-Intelligence Agency* (NRC 2006b). Identification of spatial clusters, spatiotemporal data mining and knowledge discovery, and parallel and distributed computing are on the recommended lists of topics of high priority and challenges covered in these two documents.

According to the first law of geography, "Everything is related to everything else but nearby things are more related than distant things" (Tobler 1979). Identifying

X. Shi (✉)
Department of Geosciences, University of Arkansas, 216 Gearhart Hall,
Fayetteville, AR 72701, USA
e-mail: xuanshi@uark.edu

D.A. Griffith et al. (eds.), *Advances in Geocomputation*,
Advances in Geographic Information Science,
DOI 10.1007/978-3-319-22786-3_32

spatial clusters may be cast as a machine learning process to differentiate spatial features into a certain number of classes or clusters. As a result, features within the same class are more similar, whereas features in different clusters are less similar. Although classification generally refers to supervised classification and clustering to unsupervised classification (Guo and Mennis 2009), spatial clustering is a significant function in spatial data mining and can be applied to analyze both raster or image data and vector data.

Within the context of classification and clustering approaches for spatial data mining and knowledge discovery (Lu 2000; Han et al. 2001; Jacquez 2008; Guo and Mennis 2009; Kwan et al. 2014), this research targets affinity propagation (AP) (Frey and Dueck 2007) for several reasons. The AP algorithm was introduced by *Science* in 2007, whose article has more than 2,800 citations since then. As a relatively new clustering algorithm, however, AP has not yet been widely applied in geospatial research and applications. Unlike other classification or clustering algorithms, such as the Iterative Self-Organizing Data Analysis Technique Algorithm (ISODATA), k-means, and maximum likelihood classification, AP does not specify a predefined arbitrary number of clusters in advance. Rather, it derives the number of clusters as a result. Furthermore, according to Frey and Dueck (2007, 974), "Unlike metric-space clustering techniques such as k-means clustering, affinity propagation can be applied to problems where the data do not lie in a continuous space. Indeed, it can be applied to problems where the similarities are not symmetric and to problems where the similarities do not satisfy the triangle inequality." For this reason, AP can be applied in a broader domain of science research, such as bioinformatics, image and signal processing, and text mining (Frey et al. 2005; Frey and Dueck 2007), while in geospatial research, such a cluster analysis approach can be applied to raster or image data for classification and segmentation, vector geometric data for point clustering and pattern analysis, and text mining based on attribute values.

Figures 1 and 2 display the clusters of 6,271 physicians in Fulton County and DeKalb County in Georgia by different approaches. Figure 1 is a kernel density raster that groups the points into clusters highlighted by the density. However, such an approach cannot quantify the relationship between each point and its cluster center. Figure 2 is a cluster map generated by an AP calculation in R. Each cluster center is an exemplar of the connected point features.

Although AP has significant potential in geoinformatics for the identification of spatial clusters and other research and applications, such as spatial data resampling, spatial filtering, and pattern analysis, its adoption and expansion in geospatial science research has been seriously constrained by its capability to handle big data. The size of sample data provided online[1] ranges from 400 to 17,770 points, pixels, or units. A review of AP applications in remote sensing (Xia et al. 2009; Yang et al. 2010; Napoleon et al. 2012; Chehdi et al. 2014) reveals that the size of those sample

[1]Sample AP datasets: http://www.psi.toronto.edu/affinitypropagation/vsh/.

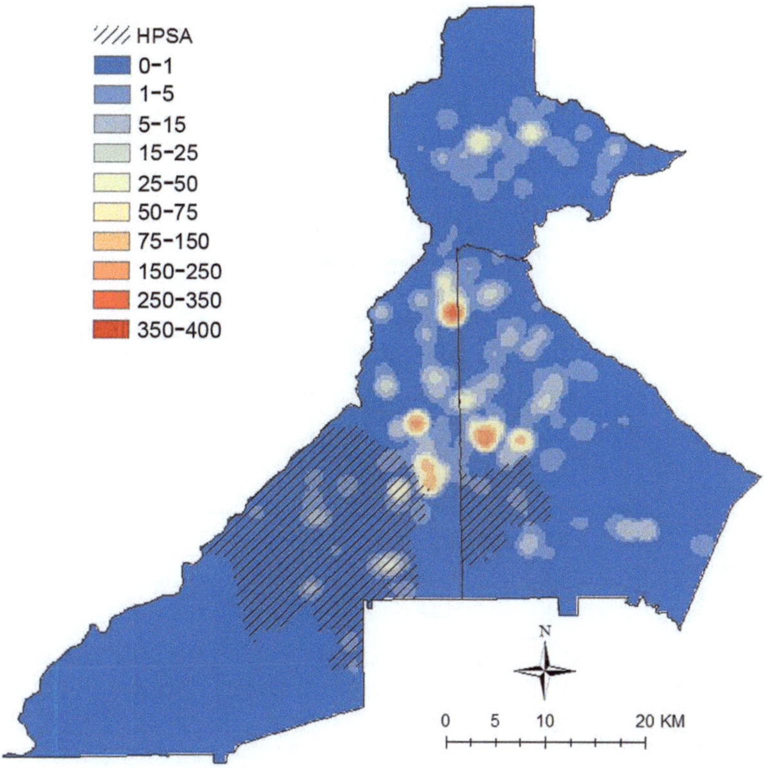

Fig. 1 Kernel density by ArcGIS

images is less than 100-by-100 pixels. If AP cannot process large amounts of data, it can hardly be adopted in geospatial applications.

This chapter introduces a current initiative that tries to parallelize the AP algorithm applicable on the modern graphics processing unit (GPU) to explore the potential of increasing its scalability and performance for big geospatial data. In the following sections, the AP algorithm is first reviewed to help understand its computational constraints. Next, the key issues in the parallelization of AP are discussed, followed by an introduction to the implementation of AP with a GPU. The scalability and performance comparisons are conducted for different datasets. Future directions of this endeavor are summarized in the conclusion.

2 The Affinity Propagation Program

The AP algorithm was first introduced by *Science* in 2007 (Frey and Dueck 2007). Implementing it requires a similarity matrix S containing n × (n − 1) records of the negative values of the distances between each point to all other points. The other

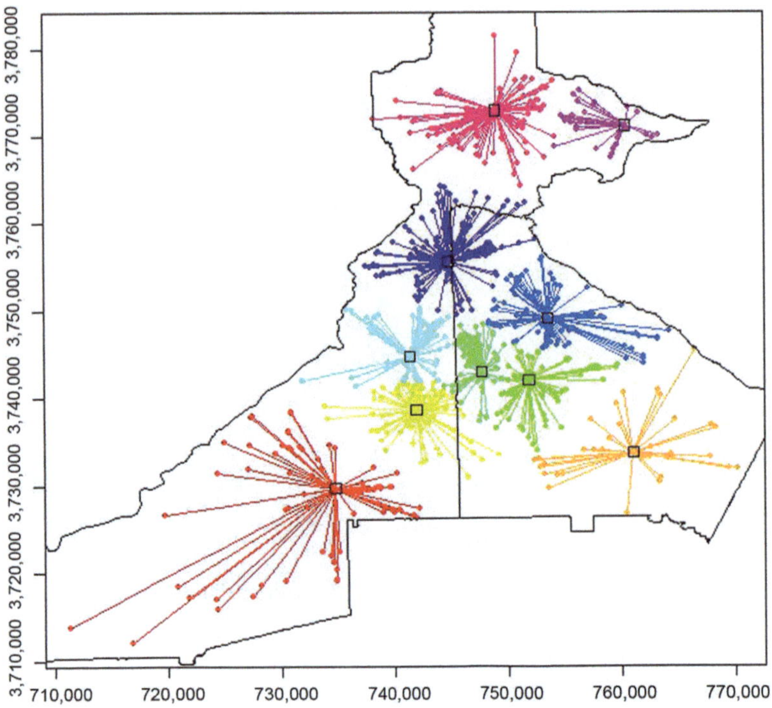

Fig. 2 AP clusters by R

input data contain the preference values for the n input points. By default, the preference value is the median value of the distance contained in the similarity matrix. The similarity matrix S describes how each data point is presented to be the exemplar or cluster center, while data points with higher preference values could be selected as cluster centers or exemplars. In this case, the preference value determines the number of identified clusters. In AP, all data points are considered equally as potential exemplars or cluster centers. For this reason, the preference values are initialized to a common value, which usually is the median distance in the similarity matrix. In general, AP is an optimization process maximizing the similarity or minimizing the total sum of intracluster similarities.

The number of clusters eventually emerges by iteratively passing messages between data points to update two matrices, A and R (Frey and Dueck 2007). The "responsibility" matrix R has values r(i, k) that quantify how well suited point k is to serve as the exemplar for point i relative to other candidate exemplars for point i. The "availability" matrix A contains values a(i, k) representing how "appropriate" point k would be as an exemplar for point i, taking into account other points' preferences for point k as an exemplar. Both matrices R and A are initialized with all zeros. The AP algorithm then performs updates iteratively over the two matrices. First, "Responsibilities" r(i, k) are sent from data points to candidate exemplars to indicate how strongly each data point favors the candidate exemplar over other candidate exemplars. "Availabilities" a(i, k) then are sent from candidate exemplars to data points to indicate the degree to which each candidate exemplar is available to be a cluster center for the data point. In this case, the responsibilities and availabilities are messages that provide evidence about whether each data point should be an exemplar and, if not, to what exemplar that data point should be assigned. For each iteration in the message-passing procedure, the sum of r(k; k) + a(k; k) can be used to identify exemplars. After the messages have converged, two ways exist to identify exemplars. In the first approach, for data point i, if r(i, i) + a(i, i) > 0, then data point i is an exemplar. In the second approach, for data point i, if r(i, i) + a(i, i) > r(i, j) + a(i, j) for all i not equal to j, then data point i is an exemplar. The entire procedure terminates after it reaches a predefined number of iterations or if the determined clusters have remained constant for a certain number of iterations.

3 Computation Constraints in the AP Program

Although AP has obvious advantages in comparison to many other approaches for cluster analysis (Frey and Dueck 2008), "Affinity propagation's computational and memory requirements scale linearly with the number of similarities input; for non-sparse problems where all possible similarities are computed, these requirements scale **quadratically** with the number of data points" (Dueck 2009, ii of Abstract). AP calculations took hours or a day to complete for some of the sample datasets discussed in Dueck's dissertation (2009).

In geospatial applications, several prior works (Xia et al. 2009; Yang et al. 2010; Napoleon et al. 2012; Chehdi et al. 2014) were able to handle only tiny datasets as prototypes to test the AP approach, because the images mentioned in these publications have a dimension of only several dozens or hundreds of pixels. The image size allowed by the AP algorithm in the MATLAB environment cannot exceed 3,000 pixels. In the case of image analytics, an image with a dimension of 100-by-100 pixels has a total of 10,000 pixels. The size of the similarity matrix is 10^8, which cannot be efficiently processed by the serial AP program. A single tile of a high-resolution image can easily contain $10^8 \sim 10^9$ pixels, resulting in computation that could simply go beyond the petascale (10^{15}) or exascale (10^{18}).

The same scalability and performance constraints exist when a large number of geospatial features in vector datasets are used. Considering the irregular distribution of spatial features, AP calculation may have to go through dozens of thousands of iterations to identify the cluster centers as exemplars. Consequently, AP calculation takes a long time to complete when a computer has sufficient memory to support the calculation. If AP is to be applied in geoinformatics to resolve real-world problems, the scalability bottleneck has to be overcome. Although modern accelerator technologies have been applied widely in general-purpose scientific computation, currently, no parallel and distributed computing solution exists for AP.[2]

4 Parallelization of the AP Program

Although sample data and a C program for AP are provided online,[3] the parallelization of AP does not seem to be an intuitive and easy process. The sample data have 25 points with x, y coordinates recorded in a text file. The two input text files include a similarity file and a preference file. The similarity file contains the information about the distance for each point to all other points and thus has $25 \times 24 = 600$ records. Each row contains the identification of a given point, the identification of a corresponding point, and the negative value of the distance between the two points. The preference file contains 25 rows each with the median value of the distance contained in the similarity file.

Consequently, when the input data are read into a program, the index of the input data seems chaotic in comparison with a regular matrix. Within a regular two-dimensional matrix annotated by n-by-n dimension, looping through the matrix to complete the calculation is easy. In the C program for AP, however, data are organized in an irregular pattern as $25 \times 24 + 25$. In a more general format, this matrix is indexed by n-by-$(n - 1) + n$. To elaborate this problem, a sample of ten points is used to describe the index of the input data of $10 \times 9 + 10$ in Table 1. While the input array is indexed from 0 to 99, the value in the array is retrieved from another *indexed* array as described in Table 1; that is, [1 2 3 4 5 6 7 8 9 0 2 3 4 5 6 7 8 9 0 1 3 4 5 6 7 8 9 0 1 2 4 5 6 7 8 9 1 2 3 5 6 7 8 9 0 1 2 3 4 6 7 8 9 0 1 2 3 4 5 7 8 9 0 1 2 3 4 5 6 8 9 0 1 2 3 4 5 6 7 9 0 1 2 3 4 5 6 7 8 0 1 2 3 4 5 6 7 8 9].

For this reason, a hybrid index approach is applied in the C program. In other words, the index of the input array is based on the output of the indexed array as described in Table 1. For a given n-by-n array, the indices of the n-by-$(n - 1)$ values, for example, are derived from the indexed array. In the following two lines of the C program, the value of m is $n \times n$.

```
for(j=0; j<m-n; j++)if(r[j]>0.0) srp[k[j]]=srp[k[j]]+r[j];
for(j=m-n; j<m; j++) srp[k[j]]=srp[k[j]]+r[j];
```

Although array k has a dimension of m, the values of k are between zero and n. Although the dimension of srp is n, the index of srp is determined by the value of k indexed by m. Meanwhile, the value of srp is calculated based on different conditions when the index of m is in a different range. Because $(m - n)$ is $n \times$

---

[2]Source: http://genes.toronto.edu/affinitypropagation/faq.html.

[3]Source: http://www.psi.toronto.edu/index.php?q=affinity%20propagation.

Table 1 The index of the input data definition in the C program

	1	2	3	4	5	6	7	8	9
0		2	3	4	5	6	7	8	9
0	1		3	4	5	6	7	8	9
0	1	2		4	5	6	7	8	9
0	1	2	3		5	6	7	8	9
0	1	2	3	4		6	7	8	9
0	1	2	3	4	5		7	8	9
0	1	2	3	4	5	6		8	9
0	1	2	3	4	5	6	7		9
0	1	2	3	4	5	6	7	8	
+									
0	1	2	3	4	5	6	7	8	9

$(n - 1)$, when srp is calculated differently for a separate data range of $n \times (n - 1)$ and n, which is just $m - (m - n)$, many other similar situations in the serial C program of AP have to be handled appropriately and carefully.

Reviewing the implementation details implies that breaking the data dependency within the sequential AP program could be the *key issue* in redesigning the parallel programs. When the index is applied as $n \times (n - 1) + n$, it is difficult to parallelize the serial AP program. For this reason, we need to first reconstruct or restore the regular (n-by-n) index framework in order to parallelize the serial C program for AP. As a result, certain individual modules of the serial C program have to be divided into two modules covering the two data ranges of $n \times (n - 1)$ and n.

5 Implementation of the Parallelized AP Program with the GPU

The parallelized AP program has been developed for implementation with the GPU. The GPU device used in this study is NVIDIA Telsa K40C installed on a desktop computer equipped with an Intel quad core i7 2.80 GHz CPU. The device has a total of 2,880 CUDA (i.e., NVIDIA's Compute Unified Device Architecture) cores with a total device memory of 12 GB.

Significant algorithm redesign and reconstruction have to be explored to achieve parallelism and to derive exactly the same result as that generated from the serial AP program. The general work flow of GPU implementation of AP is summarized as the following steps:

1. Specify the types and sizes of input and output data.
2. Allocate memory on the GPU for input data, output data, and intermediate data.
3. Allocate computing resources on the GPU; that is, specify the number of threads per block and the total number of blocks.
4. Copy both input and output data from the CPU to the GPU.

5. Execute the algorithm for AP computation.
6. Copy the output data from the GPU to the CPU.
7. Free the allocated GPU memory.

Appendix 1 describes the transformed code in the CUDA/GPU program that reflects the preceding work flow for implementation. For comparison purposes, a reader can examine the original C program[4] to understand the fundamental difference.

As concluded in the prior section, breaking the data dependency within the sequential AP program could be the key issue in redesigning the parallel programs. A significant amount of the original C program has to be rewritten and transformed into the CUDA program. For example, for the previously mentioned single line of code,

$$\text{for}(j = 0; j < m - n; j{+}{+})\text{if}(r[j] > 0.0)\ \text{srp}[k[j]] = \text{srp}[k[j]] + r[j],$$

the corresponding CUDA programs, including both the host and device programs, are described in Appendix 2. In this case, this one line of original serial C code has to be replaced by more than *fifty* lines of CUDA code.

Table 2 displays the performance comparison for AP calculations by the serial C program, the AP module in the R software, and the CUDA/GPU program. Dataset 1 has 3,736 point features denoting school locations in Arkansas. Dataset 2 has 6,271 point features denoting physician locations in Fulton County and DeKalb County in Georgia. Dataset 3 has 9,378 point features denoting school locations in Texas. Dataset 4 is a three-band image of 100-by-100 pixels. Dataset 5 is the Gene expression data downloaded from Frey's online resource (see note 1). Dataset 6 has 14,850 point features denoting school locations in Arkansas, Texas, and Oklahoma. In general, the CUDA/GPU program can achieve a speedup of 12–18 times in comparison to the serial C program and is much faster than the R program when memory is sufficient. R is slower than the serial C program. The APCluster software manual explains the performance issues when handling big data in R (Bodenhofer et al. 2015).

The calculation time for AP may vary because the optimal preference value can be derived from multiple experiments. In one test with the sample data containing 3,736 points, the AP program needs to complete 23,331 iterations to converge all features into 49 clusters. The serial C program needs about 6,459 s to generate this result, whereas the CUDA program needs about 615 s to complete this task. In the case of 6,271 physicians in Fulton and DeKalb counties, Fig. 2 displays a result with only 10 clusters, while a hierarchical number of clusters can be generated by adjusting the preference values.

One other associated computing problem in AP calculations is generating the similarity matrix and then deriving the preference values. By default, the preference value is the median of the distance between each feature and all other features.

[4]Source: http://www.psi.toronto.edu/affinitypropagation/apcluster_unsupported.txt.

Table 2 Performance comparisons for the serial C, R, and CUDA/GPU programs

Dataset	AP time (seconds)			Speedup	
	C	R	GPU	C/GPU	R/GPU
D1 (3,736)	119.04	155.66	9.26	12.86	16.81
D2 (6,271)	642.19	1,034.13	36.88	17.41	28.04
D3 (9,378)	1,154.79	2,008.89	65.64	17.59	30.60
D4 (10,000)	1,303.65	2,648.01	70.93	18.38	37.33
D5 (10,000)	1,019.94	2,003.71	58.66	17.39	34.16
D6 (14,850)	2637.91	5953.03	363.81	7.25	16.36

When a large volume of spatial features is involved, the similarity matrix is generated by n × (n − 1) calculations, while the median value can be derived by sorting n × (n − 1) values. Such data preparation procedures have to be parallelized to improve the scalability and performance of the computation.

Examining the original C and CUDA programs in Appendix 1 for AP calculations reveals that a large volume of memory is required to hold the input, intermediate, and output datasets. In general, when 10 K points are used in AP calculations, 4 GB of memory are required. For 20 K points, 16 GB of memory are required, while AP calculations over 40 K points need 64 GB of memory. As a result, when sample imagery data with 20 K pixels are used for AP calculations on a K40 GPU, no result can be generated. When sample data with 14,850 points are used, AP calculations on a K40 GPU can be finished in 363.81 s but may achieve a lower speedup when the memory limit on the GPU is approached.

6 Conclusion

By breaking data dependency, a serial C program for AP calculations has been successfully transformed into a CUDA/GPU program with significant performance improvement. The proposed solution implies that adopting an appropriate data structure is critical to improving the scalability and performance of geocomputation in geoinformatics and spatial statistics when big data are involved. In the case of local indicators of spatial association (LISA) calculations (Anselin 1995), for example, PySAL[5] is reasonable to process a small dataset. Once census block group data were used for LISA calculations, PySAL crashed because of insufficient memory. The serial C program for LISA calculations would lead to the same problem if a regular matrix data structure is applied. Because LISA calculations normally work with a sparse matrix, once the compressed row storage (CRS) data structure is applied to handle the sparse matrix, LISA calculations over 220,000 polygons in the census block group data can be finished in a couple of seconds,

[5]PySAL: https://geodacenter.asu.edu/pysal.

even by a serial C program. The parallelization of LISA on the GPU is pretty straightforward because a GPU is a perfect platform for matrix calculations.

Once data dependency can be removed, the majority of the AP program can be implemented by the embarrassingly parallel approach. As a result, the proposed solution has the potential to deploy clusters of GPUs to achieve the goal of scalable AP computation over big geospatial datasets as the memory limit on a single GPU can be overcome. Once the main frame of the AP algorithm can be transformed and implemented with a GPU, the same approach can be extended to transform the CUDA program to appropriate solutions doable on Intel's new hardware; that is, the many-integrated core (MIC). Considering the potential of AP in the scientific community in general, the parallel version of AP on a GPU/MIC and clusters of GPU/MIC will have a broader impact in the future.

Acknowledgments This research was partially supported by the National Science Foundation (NSF) through the award NSF SMA-1416509 "IBSS: Spatiotemporal Modeling of Human Dynamics across Social Media and Social Networks" and the National Institutes of Health (NIH) through the award NIH 1R21CA182874-01 "Reducing Physician Distribution Uncertainty in Spatial Accessibility Research." Any opinions, findings, recommendations, or conclusions expressed in this material are those of the author and do not necessarily reflect the views of the NSF or NIH.

Appendix 1

CUDA/GPU program that reflects the preceding work flow for AP implementation in comparison with the original serial C program

```
long blockSize = 512;
long nBlocks = m/blockSize + (m%blockSize == 0?0:1);
long nBlocks_n = n/blockSize + (n%blockSize == 0?0:1);
size_t sizeM = m*sizeof(double);
size_t sizeN = n*sizeof(double);
cudaMalloc((void **) &mx1_d, sizeN);
cudaMalloc((void **) &mx2_d, sizeN);
cudaMalloc((void **) &a_d, sizeM);
cudaMalloc((void **) &s_d, sizeM);
cudaMalloc((void **) &r_d, sizeM);
cudaMalloc((void **) &i_d, sizeM);
cudaMalloc((void **) &srp_d, sizeN);
cudaMalloc((void **) &k_d, sizeM);
cudaMalloc((void **) &dec_d, sizeof(unsigned long)*n);
```

```
cudaMemcpy(mx1_d, mx1, sizeof(double)*n,
cudaMemcpyHostToDevice);
cudaMemcpy(mx2_d, mx2, sizeof(double)*n,
cudaMemcpyHostToDevice);
cudaMemcpy(k_d, k, sizeof(unsigned long)*m,
cudaMemcpyHostToDevice);
cudaMemcpy(a_d, a, sizeof(double)*m,
cudaMemcpyHostToDevice);
cudaMemcpy(s_d, s, sizeof(double)*m,
cudaMemcpyHostToDevice);
cudaMemcpy(r_d, r, sizeof(double)*m,
cudaMemcpyHostToDevice);
cudaMemcpy(i_d, i, sizeof(unsigned long)*m,
cudaMemcpyHostToDevice);
cudaMemcpy(srp_d, srp, sizeof(double)*n,
cudaMemcpyHostToDevice);
cudaMemcpy(dec_d, dec[decit], sizeof(double)*n,
cudaMemcpyHostToDevice);

while(dn==0){
      it++; /* Increase iteration index */
      for(j=0;j<n;j++){ mx1[j]=-MAXDOUBLE; mx2[j]=-
MAXDOUBLE; srp[j]=0.0;}
      CUDAmodule00 <<< nBlocks, blockSize >>> (n,
mx1_d, mx2_d, srp_d);
      CUDAmodule0 <<< nBlocks, blockSize >>> (n, a_d,
s_d, mx1_d, mx2_d);
      CUDAmodule01 <<< nBlocks, blockSize >>> (m, n,
a_d, s_d, mx1_d, mx2_d);
      CUDAmodule1 <<< nBlocks, blockSize >>> (r_d, m,

i_d, mx1_d, mx2_d, a_d, s_d, lam);
      CUDAmodule2 <<< nBlocks, blockSize >>> (n, r_d,
srp_d);
      CUDAmodule3 <<< nBlocks, blockSize >>> (m, n,
```

```
r_d, srp_d);
      CUDAmodule4 <<< nBlocks, blockSize >>> (m, n,
r_d, k_d, srp_d, a_d, lam);
      CUDAmodule5 <<< nBlocks, blockSize >>> (m, n,
r_d, k_d, srp_d, a_d, lam);
      decit++; if(decit>=convits) decit=0;
      for(j=0;j<n;j++) decsum[j]=decsum[j]-
dec[decit][j];
      CUDAmodule7 <<< nBlocks_n, blockSize >>> (m, n,
r_d, a_d, dec_d);
      cudaMemcpy(dec[decit], dec_d, sizeof(double)*n,
cudaMemcpyDeviceToHost);
      K=0; for(j=0;j<n;j++) K=K+dec[decit][j];
      for(j=0;j<n;j++)
decsum[j]=decsum[j]+dec[decit][j];
      if((it>=convits)||(it>=maxits)){
            conv=1;
            for(j=0;j<n;j++)
if((decsum[j]!=0)&&(decsum[j]!=convits)) conv=0;
            if(((conv==1)&&(K>0))||(it==maxits)) dn=1;
      }
}

cudaMemcpy(a, a_d, sizeof(double)*m,
cudaMemcpyDeviceToHost);
cudaMemcpy(r, r_d, sizeof(double)*m,
cudaMemcpyDeviceToHost);

cudaFree(dec_d);
cudaFree(r_d);
cudaFree(k_d);
cudaFree(srp_d);
cudaFree(a_d);
cudaFree(s_d);
cudaFree(i_d);
cudaFree(mx1_d);
cudaFree(mx2_d);
```

Appendix 2

The host program and device program in CUDA/GPU code

```
/****************CUDA module – host*************************/

size_t size2 = (m-n)*sizeof(double);
size_t size3 = n*sizeof(double);

k_h=(unsigned long *)calloc(m-n,sizeof(unsigned long));
r_h=(double *)calloc(m-n,sizeof(double));

for(j=0;j<m-n;j++){
               r_h[j]=r[j];
               k_h[j]=k[j];
}

cudaMalloc((void **) &r_d, size2);
cudaMalloc((void **) &srp_d, size3);
cudaMemcpy(r_d, r_h, sizeof(double)*(m-n),
cudaMemcpyHostToDevice);
cudaMemcpy(srp_d, srp, sizeof(double)*n,
cudaMemcpyHostToDevice);
blockSize = 4;
nBlocks = n/blockSize + (n%blockSize == 0?0:1);

CUDAmodule2 <<< nBlocks, blockSize >>> (n, r_d, srp_d);

cudaMemcpy(srp, srp_d, sizeof(double)*n,
cudaMemcpyDeviceToHost);

cudaFree(r_d);
cudaFree(srp_d);
```

```
/****************CUDA module – device **************************/
__global__ void CUDAmodule2(unsigned long n, double *r,
double *srp)
{
   int idx = blockIdx.x*blockDim.x + threadIdx.x;
   __syncthreads();

   if (idx<n) {
      for(int j=0;j<n;j++) {
         if(j!=idx){
            if(j==0){
               if(r[(idx-1)+j*(n-1)]>0.0){
                  srp[idx]=srp[idx]+r[(idx-1)+j*(n-1)];
               }
            }
            else{
               if(idx>j-1){
                  if(r[(idx-1)+j*(n-1)]>0.0){
                  srp[idx]=srp[idx]+r[(idx-1)+j*(n-
 1)];
                  }
                 }else{
                 if(r[(idx-1)+j*(n-1)+1]>0.0){
                        srp[idx]=srp[idx]+r[(idx-1)+j*(n-
 1)+1];
                  }
               }
            }
         }
      }
      __syncthreads() ;
   }
 }
```

References

Anselin L (1995) Local indicators of spatial association—LISA. Geogr Anal 27:93–115

Bodenhofer U, Palme J, Melkonian C, Kothmeier A (2015) APCluster: an R package for affinity propagation clustering. https://cran.r-project.org/web/packages/apcluster/vignettes/apcluster.pdf

Chehdi K, Soltani M, Cariou C (2014) Pixel classification of large-size hyperspectral images by affinity propagation. J Appl Remote Sens 8(1), 083567: 1–14

Dueck D (2009) Affinity propagation: clustering data by passing messages. Dissertation, University of Toronto

Frey BJ, Dueck D (2007) Clustering by passing messages between data points. Science 315: 972–976

Frey BJ, Dueck D (2008) Response to comment on "Clustering by passing messages between data points". Science 319:726

Frey BJ, Mohammad N, Morris QD, Zhang W, Robinson MD, Mnaimneh S, Chang R, Pan Q, Laurin N, Sat E, Rossant J, Bruneau BG, Aubin JE, Blencowe BJ, Hughes TR (2005) Genome-wide analysis of mouse transcripts using exon tiling microarrays and factor graphs. Nat Genet 37(9):991–996

Guo D, Mennis J (2009) Spatial data mining and geographic knowledge discovery—an introduction. Comput Environ Urban Syst 33(6):403–408

Han J, Kamber M, Tung AKH (2001) Spatial clustering methods in data mining: a survey. In: Miller HJ, Han J (eds) Geographic data mining and knowledge discovery. Research monographs in GIS. Taylor and Francis, London, pp 201–232

Jacquez GM (2008) Spatial cluster analysis. In: Fotheringham S, Wilson J (eds) The handbook of geographic information science. Blackwell, Oxford, pp 395–416

Kwan M, Xiao N, Ding G (2014) Assessing activity pattern similarity with multidimensional sequence alignment based on a multiobjective optimization evolutionary algorithm. Geogr Anal 46:297–320

Lu Y (2000) Spatial cluster analysis for point data: location quotients versus kernel density. University Consortium for Geographical Information Science Summer Assembly, Portland, Oregon. http://dusk.geo.orst.edu/ucgis/web/oregon/papers/lu.htm

Napoleon D, Praneesh M, Subramanian MS, Sathya S (2012) Manhattan distance based affinity propagation technique for clustering in remote sensing images. Int J Adv Res Comput Sci Softw Eng (IJARCSSE) 2(3):326–330

National Research Council (2006a) Beyond mapping: meeting national needs through enhanced geographic information science. National Academies Press, Washington, DC

National Research Council (2006b) Priorities for GEOINT research at the National Geospatial-Intelligence Agency. National Academies Press, Washington, DC

Tobler W (1979) Cellular geography. In: Gale. S, Olsson G (eds) Philosophy in geography. Reidel, Dortrecht, pp 379–386

Xia HY, Chen XY, Guo P (2009) A shadow detection method for remote sensing images using affinity propagation algorithm. In: Proceedings 2009 IEEE international conference on systems, man and cybernetics, pp 3116–3121. doi:10.1109/ICSMC.2009.5346147

Yang C, Bruzzone L, Sun F, Lu L, Guan R, Liang Y (2010) A fuzzy-statistics-based affinity propagation technique for clustering in multispectral images. IEEE Trans Geosci Remote Sens 48(6):2647–2659

A Web-Based Geographic Information Platform to Support Urban Adaptation to Climate Change

Philip J. Nugent, Olufemi A. Omitaomu, Esther S. Parish, Rui Mei, Kathleen M. Ernst, Mariya Absar and Linda Sylvester

Abstract Increased frequency and intensity of extreme weather events and increasing population growth in cities bring to the forefront the need to easily evaluate risks in urban landscapes regarding critical infrastructure and vulnerable populations. In this chapter, we present an integrated framework for an urban climate adaptation tool (Urban-CAT) that will help cities plan for, rather than react to, possible risks to urban infrastructure and populations due to climate change. The core of the framework focuses on reducing risk by bringing together disparate, high-resolution data of risk indicators to characterize urban landscape and to develop resilience profiles. Additionally, the framework integrates climate and population growth data to better understand future impacts of these stressors on urban resiliency to develop effective adaptation strategies aimed at reducing socioeconomic costs associated with extreme weather events. This framework requires integration of the multitude of disparate, high-resolution data for analysis in a dynamic web environment. We address how to achieve this integration through the development of a distributed, high-performance geoprocessing engine.

Keywords Climate adaptation · Urban resilience · Raster processing · Distributed computing · Green infrastructure

P.J. Nugent (✉) · O.A. Omitaomu · E.S. Parish · L. Sylvester
Urban Dynamics Institute, Oak Ridge National Laboratory, Oak Ridge, TN 37831, USA
e-mail: nugentpj@ornl.gov

O.A. Omitaomu
e-mail: omitaomuoa@ornl.gov

O.A. Omitaomu · E.S. Parish · R. Mei · K.M. Ernst · M. Absar
Climate Change Science Institute, Oak Ridge National Laboratory, Oak Ridge, TN 37831, USA

D.A. Griffith et al. (eds.), *Advances in Geocomputation*, Advances in Geographic Information Science,
DOI 10.1007/978-3-319-22786-3_33

1 Introduction

Urban climate is changing rapidly. Therefore, climate change and its projected impacts on environmental conditions must be considered in assessing and comparing urban planning alternatives. In this chapter, we present an integrated framework for an urban climate adaptation tool (Urban-CAT) that will help cities plan for, rather than react to, possible risks. Urban-CAT is developed here as a scenario planning tool that is locally relevant to existing urban decision-making processes.

Cities have an opportunity today to become more resilient to climate change through changes made to urban infrastructure. Comprehensive characterization of a complex urban landscape and its critical infrastructure is newly possible as a result of recent simultaneous advances in computing and the collection and integration of large disparate datasets. Higher-resolution earth system models are advancing to the point of being able to directly characterize future climate conditions at scales vitally needed by urban decision makers. Providing cities with climate projection information now, in the form that they need most to identify key vulnerabilities and effectively allocate funding, may help reduce the vulnerability of our cities to extreme weather events over the next 30 plus years.

Addressing the question of urban system resilience to climate change is particularly urgent for several reasons. By 2030, over 60 % of the world's population will live in cities (WHO 2014). Although cities have proven to be extraordinarily resilient, the recent frequency and intensity of extreme weather events worldwide have raised concerns about their future vulnerability (IPCC 2014). Climate change impacts are expected to vary by location (IPCC 2014; EPA 2014), and provision of scientifically based tools for understanding and evaluating these impacts in conjunction with growing urban populations is critical to the development of adaptation strategies designed to avoid increasing socioeconomic costs of severe weather-related damages to urban landscapes (Preston 2013). These tools will be especially beneficial to mid-sized cities that currently house nearly half of all urban dwellers worldwide (WHO 2014).

The prelude for much of the current discussion regarding climate change and adaptation strategies stems from the Intergovernmental Panel on Climate Change (IPCC) 2007 *Fourth Assessment Report*, which outlines the impact on climate described by 23 different global climate models run with varying greenhouse emission scenarios (Quay 2010). In addition to addressing the threat of climate change, the IPCC 2007 report enumerates several adaptation and mitigation strategies, and discusses socioeconomic impacts associated with these strategies (IPCC 2007). Significant research exists that is aimed at building a conceptual framework of vulnerability and understanding how that concept relates to landscape and the extreme events that impact the landscape (Füssel and Klein 2006; Füssel 2007). Many emerging methods and frameworks for addressing climate change adaptation focus on identifying, ranking, and evaluating vulnerability and resilience metrics, particularly in iterative and scenario-based frameworks. Iterative approaches based on anticipatory learning rely on monitoring successes, failures, and reassessment throughout the adaptation process to make the most appropriate decision moving forward (Tschakert and Dietrich 2010).

As a first step toward achieving scalable, comprehensive urban climate change resilience planning, we are developing a web-based geographic information platform called Urban-CAT that will help cities plan for, rather than react to, possible risks. This is a scenario planning tool with local relevance. While cities already may have sophisticated tools to evaluate current site-specific scenarios, they lack (1) tools that scale site-specific conditions to neighborhood and citywide scales, and (2) credible climate data projections and population growth data to project future changes to an urban landscape. According to a recent global survey of 468 cities, changing storm water runoff and storm water management requirements are the most widely anticipated urban climate change impacts (Carmin et al. 2012). The 2014 IPCC identified changes to urban drainage systems as a key adaption issue for North America. Consequently, the proposed Urban-CAT framework uses storm water management as an application area. Because this is an adaptation tool, we also use green infrastructure (GI) as an adaptation strategy for storm water management.

2 The Urban-CAT Framework

The Urban-CAT tool (Fig. 1) is designed to support the development of resilient urban solutions. Its capabilities will include an advanced visualization platform to support decision making, access to future climate scenarios and environmental

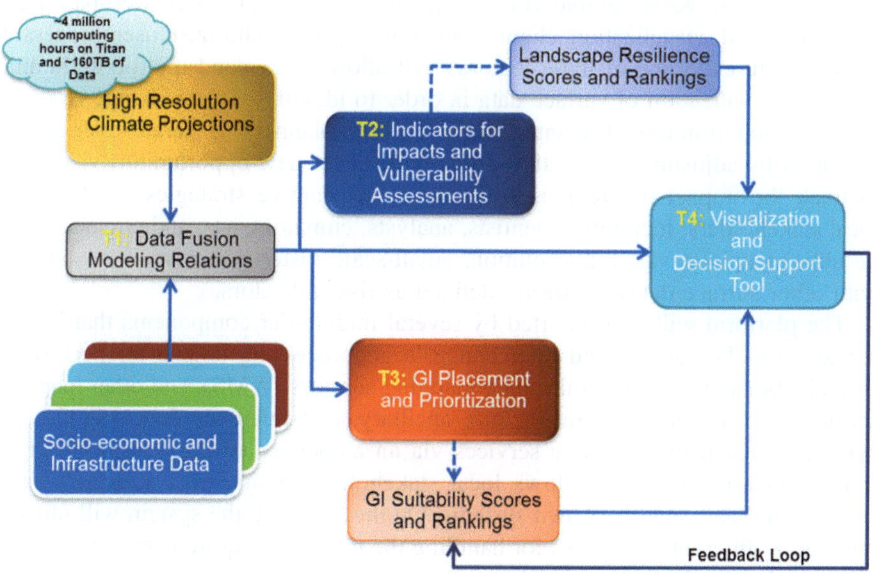

Fig. 1 Methodology overview

modeling results tailored for urban planning, connectivity to a multitude of data sources that promote assessment and comparison of local project scenarios under different climate conditions, and better insights into local effects of climate change through a scenario management capability for testing and comparing planning alternatives.

2.1 Framework

Prior to development of the visualization platform, several data development and modeling tasks have to be completed that feed into the platform capabilities. The initial task is to collect socioeconomic, infrastructure, and climate data, and to standardize these data to a common grid framework, as indicated by T1 in Fig. 1. The resulting data cube will help characterize the complex urban landscape and will feed into the effort to define vulnerability indicators, indicated by T2 in Fig. 1. The platform aggregates these developed indicators to construct resilience profiles for a surface and, conversely, to be able to identify areas at risk. Part of the framework includes the ability to evaluate GI emplacement strategies iteratively by updating the resilience profile of a landscape based on proposed emplacement options, as indicated by T3 in Fig. 1. This evaluation allows a scenario-based approach to planning that helps prioritize GI placement options. All of these tasks feed into the visualization platform aimed at helping stakeholders plan climate adaptation strategies.

The visualization platform is designed and developed as a multilayered system. The system will be accessible via a simple-to-use yet feature-rich web-based collaboration and visualization client, which will give authorized users access to system features. Some of these features will allow on-demand multi-criteria analysis and visualization of surface data in order to identify climate change risks and adaptive opportunities. The interface will allow planners to iterate over several scenarios by adjusting criteria thresholds to discover new opportunities visually or to track the impact of previously implemented adaptive strategies. Additionally, local stakeholders, including scientists, analysts, climatologists, and engineers, will be able to upload variable-resolution, small-scale raster information for analysis with other surface datasets already defined as risk indicators.

The platform will be supported by several middle-tier components that include services for data access and interchange, a geoprocessing engine for performing map algebra and serving results, web mapping services (WMS), and other map data services for visualizing infrastructure, ancillary, and reference data (Fig. 2). The system also will expose these services via an authenticated application programming interface (API) that allows local stakeholders to integrate standards-based services and data into their own systems. On the backend, the system will utilize a number of different data stores for handling the technical requirements necessary to support the application, including application persistence, data retrieval and ingestion, and geoprocessing tasks.

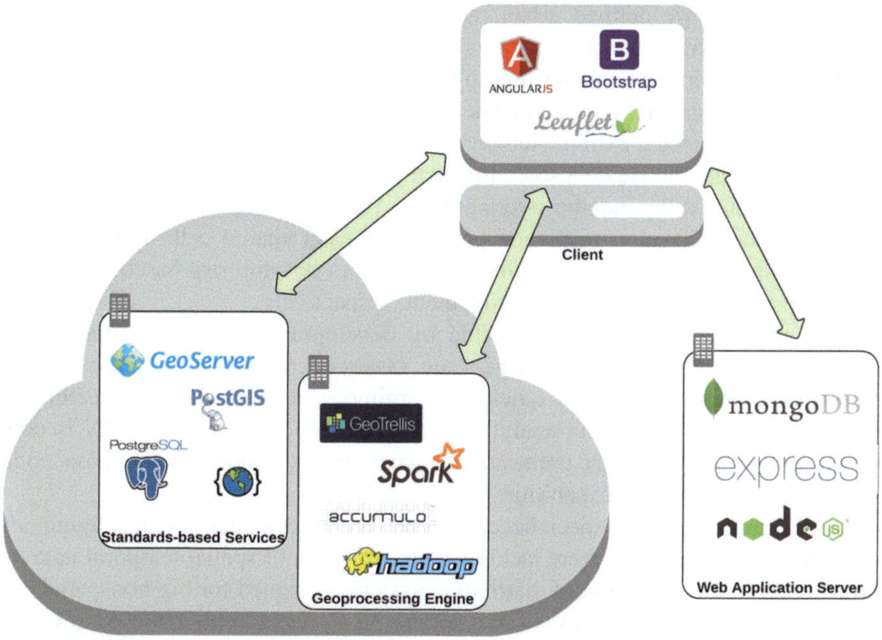

Fig. 2 The Urban-CAT platform

The framework is built on the premises of being deployable to various cloud environments and of exhibiting a service-oriented architecture for interoperability with multiple data sources and geoprocessing tasks via standards-based constructs. Additionally, the framework is highly configurable in that the stack may be rebranded per stakeholder organization and configured with a predefined area of interest and relevant datasets and processes.

2.2 Methods

Using a common spatial grid, we meshed downscaled and bias-corrected climate data for both historical (1960–2005) and future (2010–2050) periods (Ashfaq et al. 2014; Mei et al. 2013) with land use/land cover information, topography demographics, sewer pipe layouts, and social media accounts of local flooding events, among other sources, to characterize effectively the complex Knoxville, Tennessee, urban landscape and its water infrastructure. This integration helps to identify areas vulnerable to flooding and to discriminate by system exposure, sensitivity, and stress, among other risk factors. To integrate approximations of both adaptive capacity and the adaptive process into the tool, a set of indicators was developed and used to quantify each spatial grid.

For this tool, we defined urban resilience as a measure of eight components (Ross 2013)—climate, social, community, capital, economic, institutional, infrastructure, and ecological—using multiple indicators from different sources, including land cover/land use, imperviousness, slope, demographics, projected extreme precipitation, projected extreme temperature, and floodplain areas. These indicators may be aggregated to create a score for each grid cell. In turn, the scores are used to rank the spatial cells and the overlapping urban areas. The ranking was subsequently used to develop resilience profiles for each spatial cell. The developed indicators and resilience profiles are jointly used to develop risk-based approaches for storm water and floodplain management, respectively.

This methodology is made possible by developing a scalable geoprocessing engine capable of performing large-scale raster operations at web speeds. The architecture employed is one quickly emerging as a promising architecture for dealing with big spatial data analytics and visualization, and generically is composed of a cluster manager orchestrating work across several computation nodes that have geoprocessing capabilities.

Until fairly recently, specialized systems and techniques for managing and analyzing big spatial data were lacking. The explosion of spatial-temporal data has brought about the extension of distributed systems designed for big nonspatial data to handle spatial data. Many of the emerging systems follow the MapReduce paradigm employed in the Hadoop ecosystem and Spark-based computing clusters (Eldawy and Mokbel 2015). Additionally, advances in techniques for spatial data processing have prompted the need for improvements in data storage and indexing for spatial data. Traditional spatial indexing is being applied to newer nonrelational databases for storing and accessing spatial data (Xie et al. 2014). Also, new spatial indexing techniques based on space-filling curves allow multidimensional datasets to be transformed into a one-dimensional index for partitioning across distributed databases (Fox et al. 2013).

The initial implementation of the geoprocessing engine (Fig. 3) consists of a Spark cluster compute engine for handling distributed processing and a YARN (yet another resource negotiator) cluster manager utilized in the Hadoop ecosystem. An Accumulo columnar database manages how the data are chunked and distributed, and how they are accessed quickly via indexing on a Hadoop Distributed File System (HDFS). This distributed architecture, coupled with the GeoTrellis geospatial library, constitutes the core of the engine. GeoTrellis provides Spark integration for distributed raster processing while providing very fast raster input/output (IO), robust operations for manipulating raster data, and utilities for providing all these capabilities via standards-based services and RESTful ([representational state transfer]ful) endpoints. In future implementations of the engine, certain components of the system could be exchanged if other technologies are found to perform better, such as using Apache Mesos for cluster management. The initial implementation of this platform is a single instance but can be easily scaled by adding additional Spark compute nodes.

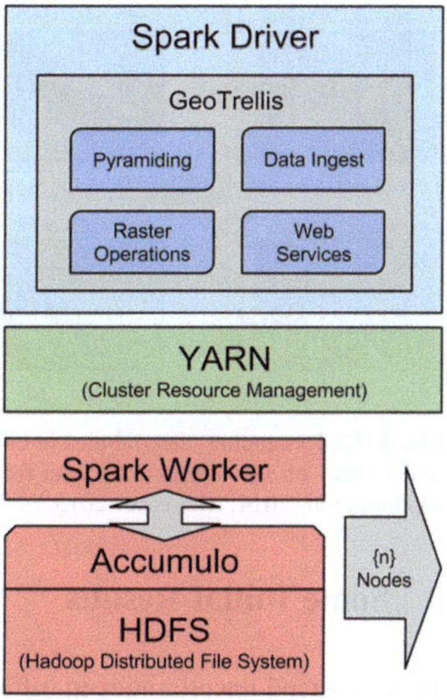

Fig. 3 Initial implementation of the geoprocessing engine

The coarse-grained, distributed processing framework implemented in the geoprocessing engine lends itself very well to the surface analysis being performed within the Urban-CAT framework. Being able to chunk, standardize, and ingest data quickly in a distributed environment as well as perform distributed analysis on several raster datasets with potentially differing resolutions, albeit common grids, is necessary for our purposes; to make this system scalable and flexible for several scenarios and operational in a web environment is essential to making this framework as valuable and accessible as possible to more medium and small stakeholder organizations.

Developing a geoprocessing engine that can quickly ingest large, disparate datasets and combine these relevant indicators to compute meaningful indexes and statistics across a surface would permit landscape resilience profiles to be created for any number of climate change scenarios. Indicators developed by knowledgeable local stakeholders could be consumed quickly by the framework and combined per best practices and judgment of stakeholders. Aside from floodplain management, other resilience profiles that could be implemented include metrics of energy demand due to rainfall and heating/cooling indicators and air quality by assessing such indicators as ground-level ozone and particle pollution. Being able to measure and account for future energy demand and air quality would pay large dividends in curtailing economic and social costs by identifying populations and infrastructure potentially at risk due to changing climate.

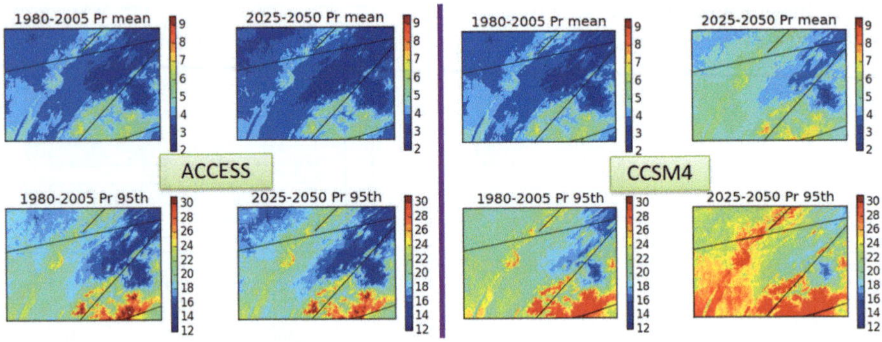

Fig. 4 Understanding uncertainties in climate models. The data from the ACCESS model are the four images on the *left*, and the data from the CCSM4 model are the four images on the *right* (Ashfaq et al. 2014; Mei et al. 2013)

3 Some Initial Results

To understand uncertainties in different climate models, we plan to provide climate projection data from major climate models. These data will provide opportunities for users to experiment with different models and begin to understand how the impacts for their area of study vary from one model to another. As an illustration, Fig. 4 shows mean and 95th percentile statistics for a historical time frame (1980–2005) and a future time frame (2025–2050) calculated from precipitation data generated by runs of two climate models (ACCESS and CCSM4). According to these results, using data from only one of these models underestimates or overestimates the projected impacts. However, using data from both models provides opportunity for quantifying the uncertainty bounds for the impact analysis.

In addition, Fig. 5 shows the population data for 2010 as well as the population projection data for the same area—Knox County, Tennessee. This visualization provides users with an opportunity to begin to understand the cost of impacts with respect to population growth. While these population projections are based on a business-as-usual scenario, understanding that these population numbers may

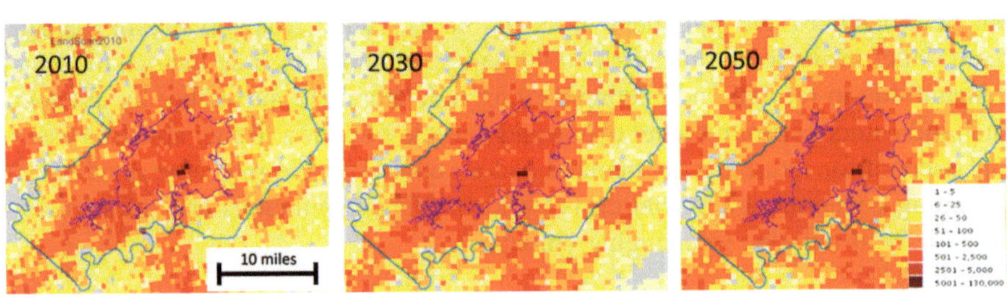

Fig. 5 Population data for 2010 and forecasts for 2030 and 2050 based on a business-as-usual population projection modeling approach

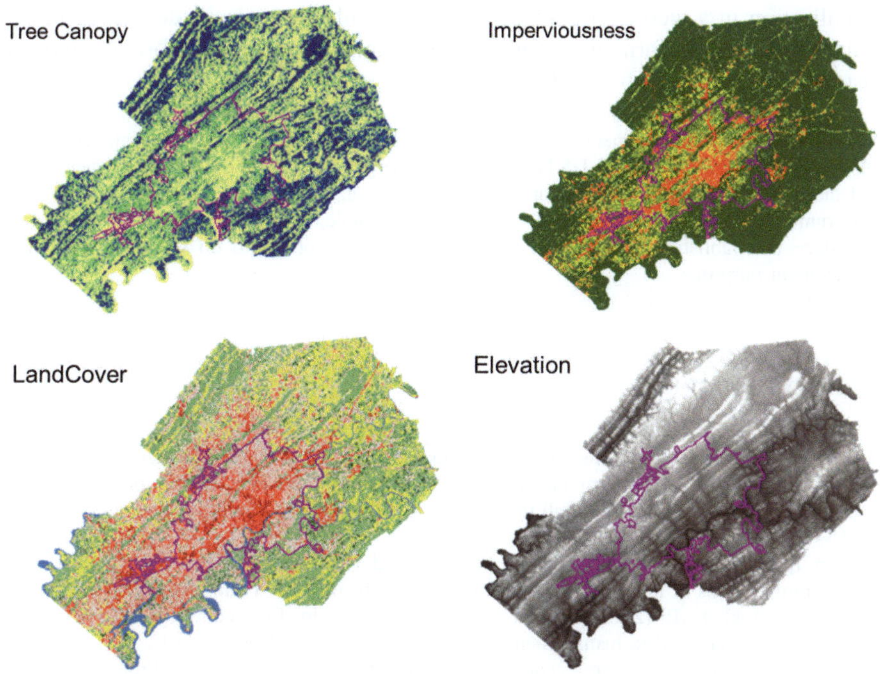

Fig. 6 Data representing other environmental indicators

change due to additional factors, such as impacts of sea-level rise in the coastal areas, is important.

In addition to socioeconomic indicators, data exist that are used to quantify factors that could provide additional insights about vulnerability and risk. Some of these data are shown in Fig. 6 and include elevation, tree canopy, imperviousness, and land cover data. With landscapes largely varying by city, some combinations of all these data layers, in conjunction with stakeholder knowledge, will provide new insights for decision makers.

4 Conclusion

With an increase in the frequency and intensity of extreme weather events and the continuous growth of urban populations, the need of city planners to quickly and easily identify risks to infrastructure and population and to plan climate adaptation strategies has become more pertinent. The Urban-CAT framework helps planners identify vulnerabilities of urban landscapes as well as plan for the future by identifying stressors on urban resiliency via the analysis of population and climate projections. The fusion of these varying datasets for analysis and visualization in a web environment is possible only through the development of a distributed, high-performance geoprocessing engine capable of performing dynamic raster processing at web speeds.

Acknowledgements This manuscript has been authored by UT-Battelle, LLC, under contract DE-AC05-00OR22725 with the US Department of Energy. The United States Government retains, and the publisher by accepting the article for publication, acknowledges that the United States Government retains, a nonexclusive, paid-up, irrevocable, worldwide license to publish or reproduce the published form of this manuscript, or allow others to do so, for United States Government purposes.

References

Ashfaq M, Rastogi D, Mei R, Kao SC, Naz BS, Gangrade S (2014) Ultra high-resolution ensemble projections of the near-term climate change over the U.S. In: Abstract presented at the American Geophysical Union fall meeting 2014 Fall Meeting, AGU, San Francisco, CA, Dec 15–19, Abstract #A53Q-02

Carmin J, Nadkarni N, Rhie C (2012) Progress and challenges in urban climate adaptation planning: results of a global survey. Massachusetts Institute of Technology, Cambridge, MA

Eldawy A, Mokbel MF (2015) The era of big spatial data. In: Proceedings of the international workshop of cloud data management CloudDM 2015, Seoul, Korea, April 2015

EPA (Environmental Protection Agency) (2014) United States Environmental Protection Agency (EPA) Green infrastructure website. http://water.epa.gov/infrastructure/greeninfrastructure. Accessed 10 Mar 2014

Fox A, Eichelberger C, Hughes J, Lyon S (2013) Spatiotemporal indexing in non-relational distributed databases. In: Proceedings of the IEEE international conference on big data, pp 291–299

Füssel HM (2007) Vulnerability: a generally applicable conceptual framework for climate change research. Glob Environ Change 17(2):155–167

Füssel HM, Klein RJ (2006) Climate change vulnerability assessments: an evolution of conceptual thinking. Clim Change 75(3):301–329

IPCC (Intergovernmental Panel on Climate Change) (2007) Climate change 2007: the physical science basis. Contribution of Working Group I to the fourth assessment report of the Intergovernmental Panel on Climate Change. Cambridge University Press, Cambridge

IPCC (Intergovernmental Panel on Climate Change) (2014) Climate change 2014: impacts, adaptation, and vulnerability: summary for policymakers. Contribution of Working Group II to the fifth assessment report of the Intergovernmental Panel on Climate Change. Cambridge University Press, Cambridge

Mei R, Rastogi D, Ashfaq M (2013) Land-atmosphere interactions in the high-resolution regional climate ensemble projections over the United States. In: Abstract presented at the American Geophysical Union fall meeting 2013 Fall Meeting, AGU, San Francisco, CA, Dec 9–13, Abstract #A33E-0294

Preston BL (2013) Local path dependence of U.S. socioeconomic exposure to climate extremes and the vulnerability commitment. Glob Environ Change 23:719–732

Quay R (2010) Anticipatory governance: a tool for climate change adaptation. J Am Plann Assoc 76(4):496–511

Ross A (2013) Local disaster resilience across the Gulf Coast: intersecting capacities for and perceptions of resilience. In: Proceedings of the THC-IT-2013 conference and exhibition, University of Houston, Houston, 2 Aug 2013

Tschakert P, Dietrich KA (2010) Anticipatory learning for climate change adaptation and resilience. Ecol Soc 15(2):11

WHO (World Health Organization) (2014) Situation and trends in key indicators. In: Global Health Observatory: urban population growth. http://www.who.int/gho/urban_health/situation_trends/. Accessed 2 May 2014

Xie X, Xiong Z, Hu X, Zhou G, Ni J (2014) On massive spatial data retrieval based on Spark. In: Proceedings of web-age information management. Springer, Berlin, pp 200–208

A Fully Automated High-Performance Image Registration Workflow to Support Precision Geolocation for Imagery Collected by Airborne and Spaceborne Sensors

Devin A. White and Christopher R. Davis

Abstract Deriving precise coordinates from airborne and spaceborne imagery, with uncertainty estimates, is very challenging. Doing so is significantly more difficult when imagery is coming from one or more sensors that have questionable and/or incomplete photogrammetric metadata. Before precision geolocation activities can take place, that metadata must be complete and consistent such that images are correctly registered to one another and the Earth's surface. This chapter describes an automated, high-performance image registration workflow that is being built at Oak Ridge National Laboratory to meet this need and focuses on the core concepts and software libraries underlying its creation. Highly encouraging initial system performance metrics are included as well.

Keywords Image registration • Photogrammetry • Computer vision • Uncertainty propagation • High-performance computing

1 Introduction

Deriving precise coordinates from airborne and spaceborne imagery, with uncertainty estimates, is a very challenging and time-consuming task. That task is significantly more difficult to accomplish when the imagery comes from one or more sensors that have questionable and/or incomplete photogrammetric metadata. Before precision geolocation activities can take place, photogrammetric metadata

D.A. White (✉) · C.R. Davis
Geographic Information Science & Technology Group, Oak Ridge National Laboratory,
1 Bethel Valley Road Knoxville, Oak Ridge, TN 37831, USA
e-mail: whiteda1@ornl.gov

C.R. Davis
e-mail: daviscr@ornl.gov

D.A. Griffith et al. (eds.), *Advances in Geocomputation*,
Advances in Geographic Information Science,
DOI 10.1007/978-3-319-22786-3_34

must be complete and consistent such that images are correctly registered to one another and the Earth's surface (McGlone and Lee 2013; Mikhail et al. 2001).

For projects with a limited number of images from a small number of sensors, a manual registration solution executed through commercially available desktop software is appropriate and recommended. However, what happens when a researcher is dealing with a very large and rapid stream of images from a wide variety of sensors with differing modalities and wildly varying photogrammetric metadata? An automated, high-performance image registration workflow is being built at Oak Ridge National Laboratory to meet this need, similar in some respects to the geospatial image registration and alignment with features (GIRAF) framework that was proposed by the University of Arkansas (Apon 2010). This chapter describes the core concepts underlying its creation and discusses all of the steps in the automated workflow—including the software being used to accomplish them. Some highly encouraging initial system performance metrics are provided as well.

2 Core Development Concepts

An automated workflow is being built around several core concepts that are expressed through the following actions: (1) develop required applications using only open source, in-house, or government-furnished software, (2) leverage well-established photogrammetric and computer vision techniques to reduce risk, (3) expose all components as services that can communicate with one another, (4) use high-performance computing architectures and paradigms wherever possible, (5) strictly adhere to and take full advantage of metadata standards for the National Imagery Transmission Format (NITF) to enable interoperability, and (6) keep the system as flexible as possible through extensive use of plugin frameworks. These operational concepts are at the heart of the workflow described subsequently and drive every aspect of the system's creation.

3 Registration Workflow

The automated workflow consists of six steps: preprocessing, trusted source selection, global localization, image registration, sensor model resection and uncertainty propagation, and enhanced metadata generation, as portrayed by Fig. 1. Each step builds on the preceding one and is discussed in turn. Table 1 summarizes where various software libraries are used in each step of the workflow and its companion data service.

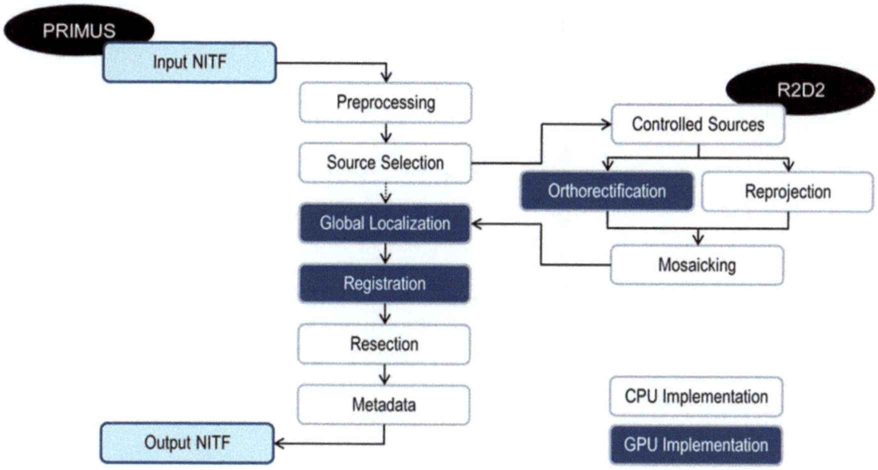

Fig. 1 The automated image registration workflow and its companion data service, highlighting the portions that take advantage of CPU- and/or GPU-based parallelization

3.1 Preprocessing

Preprocessing is a critical step in the automated image registration workflow. Beyond ensuring that the supplied image can be handled by the workflow (e.g., it is in the correct format and was collected by a supported sensor), it scans the image for any number of potential issues that might prevent a successful registration and attempts to mitigate them. Some issues such as sensor artifacts, which can be filtered out or smoothed over, are more easily dealt with than others such as clouds, water, and shadows. Solutions for at least detecting and masking regions containing these notoriously problematic items currently are under development. The integration of automated radiometric calibration and atmospheric compensation algorithms are planned for the near future, not only to help with detecting and masking but also to aid in downstream exploitation of the imagery by standard spectral analysis software such as ERDAS, ENVI, and Opticks.

No automated solution is perfect, but the preprocessing step is essentially a "triage" pass at the image. Assuming an image passes triage, relevant spatial and photogrammetric metadata are extracted from the file, a rigorous sensor model is built, and the image is subjected to dynamic range adjustment in order to make interesting features more visible during the subsequently described global localization and registration processes.

Table 1 The core software libraries used in the workflow and its companion data service

Software library	Registration workflow							Data service			
	Preprocessing	Source selection	Global localization	Registration	Resection	Metadata generation	Output	Controlled sources	Reprojection	Ortho	Mosaicking
NITRO	X						X	X			
GDAL	X	X	X	X		X	X	X	X	X	X
Proj.4	X	X	X	X		X	X	X	X	X	X
PostgreSQL + PostGIS		X						X			
OpenCV			X	X							
CSM	X	X	X	X	X	X		X		X	
MSP						X					
CUDA			X	X						X	
OpenMP	X	X	X	X		X	X	X	X	X	X

3.2 Trusted Source Selection

Trusted source selection is the process by which the initial estimate of an image's location, based on available spatial and photogrammetric metadata, is used to determine which parts of which sources of controlled data are used to aid in the registration process. Controlled sources consist of well-characterized orthoimagery, elevation, and Earth surface deformation (geoid). We constructed a PostgreSQL + PostGIS database, which consists of a series of tables that contain pointers to repositories of tiled versions of the sources as well as polygon footprints and associated horizontal and vertical (where applicable) spatial uncertainty information. These tables are populated by running a custom-built parallelized file crawler and metadata extractor on the repositories. That C++ application is OpenMP-enabled, parallelized, and cross-platform; it heavily relies upon GDAL and Proj.4.

Using a spatial bounding box query to the database, buffered by a tuneable percentage to account for incorrect initial image placement, a list of pointers to specific tiles are returned for each source, which then are used to build mosaics and composite three-dimensional (3D) uncertainty matrices on the fly. Those mosaics and matrices are inputs into the next several steps. Currently, MongoDB and Apache Spark are being evaluated as database alternatives, but so far PostgreSQL + PostGIS performance has not been an issue.

3.3 Global Localization

Global localization is the process by which a reduced resolution version of a problematic image is matched against a reduced resolution control orthoimage to quickly ascertain a reasonable guess of actual location and to do a coarse update of available geopositioning metadata (a shift in longitude/latitude map location). Then the rough guess is used to reduce the footprint of the control data and thus restrict the search space during full-resolution image registration. First, the problematic image is orthorectified using an OpenMP and CUDA-enabled cross-platform C++ application that leverages GDAL, Proj.4, and the elevation and geoid control data returned during trusted source selection (McGlone and Lee 2013; Mikhail et al. 2001). Then, it is matched against the control orthoimage using one of three registration techniques called through a plugin framework built on top of OpenCV: patch-based normalized cross-correlation (NCC) (Brunelli 2009), point-based SIFT or ORB (Lowe 1999; Rublee et al. 2011), or patch-based normalized mutual information (NMI) ported from the Insight Toolkit (Wells et al. 1996)—depending on image modality. Hardware acceleration via CUDA is employed extensively to reduce registration processing time.

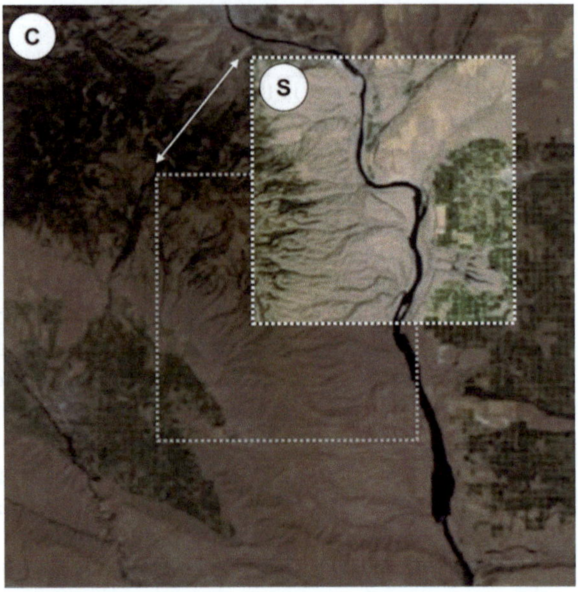

Fig. 2 A notional example of global localization. The source image (S), which was originally geolocated at the center of the control image (C), is moved across the control image until a reasonable match is found. The shift in location from its original position is stored for use in subsequent steps

The coarse update is operationalized by wrapping the sensor model in a spatial translator that removes the computed map coordinate offsets in longitude and latitude from a 3D ground location before asking the model for an image location (ground-to-image direction) and applies them to a computed 3D ground location associated with a given image location (image–to-ground direction). In other words, the model still thinks it is in its original location but is able to interact with elevation data associated with the new location, and for the rest of the workflow, the image acts as if it has been shifted to the new location. This type of interaction avoids prematurely altering core photogrammetric metadata like platform location and orientation, which could lead to sensor model instability during resection. Figure 2 provides a notional example of what takes place during global localization.

Both the resection and orthorectification processes rely upon the presence of a sensor model. To keep this element generic, the Community Sensor Model plugin framework is used. This framework provides a common set of widely used photogrammetric functions exposed through a single application programming interface (API) (National Geospatial-Intelligence Agency 2015). An application using this framework does not have to know or care about the sensor it is dealing with.

3.4 Image Registration

Image registration is a significantly more complex and computationally intensive version of global localization. If necessary, the spatial resolution of one or both of two images is adjusted so that both begin on an even footing. Then image pyramids are built for both so that matching can take place at multiple resolutions to boost the success rate.

The registration process deviates from global localization in several respects. First, due to potentially large differences in viewing geometry between a problematic image and its control orthoimage, the latter is projected into the coordinate space of the former using the sensor model, elevation data, and geoid data in concert with an OpenMP and CUDA-enabled cross-platform C++ application. This turns a 3D problem into a two-dimensional (2D) problem through perspective matching (McGlone and Lee 2013; Mikhail et al. 2001). Second, both images are partitioned into tiles to support faster matching through distributed computing and hardware acceleration, as well as to make the matching problem local instead of global. Third, the end goal of the matching process is to produce a large set of match points that can be used during resection; thus, additional work must be done when employing patch-based instead of point-based matching algorithms. For the former, interest point operators like Harris must be used within a pair of tiles to find common points (Harris and Stephens 1988). For the latter, both brute force and FLANN (fast library for approximate nearest neighbors) matchers are available (Muja and Lowe 2009). The registration process leverages the same hardware-accelerated image-matching capabilities employed during global localization. Figure 3 illustrates a notional example of what takes place during the image registration process.

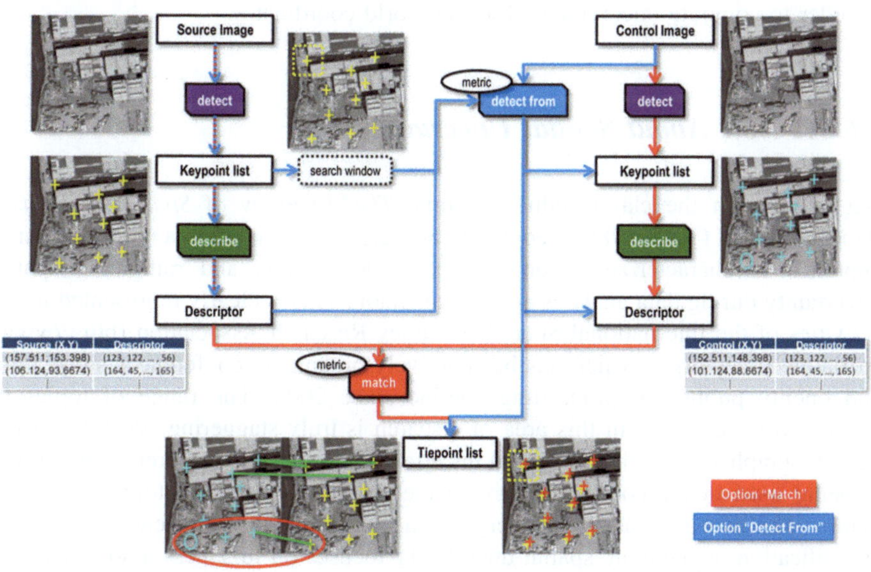

Fig. 3 A notional example of the image registration process. Interesting points are detected in the source (problematic) image and then are either matched against an independently generated set of interesting points in the control image ("Match" metric option, *bottom left*, using FLANN or similar) or searched for within a given radius of the same sample/line coordinates in the control image ("Detect From" metric option, *bottom right*, using NCC or a similar technique)

Once a large match point list has been generated across all resolutions, the sample/line coordinates of those points in the problematic image and the longitude/latitude/height above ellipsoid (HAE) coordinates of those points in the control image (calculated by projecting them back onto the ground via the sensor model) are supplied to the resection process to more accurately refine the sensor model's behavior. Using rigorous sensor models throughout the registration process, within a hardware-accelerated environment, is, to our knowledge, a novel approach.

3.5 Sensor Model Resection and Uncertainty Propagation

Resection involves the application of unified least squares (ULS) to make changes to a small set of sensor model parameters so that the model has the best agreement possible with a set of user-supplied control points (McGlone and Lee 2013; Mikhail et al. 2001). These points usually are generated through a manual selection process to ensure high quality, but a spatially bounded and localized matching process that is automated produces enough viable points that resection can take place. This is the same approach used to create structure from motion (Hartley and Zisserman 2003). The resection process generates parameter-specific uncertainty estimates as a by-product, and these estimates can be translated from sensor space to ground space in order to compute uncertainties for real-world coordinates.

3.6 A Note About Spatial Uncertainty

Beginning with the classic edited volume *The Accuracy of Spatial Databases* (Goodchild and Gopal 1989), geography as a discipline has taken a more structured approach to characterizing, communicating, leveraging, and mitigating spatial uncertainty during data analysis and visualization. This is clearly represented in the activities of the International Spatial Accuracy Research Association (http://www.spatial-accuracy.org/), which has been actively organizing conferences since 1996 and openly publishing proceedings online since 2004. The range of topics of interest to geographers in this area of research is truly staggering. With that said, how geographers conceive of spatial uncertainty is somewhat different from how it is being used in the context of this chapter, which is more photogrammetric in nature. The two approaches are very much complementary. In general, and as a simplification, geographic spatial uncertainty focuses on fuzziness associated with derived products like elevation models, image classification results, and workflows composed of many processes chained together that could have a range of possible inputs and outputs (e.g., Monte Carlo analysis). This fuzziness can come in the form of actual map location, product content (i.e., how much one can trust a computed value associated with a particular map location), or both. How one then effectively combines/fuses derived products with different types and levels of uncertainty becomes a paramount concern, one that geographers currently are trying to address constructively. It is a very challenging problem.

Photogrammetric spatial uncertainty focuses on the fuzziness that influences computed ground locations based on selected image locations (and vice versa), plus how that model interacts with 3D terrain (which has its own spatial uncertainties), as facilitated through the use of a rigorous sensor model that describes the complex nonlinear mathematical relationship between the two coordinate spaces. Even simple models, like the one for a frame camera, can have more than a dozen input terms—and each of them has an associated uncertainty that must be accounted for (National Geospatial-Intelligence Agency 2011). Many of these terms have uncertainties that relate to one another in very complex ways, so variance-covariance matrices are used extensively to more fully understand their behavior.

Simply put, no one single set of spatial uncertainty values is associated with an image that is still in its original viewing geometry; rather, a range of potential uncertainties are associated with such an image. As one attempts to determine the "correct" ground location associated with a pixel in an image, a cone of uncertainty is projected onto the ground, and the exact morphology of it can vary by location in the image and by elevation, leading to different spatial uncertainty values in different physical locations. More compact representations of this cone have been developed and are widely used, like the linear and circular error at 0.90 probability (known as LE90 and CE90) and the bias and random error terms in RPC00B (National Geospatial-Intelligence Agency 2012). However, a significant amount of information is lost in the process of making the spatial uncertainty information easier to use by moving it from sensor space into ground space, essentially reducing a highly variable, high-dimensional model to a fixed representation in 3D.

In addition to aiding in the computation of ground and image coordinates, photogrammetric uncertainty information also helps to bound the preceding unified least squares resection problem by indicating how much specific model terms can vary as they are being adjusted through the use of ground control points, which have 3D spatial uncertainties associated with them, too (McGlone and Lee 2013; Mikhail et al. 2001). Accounting for all of this uncertainty is critical for a successful resection.

3.7 Enhanced Metadata Generation

The remaining challenge is capturing the behavior of the adjusted sensor model, with its improved geolocation accuracy and associated photogrammetric spatial uncertainty estimates, and storing it in a standardized way that can be leveraged by a large number of desktop applications—all while leaving the original metadata intact to preserve its pedigree. To accomplish this, imagery is read from and written to NITF, a tightly regulated hierarchical format that permits the inclusion of extensive metadata and multiple images in the same file (Department of Defense 2006).

Standardized metadata containers exist to support many types of information, but the ones of chief interest to our project are IGEOLO (image geographic location), BLOCKA (image block information, version A), RPC00B (rapid positioning capability, version B), RSM (replacement sensor model), and SENSRB (general electro-optical sensor parameters, version B). The first two describe a four-vertex polygon that covers the footprint of an image in a real-world coordinate system, with the latter being able to do so with higher fidelity (Department of Defense 2006, 2011a). RPC00B and RSM are two forms of replacement sensor models, which for the purposes of this chapter are generic sensor models built by fitting observations from a rigorous model to a set of polynomials that have either a fixed (RPC) or variable (RSM) number of terms (Tao and Hu 2001; Department of Defense 2011a, b). Both can communicate spatial uncertainty, but RSM does so more robustly. With that said, more software packages support RPC, so both are generated. SENSRB provides a standardized set of containers where photogrammetric metadata can be stored and used by downstream applications (Department of Defense 2011c). In some cases, native metadata must be converted to SENSRB's standardized versions, which greatly improves interoperability.

Underlying all of the NITF components of the project is an open source library called NITRO, which can handle all aspects of reading and writing these very complicated files. We accelerated performance for multithreaded and distributed architectures using OpenMP, including the I/O steps, which can take a significant amount of time for large images. We call this new high-performance version Glycerin.

4 Initial System Performance Metrics

Initial tests of system performance, with respect to overall processing time, have been conducted on a high-density HPC node. This node, a Dell PowerEdge C4130 Rack Server, is configured with dual Intel Xeon E5-2670 v3 2.3 GHz central processing units (CPUs; 48 logical cores in total), 256 GB of RAM, dual 200 GB SSDs, and four NVIDIA Tesla K80 cards. Each K80 is composed of two K40s, so that eight graphic processing units (GPUs) are available. Tests were run using only one GPU to simulate real system resource availability, where eight image registration problems can run simultaneously, with each one being assigned a single GPU. Test imagery consisted of JPEG2000-compressed WorldView-2 scenes, each approximately 35,000 pixels on a side. Figure 4 highlights test results throughout the system development process, with the most recent ones appearing on the right. The initial parallelized implementation was not processing images as fast as expected, so additional resources were dedicated to refactoring and optimizing it. At present, the node is now capable of processing eight images simultaneously, every fifteen seconds, and we are continuing to refactor and optimize before eventual deployment on a multimode system. Although the CPU-only option outperforms the GPU option, it requires the full dedication of all non-GPU computing resources

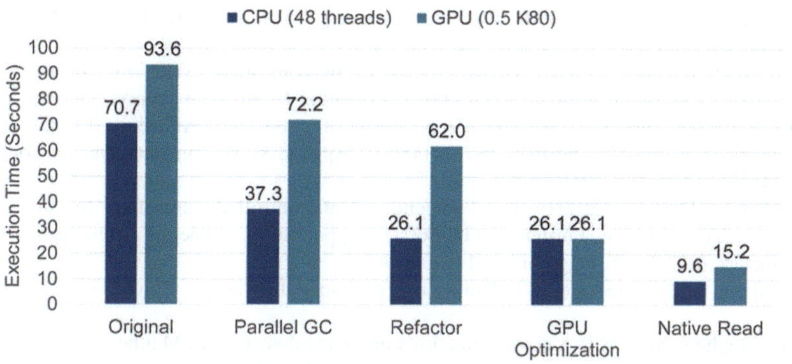

Fig. 4 The evolution of system performance, with oldest results on the *left* and most recent results on the *right*. Several rounds of parallelization, refactoring, and optimization have taken place. "Native read" refers to the more efficient handling of input imagery by keeping it for as long as possible in its native type as it moves through the workflow

on a node. That means only one image can be processed every ten seconds, compared to eight every fifteen seconds. The latter outcome is a far more scalable and responsive approach.

5 Conclusion

When precision geolocation has to be done on a very large scale and at a very rapid pace, leveraging imagery that may have only a limited amount of photogrammetric metadata, traditional manual solutions break down. An automated, high-performance registration workflow is being built to address this need. It is adhering to an open source philosophy, using well-known techniques and following well-established standards to ensure the maximum possible level of utility. Early stage performance metrics for it are very encouraging.

Acknowledgements Prepared by Oak Ridge National Laboratory, PO Box 2008, Oak Ridge, Tennessee 37831-6285, managed by UT-Battelle, LLC for the United States Department of Energy under contract number DEAC05-00OR22725.

References

Apon A (2010) GIRAF: geospatial image registration and alignment with features. In: Paper presented at the SC10, New Orleans, LA

Brunelli R (2009) Template matching techniques in computer vision: theory and practice. Wiley, New York

Department of Defense (2006) National Imagery Transmission Format Version 2.1 (MIL-STD-2500C)

Department of Defense (2011a) The compendium of controlled extensions for the National Imagery Transmission Format (STDI-0002), Appendix E: airborne support data extensions

Department of Defense (2011b) The compendium of controlled extensions for the National Imagery Transmission Format (STDI-0002), Appendix U: replacement sensor model (RSM) tagged record extensions

Department of Defense (2011c) The compendium of controlled extensions for the National Imagery Transmission Format (STDI-0002), Appendix Z: general electro-optical sensor parameters (SENSRB) tagged record extension

Goodchild M, Gopal S (eds) (1989) The accuracy of spatial databases. Taylor and Francis, New York

Harris C, Stephens M (1988) A combined corner and edge detector. In: Matthews MM (ed) Fourth Alvey vision conference. Manchester, UK, pp 147–151 (1988)

Hartley R, Zisserman A (2003) Multiple view geometry in computer vision. Cambridge University Press, Cambridge, UK

Lowe DG (1999) Object recognition from local scale-invariant features. In: Proceedings of the seventh IEEE international conference on computer vision, Kerkyra, Greece, 1999, pp 1150–1157

McGlone JC, Lee GYG (eds) (2013) Manual of photogrammetry, 6th edn. American Society for Photogrammetry and Remote Sensing, Bethesda, MD

Mikhail EM, Bethel JS, McGlone JC (2001) Introduction to modern photogrammetry. Wiley, New York

Muja M, Lowe DG (2009) Fast approximate nearest neighbors with automatic algorithm configuration. In: Paper presented at the international conference on computer vision theory and applications, Lisboa, Portugal

National Geospatial-Intelligence Agency (2011) Frame sensor model metadata profile supporting precise geopositioning (NGA.SIG.0002_2.1)

National Geospatial-Intelligence Agency (2012) Generation and application of RPC uncertainty parameters

National Geospatial-Intelligence Agency (2015) Community sensor model (CSM) technical requirements document version 3.0.2 (NGA.STND.0017_3.0.2)

Rublee E, Rabaud V, Konolige K, Bradski GR (2011) ORB: an efficient alternative to SIFT or SURF. In: 2011 IEEE international conference on computer vision, Barcelona, Spain, 2011, pp 2564–2571

Tao CV, Hu Y (2001) A comprehensive study of the rational function model for photogrammetric processing. Photogram Eng Remote Sens 67(12):1347–1357

Wells WMI, Viola P, Atsumi H, Nakajima S, Kikinis R (1996) Multi-modal volume registration by maximization of mutual information. Med Image Anal 1(1):35–51

MIRAGE: A Framework for Data-Driven Collaborative High-Resolution Simulation

Byung H. Park, Melissa R. Allen, Devin White, Eric Weber, John T. Murphy, Michael J. North and Pam Sydelko

Abstract Information about how human populations shift in response to various stimuli is limited because no single model is capable of addressing these stimuli simultaneously, and integration of the best existing models has been challenging because of the vast disparity among constituent model purposes, architectures, scales, and execution environments. To demonstrate a potential model coupling for approaching this problem, three major model components are integrated into a fully coupled system that executes a worldwide infection-infected routine where a human population requires a food source for sustenance and an infected population can spread an infection when it is in contact with the remaining healthy population. To enable high-resolution data-driven model federation and an ability to capture dynamics and behaviors of billions of humans, a high-performance computing agent-based framework has been created and is demonstrated in this chapter.

Keywords Model coupling · Agent-based simulation · High-performance computing

B.H. Park (✉) · M.R. Allen · D. White · E. Weber
Oak Ridge National Laboratory, PO Box 2008, Oak Ridge, TN 37831, USA
e-mail: parkbh@ornl.gov

M.R. Allen
e-mail: allenmr@ornl.gov

D. White
e-mail: whiteda1@ornl.gov

E. Weber
e-mail: weberem@ornl.gov

J.T. Murphy · M.J. North · P. Sydelko
Argonne National Laboratory, 9700 S. Cass Avenue, Argonne, IL 60439, USA
e-mail: jtmurphy@anl.gov

M.J. North
e-mail: north@anl.gov

P. Sydelko
e-mail: psydelko@anl.gov

D.A. Griffith et al. (eds.), *Advances in Geocomputation*,
Advances in Geographic Information Science,
DOI 10.1007/978-3-319-22786-3_35

1 Introduction

Various drivers—economic, environmental, technological, and social—cause human populations to shift, changing the topology of urban infrastructure. These changes create emerging vulnerabilities that, in anticipation of their consequences, require integrated approaches and high-resolution data for local decision makers. Such information currently is limited, because no one model can address these drivers simultaneously, and integration of the best existing models has been challenging because of the vast disparity among constituent model purposes, architectures, scales, and execution environments. The Foresight Initiative, a collaborative research effort led by the National Geospatial-Intelligence Agency and supported by Arizona State University and several national laboratories, is attempting to find solutions to these problems in order to bring the rich and complex world of computational modeling to bear on matters of national and international significance. One particular challenge of constructing such high-resolution data-driven model federations is simulating billions of actors and their dynamics and behaviors in response to various scenarios. Thus, bringing high-performance computing (HPC) capabilities into these studies is imperative. To demonstrate a possible method for addressing this challenge, we combined three major model components into a two-way-coupled system that executes a worldwide susceptible and infected (SI) epidemic simulation with a resource competition component built in—one that has the long-term potential to scale to HPC systems.

A variety of methods for implementing the individual components of this set (Collier and North 2013; McKee et al. 2015) have been implemented successfully, and a coupling for other purposes has been accomplished (e.g., Bond-Lamberty et al. 2014; DiVittorio et al. 2014). Unfortunately, no individual model or modeling framework has fully addressed simultaneous influences among these particular components. In the sections that follow, we introduce the *Modeling Interactive Repast and GCAM Ensemble* (MIRAGE), which is a framework for collaborative high-resolution simulations. We then detail the datasets, models, and procedures used for model intercommunication; the experiment that was performed using MIRAGE; the results that have been obtained to date; and our conclusions and recommended next steps.

2 Methods and Data

This section presents two versions of the MIRAGE framework. The first is MIRAGE 1.0, in which an infection model, created as an extension of an instructive model included in a parallel agent-based modeling environment, is coupled with output from an integrated assessment model and initiated with a spatially explicit population dataset. The second is a proposed MIRAGE 2.0, in which the framework is further coupled with a hydrological model so that water availability can affect agents in the simulation. This version of the framework receives world population data, streamflow, and runoff from external sources, and then implements a river routing model using elevation and flow direction data to determine water supply for food production.

As shown in Fig. 1, three components comprised the initial MIRAGE framework: LandScan, GCAM (Global Change Assessment Model; Edmonds and Riley 1985), and an infection model running in the Repast for High Performance Computing (Repast HPC; http://repast.sourceforge.net) environment. GCAM and the infection model within Repast work in a closed loop, repeatedly feeding their outputs as inputs to the other model. Such a cycle of runs iterates until the results mature, when the final outputs are generated. Each model requires a different file type for its input and output; therefore, conversions among the models are made via Python scripts. Conversions from GIS rasters, csv, and NetCDF to HDF5 are required as well as conversions back to csv from HDF5.

The LandScan 2013 global population dataset is the initial input to the infection and GCAM models. LandScan, at approximately 1 km resolution, was created using a multivariable dasymetric modeling approach along with spatial data and imagery analysis technologies to disaggregate census counts within administrative boundaries at the town, county, and state levels (http://web.ornl.gov/sci/landscan/landscan_documentation.shtml).

GCAM, version 4, is a dynamic recursive serial model, written in C++ with continuing development by Pacific Northwest National Laboratory, which includes as drivers representations of the global economy, land use, agriculture, an energy system, and climate. In GCAM, potential gross domestic product is computed based

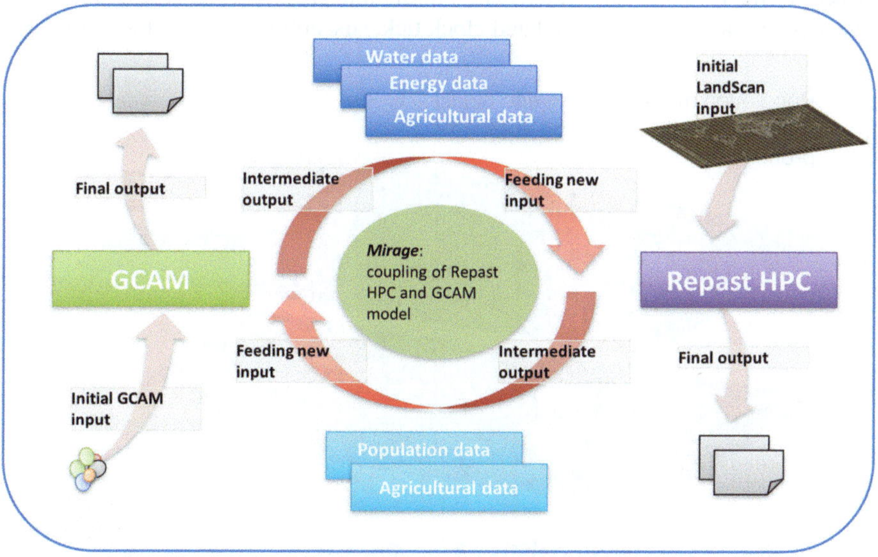

Fig. 1 The MIRAGE 1.0 framework

on labor productivity and population size in each of 32 regions spanning the globe. GCAM integrates energy, the terrestrial carbon cycle, agriculture, the land market, and forestry computations across 151 agro-ecological zones to calculate simultaneous market-clearing prices under predetermined policy scenarios at five-year time intervals (Bond-Lamberty et al. 2014; Voisin et al. 2013). For our experiment, we extracted corn production from the model at the 32-region resolution.

The infection model is implemented using the Repast HPC library. Repast HPC is a free and open source agent-based modeling and simulation (ABMS) library developed at Argonne National Laboratory and written in C++ using the message-passing interface (MPI) for high-performance distributed computing platforms. It implements a dynamic discrete-event scheduler with conservative synchronization. That is, the user schedules events to occur at a specific tick, and the ticks determine the relative order in which the events occur. Any given Repast HPC simulation is composed of agents, one or more contexts containing these agents, and zero or more of three types of projections: a grid, a continuous space, or a network. Agents inhabit a parallel multiprocess environment in which memory is not shared across processes. However, the agents themselves are distributed across processes, and each process is responsible for the agents local to that process. In the MIRAGE framework, the infection model within Repast HPC reads LandScan data and places the human population in appropriate grid cells. As shown in Fig. 2, an MPI-process of the Repast model is in charge of simulating actors in a block of cells. MPI processes collectively handle the migrations of populations across adjacent blocks.

The infection model extends a pedagogical and demonstrative Repast HPC model in which susceptible and infected humans are placed in a grid space, with the former trying to avoid the latter. The simulation progresses as the global clock advances forward. At each simulated clock tick, susceptible humans move to one of

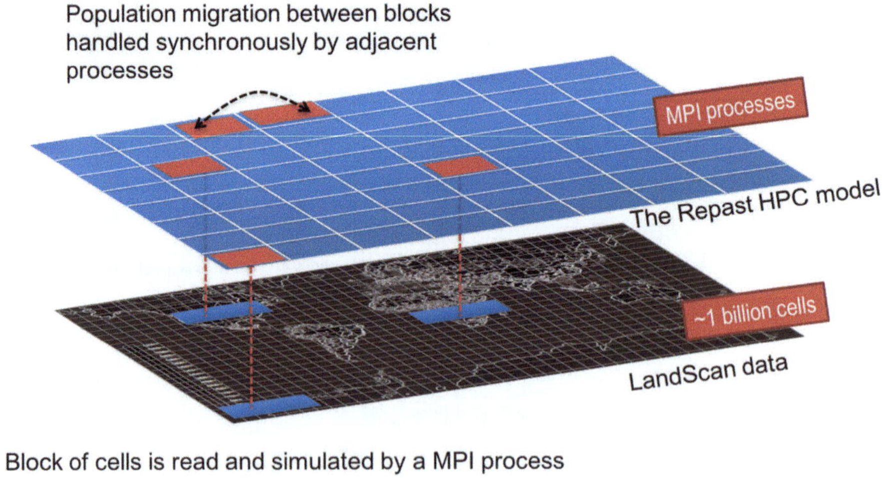

Fig. 2 The Repast HPC model with LandScan data

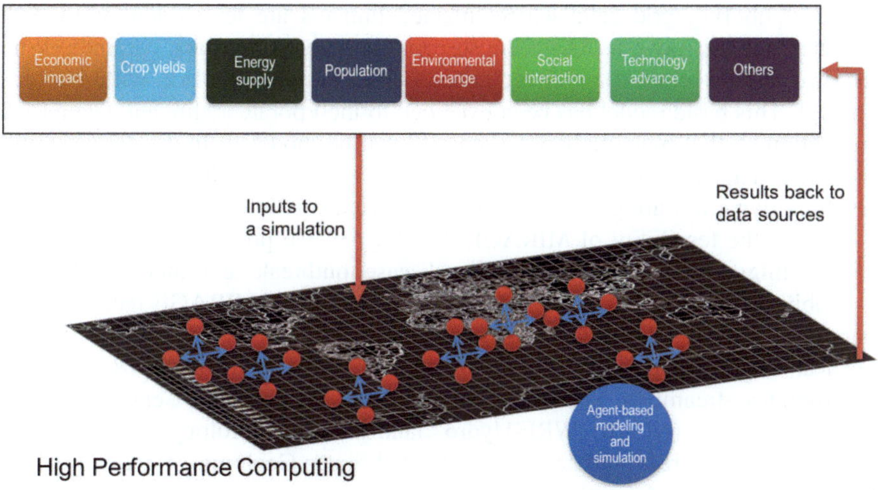

a Multiple high-resolution data from various drivers into a large-scale agent-based simulation model

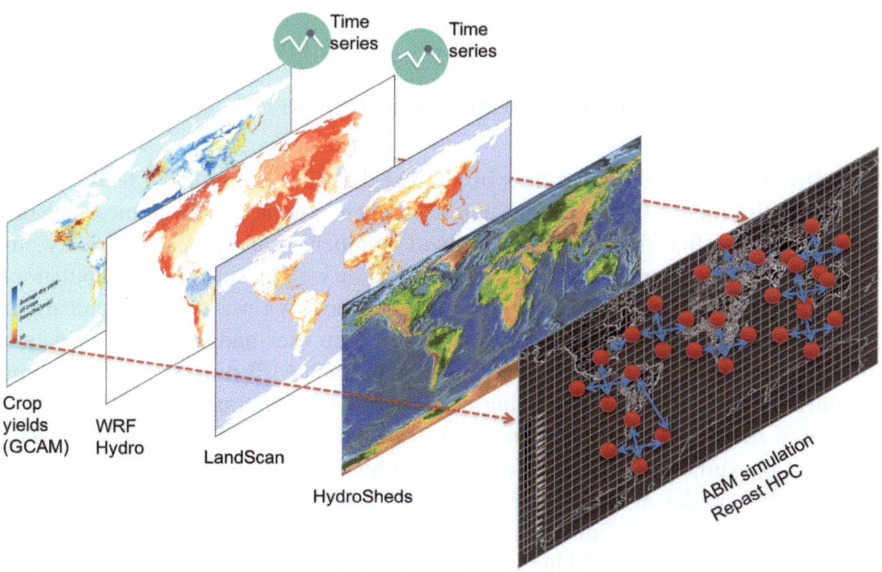

b Constituent models integrated into MIRAGE to simulate a global population migration scenario

Fig. 3 **a** MIRAGE 2.0 framework is designed to federate multiple data and send feedbacks to the data sources; **b** MIRAGE evaluates a global population migration scenario integrating four models

their neighboring grid cells where infected humans are least common, and the infected humans move to where the susceptible humans are most common. When the two groups come into direct contact, infection takes place within a short interval. This basic model has been extended to incorporate additional data into the simulation so that more elaborate scenarios can be modeled. In the demonstration presented here, corn is introduced as a resource for susceptible humans, and the resulting dynamics are generated through infected/susceptible/corn interactions.

To test the feasibility of MIRAGE 2.0 (Fig. 3), we propose a worldwide population migration scenario driven by disease outbreak and modified by food availability and climate conditions. For this application, MIRAGE receives world population data from LandScan at 1 km resolution. A river routing model developed at the Oak Ridge National Laboratory and integrated into MIRAGE, then, incorporates streamflow and runoff values from the National Center for Atmospheric Research (NCAR) WRF-Hydro[1] land surface hydrology model, and elevation and flow direction data from the United States Geological Survey's (USGS) HydroSheds data to determine water supply. Food availability is input to the system from the GCAM, and water resources for agricultural, municipal, manufacturing, and energy needs are allocated via an agent-based water management model (e.g., Akhbari and Grigg 2013; Galan et al. 2009) within the Repast HPC model.

3 Model Execution and Work in Progress

For MIRAGE 1.0, the initial inflow of GCAM data (i.e., corn yields) feeds into the Repast HPIC model that, in turn, simulates agent interaction dynamics and produces outputs after finishing a predefined number of iterations. Currently, the reading of infected populations and corn yields, as well as susceptible populations, is done through files. For consistency and efficiency, we use the hierarchical data format version 5 (HDF5) as the format for all data files and represent all data as numbers in grid cells stored as HDF5 matrices. Each MPI process reads the input for only its portion of the data using collective MPI input/output (MPI-IO).

With a resolution of 1 km, LandScan distributes the initial susceptible population of approximately 7 billion over a 21,600-by-43,200 grid of cells. Running the Repast HPC model at the full resolution of LandScan is challenging, even for leading HPC facilities. Thus, while we are planning full-scale runs on the Titan supercomputer at the Oak Ridge Leadership Computing Facility, we evaluated the model at a reduced resolution. More specifically, LandScan resolution was reduced to 360-by-720 cells, each of which spans a half degree in both longitude and latitude. Because GCAM produces output for 32 regions of the earth, GCAM output needs to be distributed over the 360-by-720 grid cells.

[1]Denotes the Weather Research and Forecasting Model Hydrological modeling extension software package.

Table 1 Procedures for mapping data from the GCAM and the Repast HPC model

1. For each admin_1 subcountry, a boundary polygon in GeoJson format is created in terms of longitude and latitude. The polygon data along with other properties of the subcountries are then stored in MongoDB (Admin_1 data were downloaded from *Natural Earth.*)
2. For each of the grid cells, a polygon is created in terms of longitude and latitude coordinates, and MongoDB is queried for all admin_1 subcountries whose boundaries intersect with that of the cell
3. Using the information obtained in step 2, human populations in the grid cells are aggregated at the admin_1 subcountry level and then at the admin_0 country level
4. Each GCAM output quantity for a region is divided into admin_1 subcountries proportional to their population sizes
5. GCAM output quantities allocated for an admin_1 subcountry level are evenly distributed over the cells

Table 1 describes the allocation process.

To facilitate understanding of a simulation and its progression over time, a visual feedback system was designed. The system uses a web-based interface through which users can retrieve their simulation results and monitor the trajectory of parameters of interest over time. Currently, susceptible/infected population changes, at the subcountry (i.e., administrative level one, or "admin_1") level, can be monitored and visualized at the same resolution. For this, the Repast HPC model produces snapshots of population for each grid cell at a regular interval. These grid-based data are mapped to the admin_1 level resolution following a procedure

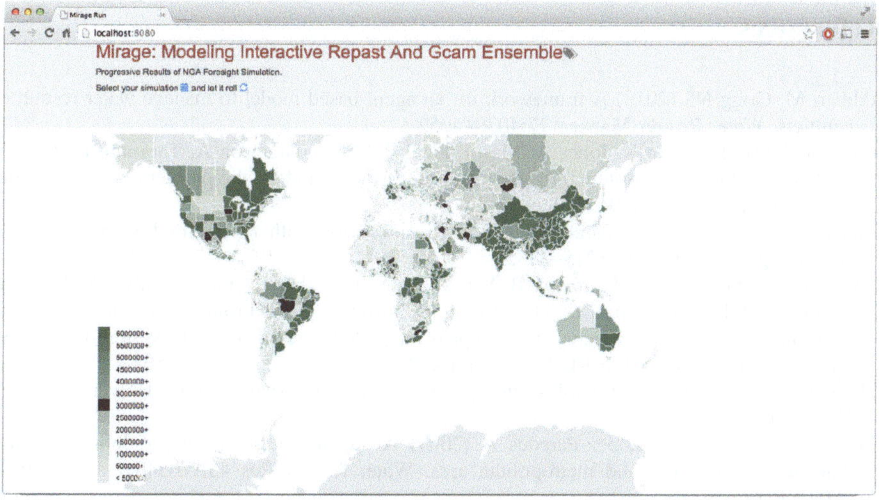

Fig. 4 Visual feedback from the MIRAGE framework

similar to the preceding one. The result is stored in a MongoDB database. The visualization front end is implemented using the Data-Driven Documents (D3) and Node.js JavaScript libraries. Figure 4 shows a screen shot of the visual feedback.

4 Conclusion and the Next Steps

This project represents a first pass at integrating agent-based modeling, global change assessment modeling, and geospatial data with an eye toward evaluating global change scenarios and their possible impacts. We show that a reasonable framework can be built that allows us to combine serial and parallel models that output different file types and incorporate data at different spatial scales to produce a result that provides new information given an established initial parameter set. The model includes two-way coupling and the potential to exchange, in a scientifically robust way, additional parameters among the currently coupled and planned models.

Acknowledgements This manuscript has been authored in part by UT-Battelle, LLC under Contract No. DE-AC05-00OR22725 with the US Department of Energy. This manuscript also has been authored in part by UChicago Argonne, LLC, Operator of Argonne National Laboratory ("Argonne"). Argonne, a US Department of Energy Office of Science Laboratory, is operated under contract number DE-AC02-06CH11357. The United States Government retains, and the publisher, by accepting the article for publication, acknowledges that the United States Government retains, a nonexclusive, paid-up, irrevocable, worldwide license to publish or reproduce the published form of this manuscript, or allow others to do so, for United States Government purposes. The work used the Oak Ridge Leadership Computing Facility.

References

Akhbari M, Grigg NS (2013) A framework for an agent-based model to manage water resources conflicts. Water Resour Manage 27:4039–4052

Bond-Lamberty B, Calvin K, Jones A, Mao J, Patel P, Shi X, Thomson A, Thornton P, Zhou Y (2014) Coupling earth system and integrated assessment models: the problem of steady state. Geosci Model Dev 7:1499–1524

Collier N, North M (2013) Parallel agent-based simulation with repast for high performance computing. Simulation 89:1215–1235

DiVittorio A, Chini L, Bond-Lamberty B, Mao J, Shi X, Truesdale J, Craig A, Calvin K, Jones A, Collins W, Edmonds J, Hurtt G, Thornton P, Thomson A (2014) From land use to land cover: restoring the afforestation signal in a coupled integrated assessment–earth system model and the implications for CMIP5 RCP simulations. Biogeosciences 11:6435–6450

Edmonds J, Reilly J (1985) Global energy: assessing the future. Oxford University Press, New York

Galán JM, del Olmo R, López-Paredes A (2009) An agent based model for domestic water management in Valladolid metropolitan area. Water Resour Res 45:W05401. doi:10.1029/2007WR006536

McKee J, Rose A, Bright E, Huynh T, Bhaduri B (2015) A locally-adaptive, spatially-explicit projection of US population for 2030 and 2050. Proc Natl Acad Sci 112(5):1344–1349
Voisin N, Liu L, Hejazi M, Tesfa T, Li H, Huang M, Liu Y, Leung L (2013) One-way coupling of an integrated assessment model and a water resources model: evaluation and implications of future changes over the US Midwest. Hydrol Earth Syst Sci 17:4555–4575

A Graph-Based Locality-Aware Approach to Scalable Parallel Agent-Based Models of Spatial Interaction

Zhaoya Gong, Wenwu Tang and Jean-Claude Thill

Abstract A great potential exists for mitigating the computational costs of spatially explicit agent-based models (SE-ABMs) by taking advantage of parallel and high-performance computing. However, spatial dependency and heterogeneity of interactions between agents pose challenges for parallel SE-ABMs to achieve good scalability. This chapter summarizes an application of the principle of data locality to tackle these challenges by extending a theoretical approach to the representation of the spatial computational domain. We propose and formalize a graph-based locality-aware approach to scalable parallelization of SE-ABMs. To demonstrate the applicability of this approach, two sets of experimentation are laid out and a locality-aware algorithm is designed to facilitate the study of model scalability. The results of simulation experiments illustrate the advantage of our approach to scalable parallel agent-based models of spatial interaction.

Keywords Locality awareness · Parallel agent-based models · Spatial interaction

Z. Gong (✉)
School of Geography, Earth and Environmental Sciences,
University of Birmingham, Birmingham, UK
e-mail: Z.Gong@bham.ac.uk

W. Tang · J.-C. Thill
Department of Geography and Earth Sciences,
University of North Carolina, Charlotte, USA
e-mail: WenwuTang@uncc.edu

J.-C. Thill
e-mail: Jean-Claude.Thill@uncc.edu

D.A. Griffith et al. (eds.), *Advances in Geocomputation*,
Advances in Geographic Information Science,
DOI 10.1007/978-3-319-22786-3_36

1 Introduction

Spatially explicit agent-based models (SE-ABMs) simulate dynamic interactions among contextually situated agents, and between these agents and their environments. These decentralized interactions propagate across scales to generate complex regularities that, in turn, reshape agent behaviors. As a computationally intensive approach, SE-ABMs can benefit from various parallel computing resources in the cyberinfrastructure domain in order to cope with the challenges of handling massive and heterogeneous geospatial data. However, spatial dependency and heterogeneity of agent-level interactions pose challenges for parallel SE-ABMs when the goal is to achieve good scalability (Gong et al. 2013). The challenges are as follows. First, when agents interact with each other, computing overhead is introduced by the cost of data access of one agent to others that are assigned to parallel computing tasks and by the cost of synchronization to maintain their data coherence and integrity. Second, this overhead is unevenly distributed across space because of the heterogeneous patterns of interactions that result from various geographic and social neighborhoods of individual agents. This chapter summarizes efforts to tackle these challenges by employing the principle of data locality to extend a theoretical approach to the representation of the spatial computational domain (Wang and Armstrong 2009) that was intended to guide the parallelization of computationally intensive geographical analyses. Our graph-based approach is tailored to minimize the interaction overhead in domain decomposition, and to maximize the efficient allocation of computing resources. The usefulness of this approach is demonstrated by applying it to an ABM of a spatial interaction system that simulates information exchange and the diffusion of opinion development among individual decision makers.

2 Literature Review

In this section, we first review the literature on SE-ABMs regarding their spatial properties. Then, we follow with a review of the application of the locality principle in the domain of computer science with respect to performance improvement.

2.1 Spatially Explicit Agent-Based Models

Agent-based modeling is a computational approach that simulates agents, environments, and their decentralized interactions for the investigation of dynamics in complex adaptive systems (Epstein 1999; Matthews et al. 2007). SE-ABMs are a special group of ABMs that consider explicit representations of spatially referenced agents or environments in their model components (Bian 2004; Brown et al. 2005;

Tang 2008; Stanilov 2012). In other words, spatial relationships are treated as an essential element related to agent behaviors and environmental characteristics in the agent-agent and agent-environment interactions in SE-ABMs. This type of model, closely related to investigations about geographic phenomena, has found application in a variety of fields, ranging from land cover change, urban growth, and ecology (Parker et al. 2003; Brown et al. 2005; Stanilov 2012; Grimm et al. 2006), to epidemiology, demography, and social sciences (Epstein 2009; Billari and Prskawetz 2003; Epstein 1999).

Due to the explicit representation and incorporation of spatial relationships, spatial interactions in SE-ABMs exhibit spatial dependence, spatial heterogeneity, and temporal dynamics (Parker et al. 2003; Irwin 2010). Spatial dependence arises from local interactions among spatially situated agents or between agents and their micro-environments. Spatial heterogeneity is caused by spatial variation in interaction processes and patterns across space at the scale of micro-environments. Dynamics refer to change in spatial interaction processes and patterns over time. The compounding effect of these three properties complicates the intensive computational requirements for SE-ABMs and further poses challenges for the parallelization of SE-ABMs in order to achieve scalable high performance (Crooks et al. 2008; Gong et al. 2013).

2.2 Locality of Reference

Locality of reference is one of the foundational principles in computer science (Denning 2006). It was established and developed to improve the efficiency of early virtual memory systems and led to the formation of the first coherent scientific framework for designing and analyzing dynamic memories (Denning 1970). Transforming virtual memory from an unpredictable to a self-regulated system, it has guided the (1) design of robust algorithms for memory page replacement; (2) creation of compiler code generators that can better group codes onto pages; and (3) successful prevention of thrashing, the near-complete collapse of system throughput due to heavy paging.

The locality idea has been generalized to a behavioral theory of computational processes interacting with storage systems (Denning 2006). It can be articulated through three main points: (1) for any computational process, there exists a series of locality sets of data to access, and all data access happens only within these sets; (2) locality sets can be inferred for a given computational process by measuring the distance from the process to data objects, which is a function defined in terms of space (data located close together in the address space are more likely to be used close together in time), in terms of time (recently used data are more likely to be reused in the near future), or in terms of any form of costs; and (3) the storage system will optimize throughput when locality sets are present in the high-speed memory connected to the computational process (Denning 1980).

The locality principle has found application in almost all types of storage and communication systems, such as the design of caches of all sorts, storage hierarchies, databases, logging systems, graphics display systems, and network interfaces. Well beyond these, it also has been adopted in search engines (e.g., Google) to quickly find the pages most relevant to keyword queries via caching mechanisms, in spam filters to determine e-mail messages in users' locality sets and those that are not, in recommender systems (e.g., Amazon) to recommend similar purchases based on a user's purchase history and the purchases of other similar users, and in many more other areas (Denning 2006). With its wide adoption, the locality principle has evolved into a modern model of locality that focuses on the awareness of, and meaningful response to, the context of a software system (Denning 2006), which is fundamental to the design, analysis, and performance of any software system.

The following are selected applications of the locality principle. First, indexing in filesystems and databases, by using B-tree, R-tree, or space-filling curves, for instance, organizes a large number of files and data on disks to improve spatial locality and minimize data access time for both linear and multidimensional information (Bayer and McCreight 1970; Lawder and King 2001; Schubert et al. 2013). Second, algorithm design critically relies on cache awareness by exploiting data locality to maximize cache performance for data structures tailored to computationally intensive algorithms (Wolf and Lam 1991; Kennedy and McKinley 1992; Kowarschik and Weiß 2003). More general cache-oblivious algorithms have been developed to perform self-tuning and achieve optimal efficiency on a multilevel memory hierarchy without prior knowledge about specifications of memories, such as cache size and cache line length (Frigo et al. 1999; Demaine 2002; Günther et al. 2006). Third, to achieve scalable high performance, parallelization of software programs and algorithms emphasizes increasing spatiotemporal data locality in computational threads/processes as a means to overcome the communication overhead caused by irregular memory reference patterns and, at the same time, to balance loads among computing units (Acar et al. 2002; Wu et al. 2011). Strategies, such as a data layout transformation, locality-aware work stealing, a topology-aware load balancer, and intelligent data migration, were devised to build context awareness into concurrent programs that can adapt to a variety of single-node parallel platforms—for example, CPU-GPU[1] and NUMA (non-uniform memory access) architectures (Shaheen and Strzodka 2012; Pilla et al. 2014; Agarwal et al. 2015). Topology awareness has become essential for applications involving networks of parallel computers to reconcile resource allocation and scheduling with underlying computing, storage, and network topologies such that communication costs are minimized, and data and job locality are maximized (Xu et al. 2003; Li et al. 2012; Jeannot et al. 2013).

[1]CPU denotes central processing unit, and GPU denotes graphics processing unit.

3 A Locality-Aware Approach

In this section, we adapt the principle of locality to the context of agent-based modeling. Given this conceptualization, we propose and formalize a locality-aware approach to scalable parallelization of SE-ABMs.

3.1 The Locality Principle

As stated previously, the principle of locality refers to the idea that system optimality is achieved when locality sets are present in close proximity to computational processes. A modern model of locality enables the context awareness of computation through the intertwinement of four key elements as follows: the *observers*' actions/interactions are monitored to identify their *neighborhoods* via *inference*, so that the performance of computational tasks can be *optimized* for all observers by sensing their neighborhoods and adapting to them. This model is well suited to the design of efficient parallel SE-ABMs. In the language of ABMs, each agent is an observer whose neighborhood definition can be inferred from either monitoring its dynamic interaction patterns or its static structural declaration. A neighborhood is not necessarily geographic but could be defined through social or other functional dimensions. Because interactions are most intensive between an agent and its neighborhood, the presence of all neighboring agents in its locality set is imperative to minimize the cost of data access such that the performance of the computational task related to this agent can be optimized.

3.2 Locality-Aware Computational Domain

Wang and Armstrong (2009) formalized a spatial computational domain that comprises layers of two-dimensional computational intensity surfaces mapped from a spatial domain. Each surface is represented by a grid on which each cell indicates the computational intensity at location (i, j). This grid representation features a spatially contiguous space. Therefore, neighborhoods defined in this space have to be geographically continuous and cannot accommodate a social structure that has neighboring agents distributed discretely across space. The granularity of the spatial computational domain representation is determined by the choice of cell size, which is coarsely dependent on two trade-off factors, namely, the cost of decomposition and sufficient concurrency, but still contingent on certain arbitrariness. Most importantly, no explicit representation of the interdependency exists between agents whose granularity is usually much finer than the cell size.

To extend this theoretical approach to parallel SE-ABMs, we employ the principle of locality and use graphs to represent the locality-aware computational

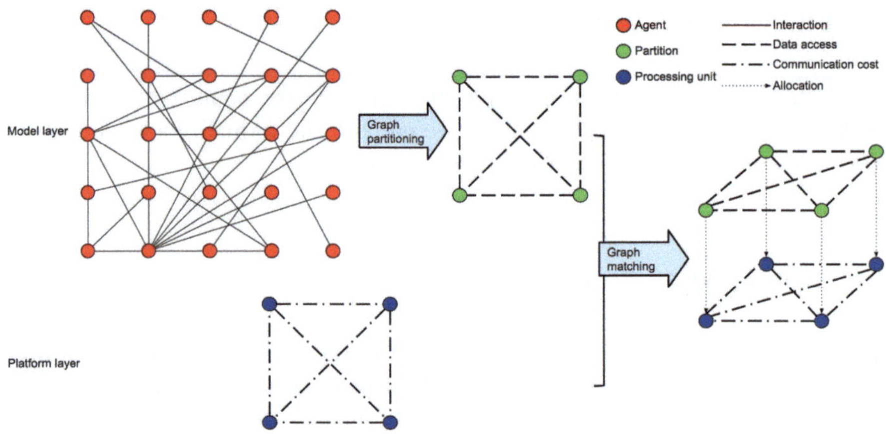

Fig. 1 A locality-aware computational domain

domain that contains two layers: model and platform (Fig. 1). In the model layer, a vertex represents an agent, and an edge represents the independency between two agents in terms of data access due to their interactions (refer to Table 1 for a comparison with a grid representation). Therefore, the space represented by a graph is topological, which enables discrete neighborhood structures. The granularity is naturally at the agent level, which avoids the uncertainty associated with predetermined cell size in a grid representation. In the platform layer, vertices represent processing units (e.g., CPUs) of a computing platform, while edges are weighted by communication (or data access) costs between processing units.

A desirable feature is that the domain decomposition produces equal partitions of the model graph where inter-partition data access is minimized. To guide this process, transformations are performed for the model layer to estimate the computational intensity of a SE-ABM. Two types of transformations can be differentiated: (1) operation-centric transformations estimate the algorithmic time complexity of the operations that process agent behaviors; and (2) data-centric transformations can be further distinguished by two specific functions. The data access function (an analog to the distance function in the principle of locality) estimates the intensity to access and transfer data. The memory function estimates the algorithmic space complexity for the memory requirement. The estimated computational intensity values are embedded in the model layer as follows: a vertex has a vector of values attached to it, indicating the estimated operation and memory intensity, whereas an edge carries a weight indicating the data access intensity between two agents.

Table 1 A comparison between grid and graph representations

Representation	Grid	Graph
Space	Spatially contiguous	Topological
Neighborhood	Continuous	Discrete
Granularity	Cell size	Agent level
Interdependency	N/A	Data access

The partitioned model layer becomes a generalized graph with partitions as vertices and data access intensity between partitions as weighted edges. To allocate model partitions to the processing units with an awareness of data locality, a problem can be formalized as matching two graphs. Given model graph $G_M = (V_M, E_M)$ and platform graph $G_P = (V_P, E_P)$, with $|V_M| = |V_P|$, the problem is to find a one-to-one mapping $f\colon V_M \to V_P$ such that an objective is optimized (e.g., minimizing total data access cost $\sum_{E_M \to E_P} E_M^W E_P^W$).

4 Design and Experimentation of Parallel SE-ABMs

In this section, we employed an agent-based spatial interaction model and laid out two sets of experimentations in order to demonstrate the applicability of the proposed locality-aware approach. Particularly, we designed and implemented a locality-aware parallel algorithm to facilitate the study of model scalability.

4.1 Agent-Based Spatial Interaction Model

To validate our expanded theoretical approach and to evaluate its effectiveness to alleviate the two stated challenges for a scalable parallel SE-ABM, we use a bounded confidence model (Weisbuch et al. 2002) as our testbed, the same one as in Gong et al. (2013). This model simulates the opinion exchange process of spatially distributed agents in a grid space. Each cell in the grid accommodates only one agent. Therefore, the grid size indicates the population size. The communication between two individual agents is bounded, because no opinion exchange occurs if the two individuals hold extremely different opinions regarding a topic. In contrast, meaningful communication occurs based on a learning function (Eq. 1), only when two individuals have sufficiently similar opinions according to a threshold ρ:

$$V_i' = \begin{cases} (1-\beta)V_i + \beta V_j & when\left|V_i - V_j\right| \le \rho \\ V_i & when\left|V_i - V_j\right| > \rho \end{cases}, \tag{1}$$

where V_i and V_j are old opinions of agents i and j, V_i' is the new opinion of agent i, and β is the learning rate.

The model is initialized with randomly assigned opinion values in the range of [0.0, 1.0] to every agent; it is executed iteratively until a consensus is reached among the majority of agents. In each iteration, every agent has one opportunity to interact with an identified neighbor. Given the setup of our model, the population of agents can be divided into two groups, A and B, regarding the computational intensity associated with each agent. Based on the estimates given by operation-centric and data-centric transformations, agents in group A with successful opinion exchange have higher computational intensity than those without opinion exchange in group B. The distributions of groups A and B agents, and the induced interaction patterns, are entirely dependent on the neighborhood definitions. Given this representative agent-based spatial interaction model, we designed two sets of experiments and conducted them on a shared-memory platform to demonstrate the applicability of our locality-aware approach.

4.2 Homogeneous Neighborhoods

The first set of experiments focuses on a homogeneous neighborhood configuration for every agent. The following distance decay function (Eq. 2) with stochasticity is used to define a circular neighborhood for agent i as the center.

$$P_{i,j} = Di, j - 1/\alpha, \tag{2}$$

where $P_{i,j}$ is the probability that a neighbor j can be selected from agent i's neighborhood, which is negatively related to the distance $D_{i,j}$ from j to agent i given the effect of the decay parameter α. The neighborhood configuration is homogeneous for all agents in that the parameter α is constant. At the model run time, a neighbor is identified by randomly generating a $P_{i,j}$ for agent i, and the search radius $D_{i,j}$ can be derived accordingly. In addition, an azimuth between 0 and 360° is randomly chosen. Combining the search radius and azimuth, a neighbor can be identified for each agent.

Although a certain level of stochasticity is involved in neighbor selection, these selections are all bounded by local neighborhoods of the same size. Furthermore, neighborhood homogeneity leads to the random distribution of groups A and B agents across the grid space, which consequently ensures a fairly homogeneous pattern of spatial interactions. Given such a pattern, the most straightforward strategy for domain decomposition is by horizontally splitting the grid into evenly sized partitions (as in Fig. 2). Because groups A and B agents are randomly distributed, the aggregated computational intensity for each partition is fairly balanced, and the inter-partition data access mostly occurs along the edges between adjacent partitions. Following Gong et al. (2013), we investigate the effects of variations in both grid space size and neighborhood size on the scalability of parallel models. In other words, given homogeneous interaction, we examine how inter-partition data access costs can be handled by our approach by comparing the locality-aware version of our ABM with the version without considering locality.

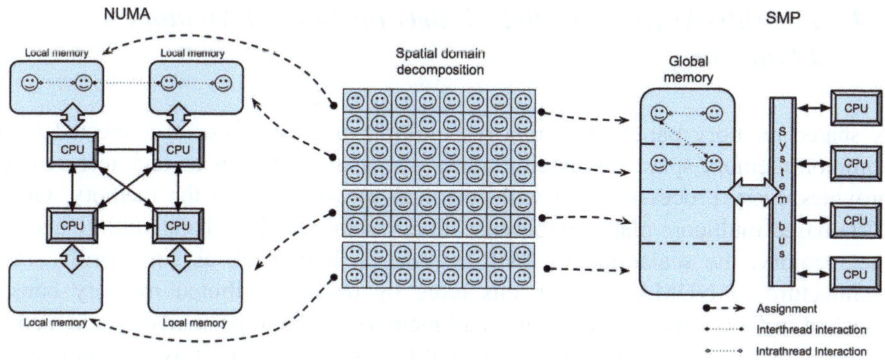

Fig. 2 A comparison of SMP and NUMA architectures regarding regular partitioning

4.3 Heterogeneous Neighborhoods

The second set of experiments features a configuration of heterogeneous agent neighborhoods, which renders spatially heterogeneous interaction patterns. Thus, instead of using spatial neighborhoods, as in Sect. 4.2, we resort to a scale-free network (Barabási and Albert 1999) constructed among spatially situated agents to form their social neighborhoods (as in Fig. 1); node degrees of such a network exhibit a heterogeneous distribution. Agents having more connections with others create larger neighborhoods. At run time, each agent randomly selects a neighbor from his/her neighborhood with which to interact. Such heterogeneous neighborhoods guarantee the generation of heterogeneous patterns of spatial interactions. However, such patterns render the domain decomposition nontrivial. Furthermore, due to the dynamics in SE-ABMs, the heterogeneity of interaction distribution may change over time. This formulation requires robust parallel strategies (e.g., quad-tree partitioning and dynamic load balancer) to automatically handle the computation decomposition and assignment (Wang and Armstrong 2009). To prevent our case study from being too complicated, we assume that the neighborhood structures for a model are predetermined and static during a model run.

According to our locality-aware approach, graph partitioning is applied for domain decomposition. Specifically, we apply multilevel k-way partitioning (Karypis and Kumar 1998b) to create contiguous and relatively balanced partitions with the amount of inter-partition data access (number of edges that straddle different partitions) minimized. However, due to the minimization objective, partitions may not be completely balanced in terms of the number of agents. Moreover, the distribution of groups A and B agents in each partition is uncertain because they are determined at run time, which becomes another possible source of load imbalance in terms of computational intensity. Different topologies of scale-free networks with the same setting were tested to examine the efficiency of our approach to guide domain decomposition and resource allocation in terms of maximizing data locality and minimizing data access costs.

4.4 *Locality-Aware Parallel Models on Shared-Memory Platforms*

A shared-memory paradigm comprises multiple processors/cores connected to a common memory space. Symmetric multi-processing (SMP) is an architecture that provides every processing unit with identical access speed to the memory. On a CPU-based multicore platform, because all cores share a fixed bandwidth to access the memory, the scalability of SMP becomes a bottleneck as cores grow. The architecture of NUMA resolves this issue by using distributed memory banks attached to CPU cores, which enables all memory banks to be shared, while a core accesses its local bank faster than nonlocal banks. Figure 2 portrays a comparison of the two platforms.

In our parallel models, we adopt multithreading parallelization, which is supported by multiprocessor/core shared-memory platforms. The common practice consists of generating the same number of threads as the number of cores and mapping these threads to cores. Accordingly, the same number of partitions is created and assigned to threads. The algorithm of our locality-aware parallel model optimized for NUMA architecture is detailed in Fig. 3. Four critical components merit highlighting. First, the specifications of decomposed subdomains (or partitions) are provided externally by strategies such as regular partitioning or graph partitioning. Second, each thread constructs and initializes its own partition of agents in its local memory and provides an interface for other threads to access its local data. Here, given homogeneous computing units (same CPU cores), we assume that only two types of data access costs exist in the platform layer, namely, inter-thread and intra-thread data access costs. Because access to local data (intra-thread) for a thread costs much less than access to remote data (inter-thread), we assume that accessing remote data owned by different threads is subjected to the same cost. Based on this assumption, allocation of partitions to computing units through graph matching does not differentiate between types of match as long as partitions belonging to threads are kept in their local memories. Such an optimization ensures that intra-partition interactions are processed at a cost of intra-thread data access, while inter-partition interactions are processed at a cost of inter-thread data access. Third, during inter-thread data access, an agent may interact with two or more agents from different partitions at the same time due to the parallel processing of threads. Mutual exclusion must be applied to prevent the data of such an agent from being simultaneously modified. However, applying mutual exclusion has the potential to cause overheads and to slow down the speedup (SP) of parallel programs. Fourth, synchronization among all threads must be performed between iterations to preserve the efficacy of a model in its parallelization. Due to the imperfect balancing between partitions, synchronization is another source of overhead.

Our parallel models are implemented using C++ and OpenMP (2011), a standard specification for multithreading programs. These models were tested on a Linux cluster with two computing nodes. Each node had 32 CPU cores (AMD

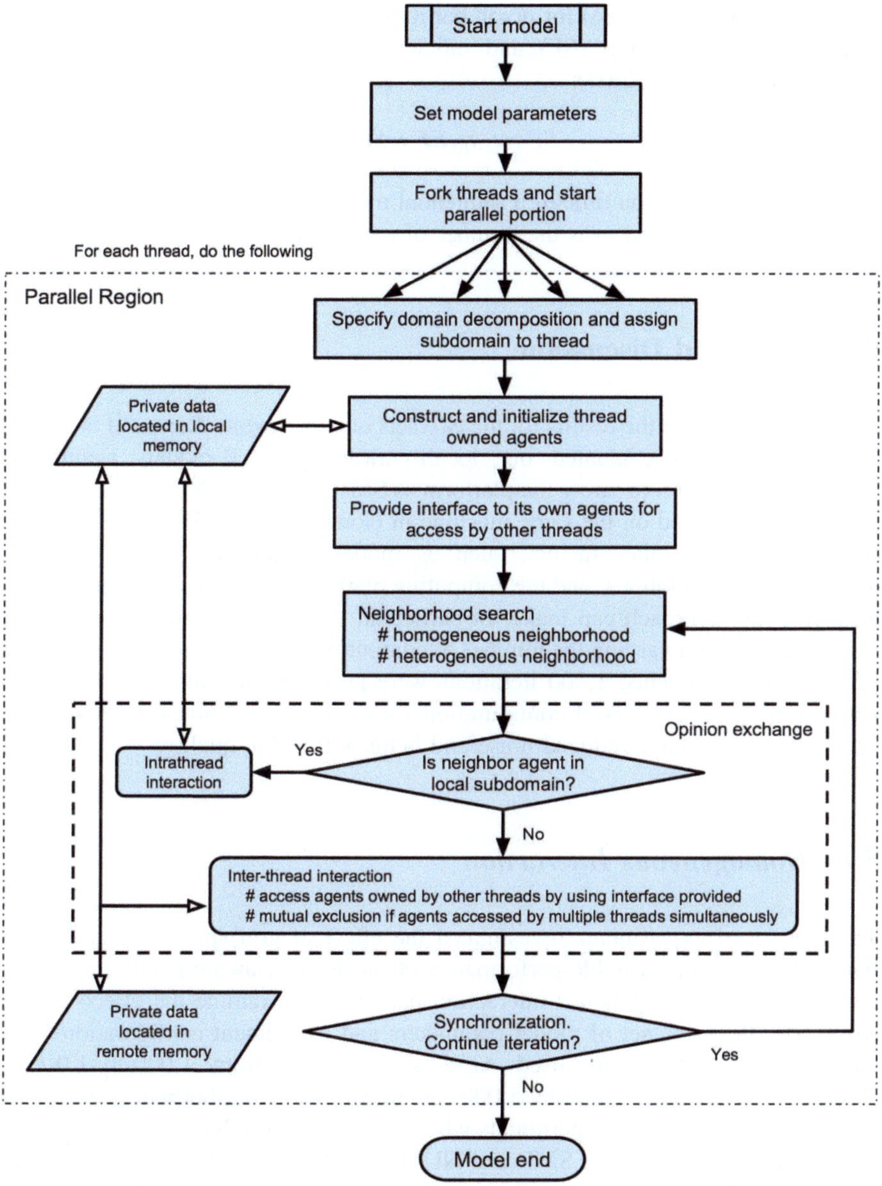

Fig. 3 Flowchart for locality-aware parallel model designed for a NUMA architecture

Opteron 2.0 GHz) and 64 GB of memory. Both nodes had an underlying NUMA architecture, although they allowed programs to operate as in SMP mode. In NUMA mode, 4 cores share a local memory of 8 GB. In addition, we used NetworkX (Hagberg et al. 2008) to generate scale-free networks and METIS

(Karypis and Kumar 1998a) for graph partitioning. To evaluate the performance of parallel models, speedup (SP) and efficiency (EF) are the two metrics used (Wilkinson and Allen 2004):

$$SP = t_s / t_p, EF = SP / m, \tag{3}$$

where t_s is the execution time of a sequential model, while t_p is the execution time of a parallel model, and m is the number of computing units.

5 Results and Discussion

This section presents the results for the two sets of experiments discussed in Sect. 4. These experiments evaluated our locality-aware parallel models against the locality-blind version (treating the platform as SMP rather than NUMA), which was developed and tested on the same machine in Gong et al. (2013). By doing so, we controlled for all other factors, such as model parameters ($\beta = 0.2$; $\rho = 0.3$), experiment configurations, and the computing platform, to elucidate the difference a locality-aware approach can make. In particular, we looked at how model performance scales when varying the number of computing cores between 2, 4, 8, 16, and 32. For all experiments, 1,000 iterations were performed for all model runs, and simulations for every model configuration were repeated 30 times to obtain averaged outputs that have reduced noise and better inferential quality.

5.1 *Homogeneous Interaction*

The first set of experiments investigated the effect of grid space size and neighborhood size on the scalable performance of our locality-aware parallel models in the presence of homogeneous interaction patterns. To examine the effects of grid space size, the same set of model parameters and experiment configurations as in Gong et al. (2013) were used. Grid size changes from 1,000-by-1,000 to 5,000-by-5,000 with a 1,000-by-1,000 increment. The neighborhood size is constant, with $\alpha = 0.9$ for all experiments when examining the effect of grid space size.

The SP and EF for both SMP and NUMA models with locality awareness are portrayed in Figs. 4 and 5. For both types of model, as more parallel computing cores are introduced, SP shows a general upward trend. However, SMP models exhibit a more severe slowing down for model SP with larger grid size as the number of cores increases. This outcome means that the grid space size has a negative impact on the scalability of SMP models. It also is indicated by the EF of SMP models (Fig. 5a). That is, as SMP models are scaled to more cores, the EF decreases in general, and its decrease becomes more pronounced with an increasing grid size. Because SMP models are not aware of data locality, partitions allocated to

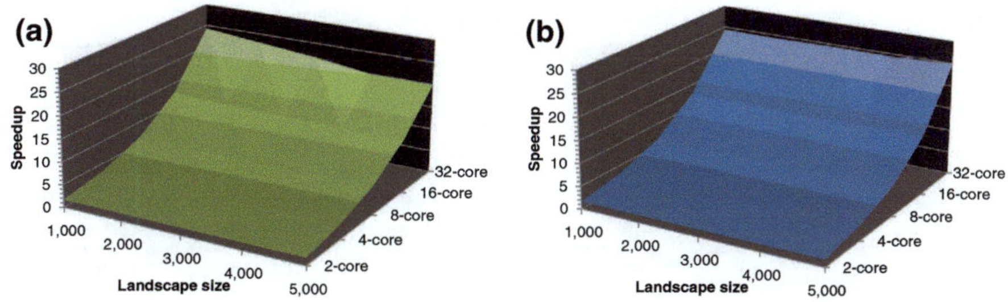

Fig. 4 A comparison of SP for **a** SMP models (Gong et al. 2013) and **b** NUMA models with locality awareness for experiments with varying grid size and core number

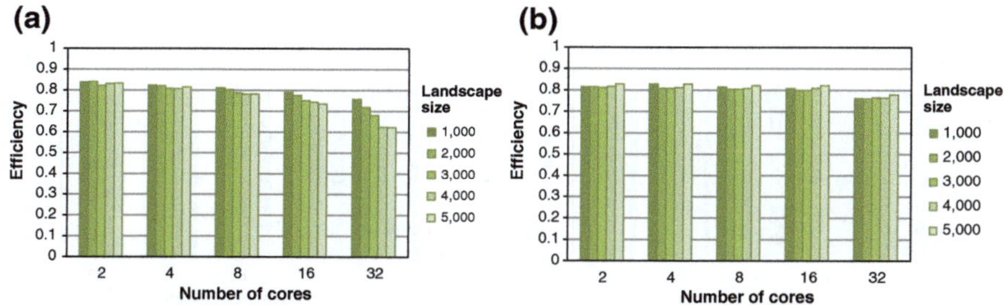

Fig. 5 A comparison of EF for **a** SMP models (Gong et al. 2013) and **b** NUMA models with locality awareness for experiments with varying grid size and core number

threads are not necessarily placed in their local memories. This mismatch between model partitions and computing units leads to a situation where many intra-partition interactions that should be processed only at a cost of intra-thread data access are subjected to inter-thread data access costs. As the number of partitions increases with the number of cores, the likelihood of this mismatch becomes much higher, which further increases overhead caused by a larger portion of intra-partition interactions subjected to inter-thread data access costs. In contrast, by optimizing data locality to allow intra-partition interactions to be processed at a cost of intra-thread data access, NUMA models eliminate the negative impact of grid size on model performance (Figs. 4b and 5b). As grid size increases, they instead exhibit a slight increasing trend (Fig. 5b). This outcome can be explained by the relatively smaller neighborhoods when grid size keeps increasing and the neighborhood size (α) is kept constant. Accordingly, the size of each partition becomes larger compared to the neighborhood size, which reduces the possibility of inter-partition interactions and the induced potential inter-thread data access costs. In addition, NUMA models achieve reasonably good scalability, which is reflected by the consistent EF of 0.8 or beyond over all experiments using 16 or fewer cores. For experiments using 32 cores, the EF values of NUMA models are below 0.8 but above 0.75, which is reasonable because as the maximum number of cores is introduced, the inter-thread data access costs reach their highest level.

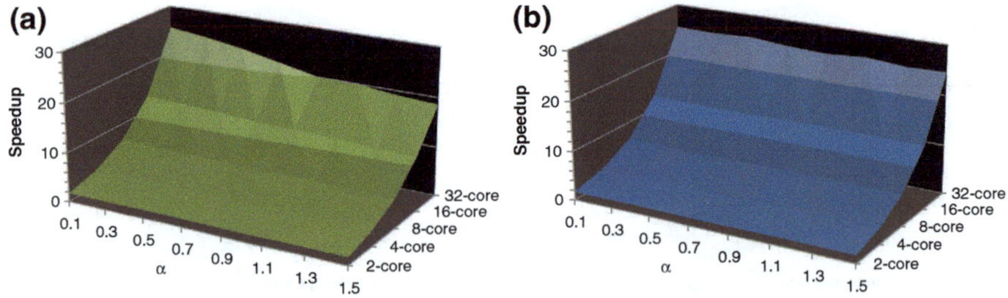

Fig. 6 A comparison of SP for **a** SMP models (Gong et al. 2013) and **b** NUMA models with locality awareness for experiments with varying neighborhood size and core number

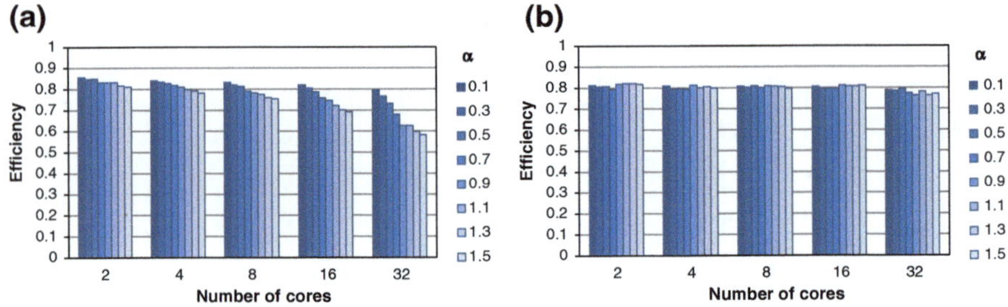

Fig. 7 A comparison of EF for **a** SMP models (Gong et al. 2013) and **b** NUMA models with locality awareness for experiments with varying neighborhood size and core number

Similarly, for the category of experiments examining the effect of neighborhood size, α changed from 0.1 to 1.5, with an increment of 0.2. The landscape size of 4,000-by-4,000 is fixed for all experiments in this category. Figures 6 and 7 show the SP and EF, respectively, for SMP and NUMA models for this category of experiment. For SMP models, neighborhood size imposes the same negative effects on the scalability of model performance. As indicated by Figs. 6a and 7a, SP slows down, and the magnitude of decrease in EF becomes more significant with increasing neighborhood size, as more cores are used. It is the same issue of mismatch between model partitions and computing threads that degrades the model performance by creating an overwhelming amount of inter-thread data access costs for intra-partition interactions. In contrast, effects of neighborhood size almost disappear in NUMA models (experiments with 32 cores in Fig. 7b show a slight decrease in the EF with increasing neighborhood size), because they control for data locality. The scalability of NUMA models is consistently good, with EF values around 0.8 over all experiments using 16 cores or less. For experiments using 32 cores, the EF is only slightly lower than 0.8, indicating a higher level of overhead with all cores fully loaded. In sum, given homogeneous neighborhood configurations, the effects of grid space size and neighborhood size on model performance can be well controlled by locality-aware parallel models. The scalability of these models is reasonably good and consistent in the presence of homogeneous patterns of spatial interaction.

5.2 Heterogeneous Interaction

The second set of experiments investigated how heterogeneous interaction patterns affect the performance and scalability of our locality-aware parallel models. Again, the results are compared to those from models with no consideration for locality. To reduce the effect of randomness in the generated social neighborhoods, networks with the same distribution of node degree were generated 30 times for 30 model runs in order to compute averaged results. Each network was fed into both NUMA and SMP models to control for the influence of neighborhood setting. Scale-free networks were generated based on the Barabási–Albert preferential attachment model with the number of nodes set to 1,000,000 (agents in 1,000-by-1,000 grid space), with three edges to attach from a new node to existing nodes.

Figure 8 depicts the SP and EF for NUMA and SMP models in the presence of heterogeneous interaction patterns. As more cores are used, the SP for NUMA models increases more rapidly than that for SMP models. Especially for experiments with 32 cores used, the SP for NUMA models is almost double that for SMP models. As a result, NUMA models exhibit scalable performance superior to that for SMP models, because optimized intra-partition interactions take only intra-thread data access costs in NUMA models. In contrast, due to their blindness to data locality, SMP models have increasing proportions of intra-partition interactions subjected to inter-thread data access costs, which deteriorate their performance gain as they scale to more computing cores. This effect also is indicated by the EF plot for SMP models (Fig. 8b). Although NUMA models perform better in terms of EF, the results also reflect a deceasing (from 0.8 to less than 0.5) trend as more cores are used. Because NUMA models have intra-partition interactions

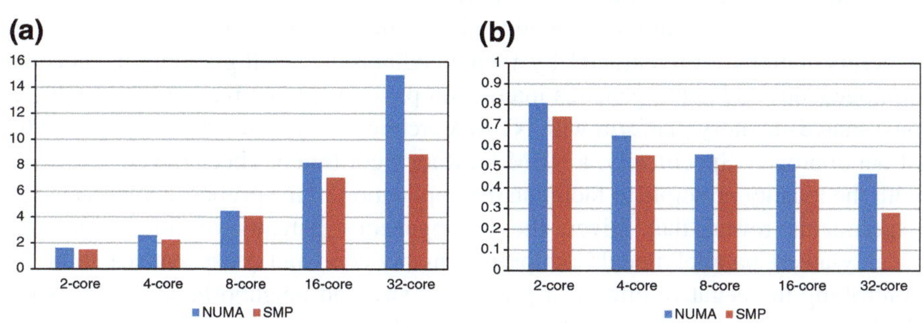

Fig. 8 A comparison of NUMA and SMP models regarding SP (**a**) and EF (**b**) in the presence of heterogeneous interaction patterns

optimized for data access, inter-partition interactions can be singled out as the main source of EF degradation. Especially with the maximum number of cores used, around 50 % of computing resources is wasted on overhead that mainly is caused by inter-thread data access costs. In addition, the overhead potentially is attributed to other sources including unbalanced partitions, mutual exclusion, and intra-thread data access costs.

However, if the inter-partition interactions are homogeneous, as in the first set of experiments, the scalability of NUMA models is not severely affected by the sole influence of increasing proportions of inter-partition interactions due to more partitions (refer to the change in EF for NUMA models with a grid size of 1,000-by-1,000 in Fig. 5b). In addition, note that the EF of NUMA models using two cores in Fig. 8b is almost the same as the EF of NUMA models with a grid size of 1,000-by-1,000 using two cores in Fig. 5b. This outcome is reasonable because the inter-partition interactions are symmetric at the aggregate level between two partitions, although heterogeneous patterns of interactions exist at the individual agent level. Once the number of partitions is larger than two, heterogeneous patterns of inter-partition interactions start to be evident at the aggregate level. As the number of partitions grows, the degree of heterogeneity becomes greater, which is negatively correlated with a decrease in EF for NUMA models using more than two cores. Therefore, the heterogeneity of inter-partition interactions can be inferred as a fundamental factor that negatively affects the scalability of NUMA models. This effect can be related to an assumption made in the design of our locality-aware algorithm. That is, inter-thread data access costs are the same from one core to the others. In other words, if data access costs are different from one core to others or if there exists a topology of inter-thread data access costs among different cores, locality awareness is not fully exploited in the current design of our algorithm.

6 Conclusions and Future Work

This chapter aims to address two challenges to scalable parallel SE-ABMs, namely, the spatial dependency and heterogeneity of agent interactions. We formalize the locality-aware computational domain with a graph-based approach. Two sets of experimentations, differentiated by their neighborhood configurations to generate homogeneous and heterogeneous interaction patterns, respectively, were conducted on a shared-memory platform with NUMA architecture to illustrate the applicability of our approach. Specifically, a NUMA-aware algorithm has been developed in line with the locality principle to facilitate the application of the proposed approach. Our experiments reveal that, given the homogeneous patterns of interactions, the locality-aware parallel models exhibit reasonably good and consistent scalability by controlling the negative effects of grid space size and neighborhood size to which the locality-blind parallel models are subjected.

In the presence of heterogeneous patterns of interaction, although locality-aware parallel models achieve better scalable performance than locality-blind versions, the experiments imply that scalability is negatively affected by the heterogeneity of spatial interaction. This finding uncovers the hot spot of performance loss in the proposed design of locality-aware algorithm. Future research should more firmly establish the extent to which interaction heterogeneity affects scalability by testing how model performance varies with the degree of heterogeneity. In addition, once the real effect of spatial heterogeneity is well understood, an operational solution can be suggested in accordance with the proposed locality-aware approach. Specifically, data locality should be fully exploited by improving existing algorithms in a manner that reflects the underlying topology of data access costs between computing units.

References

Acar UA, Blelloch GE, Blumofe RD (2002) The data locality of work stealing. Theory Comput Syst 35:321–347

Agarwal N, Nellans D, O'Connor M, Keckler SW, Wenisch TF (2015) Unlocking bandwidth for GPUs in CC-NUMA systems. In: 2015 IEEE 21st international symposium on high performance computer architecture (HPCA), Burlingame, CA. IEEE, pp 354–365

Barabási A-L, Albert R (1999) Emergence of scaling in random networks. Science 286:509–512

Bayer R, McCreight E (1970) Organization and maintenance of large ordered indices. In: Proceedings of the 1970 ACM SIGFIDET (now SIGMOD) workshop on data description, access and control—SIGFIDET '70, New York, NY. ACM Press, pp 107–141

Bian L (2004) A conceptual framework for an individual-based spatially explicit epidemiological model. Environ Plan 31:381–395

Billari FC, Prskawetz A (2003) Agent-based computational demography. Physica-Verlag HD, Heidelberg

Brown DG, Riolo R, Robinson DT, North M, Rand W (2005) Spatial process and data models: toward integration of agent-based models and GIS. J Geogr Syst 7:25–47

Crooks A, Castle C, Batty M (2008) Key challenges in agent-based modelling for geo-spatial simulation. Comput Environ Urban Syst 32:417–430

Demaine E (2002) Cache-oblivious algorithms and data structures. In: Lecture notes from the EEF summer school on massive data sets. BRICS, University of Aarhus, Denmark

Denning PJ (1970) Virtual memory. ACM Comput Surv 2:153–189

Denning PJ (1980) Working sets past and present. IEEE Trans Softw Eng SE-6, 64–84

Denning PJ (2006) The locality principle. In: Barria JA (ed) Communication networks and computer systems. Imperial College Press, London, pp 43–67

Epstein JM (1999) Agent-based computational models and generative social science. Complexity 4:41–60

Epstein JM (2009) Modelling to contain pandemics. Nature 460:687

Frigo M, Leiserson CE, Prokop H, Ramachandran S (1999) Cache-oblivious algorithms. In: 40th annual symposium on foundations of computer science, New York, NY. IEEE, pp 285–297

Gong Z, Tang W, Bennett DA, Thill J-C (2013) Parallel agent-based simulation of individual-level spatial interactions within a multicore computing environment. Int J Geogr Inf Sci 27:1152–1170

Grimm V, Berger U, Bastiansen F et al (2006) A standard protocol for describing individual-based and agent-based models. Ecol Model 198(1–2):115–126

Günther F, Mehl M, Pögl M et al (2006) A cache-aware algorithm for PDEs on hierarchical data structures based on space-filling curves. SIAM J Sci Comput 28:1634–1650

Hagberg AA, Schult DA, Swart PJ (2008) Exploring network structure, dynamics, and function using networkx. In: Varoquaux G, Vaught T, Millman J (eds) Proceedings of the 7th python in science conference (SciPy2008). Pasadena, CA, pp 11–15

Irwin EG (2010) New directions for urban economic models of land use change: incorporating spatial dynamics and heterogeneity. J Reg Sci 50:65–91 (2008)

Jeannot E, Meneses E, Mercier G, Tessier F, Zheng G (2013) Communication and topology-aware load balancing in Charm++ with treematch. In: 2013 IEEE international conference on cluster computing (CLUSTER), Indianapolis, IN. IEEE, pp 1–8

Karypis G, Kumar V (1998a) A fast and high quality multilevel scheme for partitioning irregular graphs. SIAM J Sci Comput 20:359–392

Karypis G, Kumar V (1998b) Multilevel k-way partitioning scheme for irregular graphs. J Parall Distrib Comput 48:96–129

Kennedy K, McKinley KS (1992) Optimizing for parallelism and data locality. In: Proceedings of the 6th international conference on supercomputing (ICS '92), New York, NY. ACM Press, pp 323–334

Kowarschik M, Weiß C (2003) An overview of cache optimization techniques and cache-aware numerical algorithms. In: Meyer U, Sanders P, Sibeyn J (eds) Algorithms for memory hierarchies: advanced lectures, vol 2625. Springer, Berlin, pp 213–232

Lawder JK, King PJH (2001) Querying multi-dimensional data indexed using the Hilbert space-filling curve. ACM SIGMOD Rec 30:19–24

Li M, Subhraveti D, Butt AR, Khasymski A, Sarkar P (2012) CAM: a topology aware minimum cost flow based resource manager for MapReduce applications in the cloud. In: Proceedings of the 21st international symposium on high-performance parallel and distributed computing—HPDC '12, New York, NY. ACM Press, pp 211–222

Matthews RB, Gilbert NG, Roach A, Polhill JG, Gotts NM (2007) Agent-based land-use models: a review of applications. Landscape Ecol 22:1447–1459

Parker DC, Manson SM, Ma Janssen, Hoffmann MJ, Deadman P (2003) Multi-agent systems for the simulation of land-use and land-cover change: a review. Ann Assoc Am Geogr 93:314–337

Pilla LL, Ribeiro CP, Coucheney P, Broquedis F, Gaujal B, Navaux POA, Méhaut J-F (2014) A topology-aware load balancing algorithm for clustered hierarchical multi-core machines. Future Gen Comput Syst 30:191–201

Schubert E, Zimek A, Kriegel H-P (2013) Geodetic distance queries on R-Trees for indexing geographic data. In: Nascimento MA, Sellis T, Cheng R et al (eds) Advances in spatial and temporal databases. Springer, Berlin Heidelberg, pp 146–164

Shaheen M, Strzodka R (2012) NUMA aware iterative stencil computations on many-core systems. In: 2012 IEEE 26th international parallel and distributed processing symposium. IEEE, pp 461–473

Stanilov K (2012) Space in agent-based models. In: Heppenstall AJ, Crooks AT, See LM, Batty M (eds) Agent-based models of geographical systems. Springer, Netherlands, Dordrecht, pp 253–269

Tang W (2008) Simulating complex adaptive geographic systems: a geographically aware intelligent agent approach. Cartogr Geogr Inf Sci 35:239–263

Wang S, Armstrong MP (2009) A theoretical approach to the use of cyberinfrastructure in geographical analysis. Int J Geogr Inf Sci 23:169–193

Weisbuch G, Deffuant G, Amblard F et al (2002) Meet, discuss, and segregate! Complexity 7:55–63

Wilkinson B, Allen M (2004) Parallel programming: techniques and applications using networked workstations and parallel computers, 2nd edn. Prentice-Hall Inc, Upper Saddle River, NJ

Wolf ME, Lam MS (1991) A data locality optimizing algorithm. ACM SIGPLAN Not 26:30–44

Wu B, Zhang EZ, Shen X (2011) Enhancing data locality for dynamic simulations through asynchronous data transformations and adaptive control. In: 2011 international conference on parallel architectures and compilation techniques, Galveston, TX. IEEE, pp 243–252

Xu Z, Tang C, Zhang Z (2003) Building topology-aware overlays using global soft-state. In: 23rd international conference on distributed computing systems, Providence, RI. IEEE, pp 500–508

Simulation of Human Wayfinding Uncertainties: Operationalizing a Wandering Disutility Function

Amir Najian and Denis J. Dean

Abstract This study focuses on design and implementation of an agent-based simulation model that replicates spatially disoriented walking behavior caused by decline in cognitive abilities, similar to conditions experienced by Alzheimer's patients. Results of this simulation will be used to investigate potential correlations between observable spatial patterns in walking trajectories and levels of cognitive impairment in dementia patients. Review of literature on human wayfinding behavior provides a set of operational parameters to employ in an agent-based model. The proposed mechanism of replicating spatial disorientation in this study relies on stochastic modeling of uncertainties in (1) traveled distance, (2) direction of travel toward the destination, and (3) location of landmarks in the environment. Additionally, a proposed measure of aggregate wayfinding disutility is introduced to regulate the start of spatially disoriented walking episodes in agents.

Keywords Spatial disorientation · Wayfinding · Agent-based modeling · Spatial patterns · Wandering

1 Introduction

One out of nine older Americans has Alzheimer's disease, and six out of ten of those with Alzheimer's will wander (Alzheimer's Association 2013; Stokes 1996). Experts classify wandering as a disruptive and potentially dangerous behavior, and consider it symptomatic of declining cognitive abilities (Scarmeas et al. 2009). This chapter summarizes a study design motivated by the idea of finding practical solutions to help Alzheimer's and dementia patients who wander.

A. Najian (✉) · D.J. Dean
Geospatial Information Sciences, University of Texas at Dallas, Richardson, USA
e-mail: amir.najian@utdallas.edu

D.J. Dean
e-mail: denis.dean@utdallas.edu

D.A. Griffith et al. (eds.), *Advances in Geocomputation*,
Advances in Geographic Information Science,
DOI 10.1007/978-3-319-22786-3_37

Previous studies suggest that spatial patterns exist in wandering behavior caused by dementia (Martino-Saltzman et al. 1991). Yet, these spatial patterns have not been studied analytically. Furthermore, such patterns have been recognized only indoors; to the best of the authors' knowledge, no effort has been made to identify any spatial patterns in outdoor wandering. The study proposed here tries to develop a method for analytical detection of such patterns in outdoor walking trajectories of individuals with impaired cognitive abilities. Discovering correlations between observable spatial patterns in trajectories of such individuals and the level of decline in cognitive abilities assists in identifying people with dementia who experience wandering through monitoring their walking trajectories. This identification allows for real-time intervention to avoid or at least minimize the risks imposed by dangerous wandering situations.

In this chapter, we propose an agent-based model that simulates walking trajectories of a range of pedestrians from cognitively healthy to those experiencing high levels of dementia. These trajectories will be analyzed in relation with the agents' degree of cognitive impairment. Our research goal is to identify travel patterns that are indicative of dementia. The purpose of this chapter is to present our conceptualization of this research problem coupled with the formulation of a simulation experiment to solve it.

2 Definitions

The following discussion requires clarifying the definitions of two common terms that sometimes are used interchangeably, but in this chapter have significantly different meanings. In the context of environmental cognition and cognitive psychology, *navigation* is defined as deliberate walking or moving in space (Golledge 1999). *Wayfinding* is defined as a specific type of navigation. Allen (1999, 47) defines wayfinding as "purposeful movement to a specific destination that is distal and, thus, cannot be perceived directly by [a] traveler." The important difference here is that the destination in wayfinding is specified but not directly perceivable from its starting point. This type of destination plays a central role in the method of simulating spatially disoriented walking to be used in our study.

3 The Research Problem

The principal hypothesis guiding our proposed study is that a quantitative analysis of outdoor pedestrian movement patterns can identify individuals who are wandering due to dementia. An implicit assumption of this hypothesis is that a relationship exists between spatial patterns of movement and the level of dementia in affected individuals. This contention is consistent with the approach of viewing wandering as part of a multicomponent behavior (Algase et al. 2007). In this context, the multicomponent behavior approach relates an individual's level of dementia, cognitive status, degree of wayfinding uncertainty, level of spatial disorientation, and walking patterns (Fig. 1). Our study proposes to investigate the detectability of the correlation between dementia and walking travel patterns that the multicomponent behavior approach implies must exist.

4 Background

To address the research question of this study, first a review of literature on wandering and spatially disoriented behaviors caused by cognitive impairments, specifically Alzheimer's type dementia, was conducted. The main objective was to identify quantifiable aspects of such spatial behaviors. Also, fundamental concepts of spatial orientation in human wayfinding behavior were reviewed to achieve an understanding of how spatial disorientation occurs as the result of uncertainties in cognitive mechanisms that maintain spatial orientation. Part of the main objective of this review was to arrive at a set of operational parameters that can be employed in an agent-based simulation model to replicate spatially disoriented behaviors.

4.1 Quantifying Dementia

An individual's level of dementia can be quantified using a well-established score of cognitive functionality called the mini-mental state (MMS) (Folstein et al. 1975). A MMS is measured using a 11-item questionnaire that requires approximately ten minutes to complete. The questionnaire examines cognitive functions in many areas, and a score of 24 or more (out of 30) is considered to represent normal cognitive abilities. Scores lower than 22 reflect various levels of cognitive impairment. We employ ideas from the MMS examination to quantify the degrees of dementia experienced by the agents used in our model.

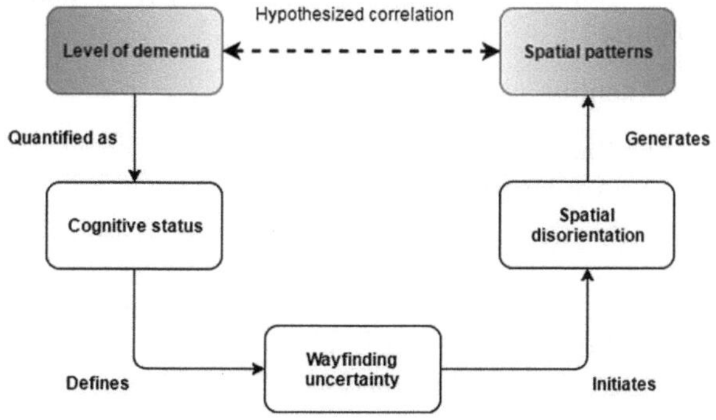

Fig. 1 A proposed method of testing the research hypothesis

4.2 Spatial Orientation and Human Wayfinding

Rieser (1999, 168) defines spatial orientation as "knowing where one is relative to objects in the surrounding environment", explains that lack of knowledge about the required directions and distances to return to a starting point or another useful destination results in disorientation. Dudchenko (2010) describes spatial disorientation as the result of failing to maintain requirements of spatial orientation, including recognition of one's surroundings, ability to decide about turning directions based on familiar landmarks, and maintaining orientation through monitoring previous experiences. He characterizes being lost as the inability to wayfind in a large-scale space. Dudchenko further states that people are more likely to lose their way if familiar landmarks are not present. Humans rely on landmarks and their *spatial abilities* to maintain spatial orientation in surrounding environments and to complete wayfinding tasks. Golledge (1999) describes spatial abilities as natural skills and memory-based abilities that form spatial knowledge in most human wayfinding activities. Golledge and Stimson (1997) provide a comprehensive list of spatial abilities, some of which are directly applicable to simulating spatial orientation in pedestrian agents. These include the ability to give and comprehend directional and distance estimates, the ability to understand network structures, and the ability to orient oneself with respect to local, relational, or global *frames of reference*. In general, two types of frames of reference exist with regard to a viewer and his/her environment. First, a viewer-centered (or self-to-object) frame of reference is based on a viewer's perspective of the world and thus is controlled by the viewer's orientation in the environment. Second, an object-centered or (object-to-object) frame of reference provides a representation of objects or places "with respect to their intrinsic axes, such as gravity and prominent landmarks" (Amorim 1999, 156). Rieser (1999) proposes that individuals employ both types of frames of reference during navigation and wayfinding.

Based on the literature about this topic, we choose a set of parameters to control the spatial orientation in pedestrian agents. These parameters can be categorized into two groups: Attention zone parameters (Torrens et al. 2012) define and control a circular area in front of an agent that represents the agent's field of view. Frame of reference parameters define an agent's ability to maintain its relative orientation to key features in the environment. The specific parameters in each category are as follows:

- Attention zone parameters
  1. Radius (r)
  2. Arc length (s)
- Frame of reference parameters
  3. Distance from origin d_o (or to destination d_d)
  4. Bearing to destination (b)
  5. Position of landmarks (p)

4.3 Wandering Behavior

Wandering is an abnormal wayfinding behavior that happens under cognitively impaired conditions caused by dementia. Wandering is caused, at least in part, by deterioration of memory and decreased capacity to think and reason, and is the result of an elderly person's impaired memory interacting with the surrounding environment (Stokes 1996). Martino-Saltzman et al. (1991) extracted spatial patterns of movement in wandering behavior among dementia patients inside a care facility. These patterns include the following four basic travel tendencies: direct, random, pacing, and lapping. Direct travel is moving from one location to another without diversion. Random travel is roundabout or haphazard moving to many locations within an area without repetition. Pacing is repetitive back-and-forth movement within a limited area. Finally, lapping is repetitive travel characterized by circling large areas. These patterns were observed in indoor travel and may or may not be observed in the outdoor environments that are the focus of this investigation.

4.4 Observation and Simulation of Human Movement and Wandering

Observational studies of wandering behavior can be categorized in two ways. The first concentrates on monitoring and detection of the basic travel patterns of wandering proposed by Martino-Saltzman et al. (1991). Algase (2003), Greiner et al. (2007), and Vuong et al. (2011) furnish examples of this type of study. The second more broadly investigates activity patterns in cognitively impaired individuals, including both spatial and nonspatial aspects of wandering, such as the ability of individuals to remember the purpose of their travels. Makimoto et al. (2008), Sposaro et al. (2010), and Shoval et al. (2008) furnish examples of this type of study.

Many studies of human walking behavior lay the groundwork for agent-based simulation modeling of spatial disorientation. Among such studies, Nasir et al. (2014) propose a method to model wandering in walking behavior using the ideas of impedance. It should be emphasized that the wandering behavior in Nasir et al.'s work is different from wandering caused by dementia, and it only characterizes stochastic fluctuations in human walking. However, the idea of defining a disutility function for disoriented behaviors (which is how Nasir et al. produced their impedance measures) is adopted in our proposed study.

Another group of authors have developed agent-based walking models that focus on simulating the behaviors of healthy individuals in various environments

(Dijkstra et al. 2006; Dijkstra 2008; Banks and Sokolowski 2009). Some of these studies have used the synthetic walking trajectories produced by their agents to evaluate the feasibility of using such trajectories to study human behavior (Torrens et al. 2011, 2012). However, to the best of the authors' knowledge, no previous study has used these approaches to investigate travel patterns of cognitively impaired individuals.

5 Methods

In our proposed agent-based model, the degree of dementia exhibited by agents is controlled by the parameters listed in Sect. 4.2. By manipulating these parameters, agents exhibit differing levels of dementia, which will be reflected in their travel trajectories. We analyze these trajectories to determine if we can deduce the level of dementia in agents that generated recognizable spatial patterns in their walking trajectories.

Our proposed study is designed as a Monte Carlo experiment, which generates a large population of pedestrian agents with various levels of cognitive abilities. These agents travel through a wide variety of simulated environments exhibiting distinct levels of network complexity, frequency and distribution of landmarks, and differing lengths of travel between origin and destination points. This experiment creates a large dataset of walking trajectories that will help to investigate hypothesized correlations through statistical analysis.

At the beginning of the simulation, a pedestrian agent is instantiated. The cognitive status of this agent, measured via the stochastic pseudo-MMS mentioned previously, is set at instantiation. Based on the agent's pseudo-MMS value, another parameter (t_0) is set stochastically. The t_0 parameter determines when the agent starts to feel uncertain about the distance and direction it has traveled. The lower the level of cognitive ability of an agent is, the shorter the uncertainty onset time will be.

Once its pseudo-MMS has been set, the temporal simulation of the agent's movement commences. Agents move along fixed travel paths (presumed to be sidewalks along roads) arranged in an interconnected network, and they travel from a fixed starting point to a fixed ending point. The number and location of landmarks throughout the network also is fixed. As agents move through a network and come to intersections where multiple paths converge, they must make decisions to choose the paths that move them toward their target. Agents with little or no cognitive impairment usually make these decisions correctly, resulting in a trajectory that more or less follows a shortest-path route from the starting to the ending point. Cognitively impaired agents have a lower probability of making correct decisions.

The likelihood that an agent makes an incorrect decision is proportional to a disutility function (d_w) unique to each agent. This disutility function is the sum of the following three uncertainty components: distance, direction, and perceived location of landmarks, including the starting and ending points. The following is the equational description of this disutility function:

$$d_w = e_{dis} + e_{dir} + e_{loc}, \quad (1)$$

where distance error (e_{dis}) is measured as the ratio of the distance traveled with uncertainty (i.e., distance traveled after t_0) to total distance traveled from the origin of the journey; direction error (e_{dir}) is measured as the ratio of change in heading toward a destination, caused by wayfinding uncertainty, to its maximum value; and location error (e_{loc}) is measured as the ratio of change in the perceived location of each landmark, to the maximum possible distance a landmark could be shifted and still remain within the region simulated in the model. This disutility measure continues to accumulate (thereby increasing the probability that a given agent makes incorrect wayfinding decisions) as the agent moves until any of the following conditions are true: (1) the agent reaches the destination of the journey, (2) the agent visits one of the prelocated landmarks in the simulation environment, or (3) the maximum time of the simulation experiment ends.

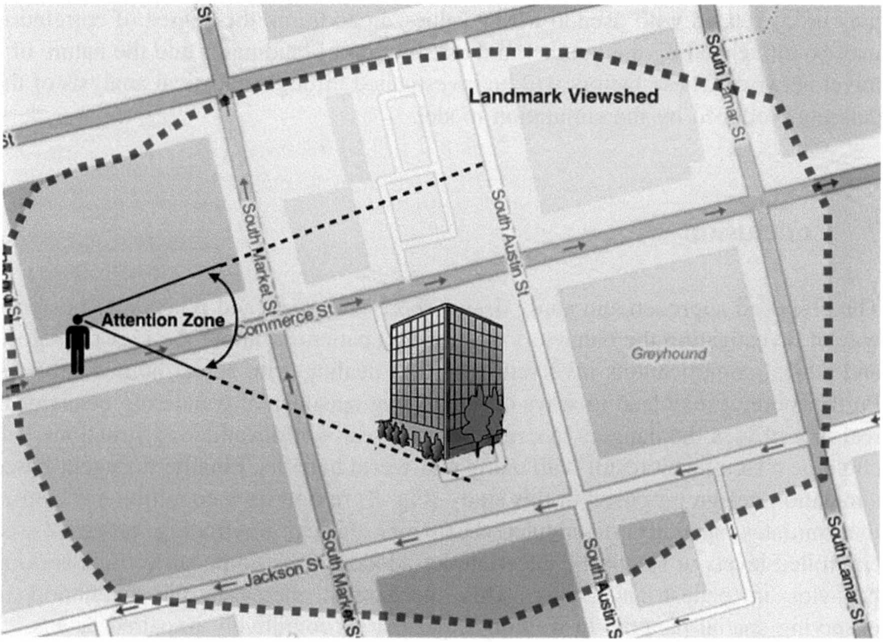

Fig. 2 An example of a viewshed and an agent's attention zone

Arriving at the destination or running out of time terminates an experiment, whereas visiting a landmark initiates a feedback mechanism that resets all three wayfinding error components to their initial values. This mechanism is triggered when an agent is inside the viewshed of a landmark, and the landmark is in the attention zone of the agent (Fig. 2). The viewshed of a landmark is defined by a nonsymmetric buffer around it. This buffer may change based on environmental characteristics that govern the visibility conditions of features in space. Two of the parameters defined previously—radius (r) and arc length (s) of the agent's attention zone—control the size of an agent's attention zone.

6 Expected Results

We expect that certain characteristics of the trajectories described by agent travel patterns are correlated with the agents' pseudo-MMS values. For example, the number of times an agent makes an incorrect wayfinding decision when it comes to an intersection (i.e., selects a route other than the one that moves it most directly toward its destination) should be correlated with its pseudo-MMS value. Other, simpler-to-measure values, such as the ratio of actual distance traveled to the shortest possible network distance between selected starting and ending points, also may be correlated with pseudo-MMS values. In addition, the degree of correlation may be influenced by the number and distribution of landmarks and the nature of a travel network. These factors will be investigated through statistical analysis of the datasets produced by the simulation model.

7 Conclusion

The proposed approach and study design represent a new and potentially powerful way of investigating the behaviors of dementia patients without the logistic, safety, and ethical complications involved in studies dealing with actual human subjects. Furthermore, it may lead to ways of identifying episodes of wandering behavior in real time. Given the dangers inherent in many real-world wandering situations, this could be a significant result with many real-world benefits. Finally, the agent-based simulation design proposed in this study (Fig. 3) resembles a cognitive mechanism that initiates spatially disoriented behaviors during wayfinding process-based controlled levels of stochastic uncertainties added to least-impedance route-seeking behavior in pedestrian agents. This mechanism increases the likelihood of observing spatial patterns in walking trajectory of cognitively impaired agents.

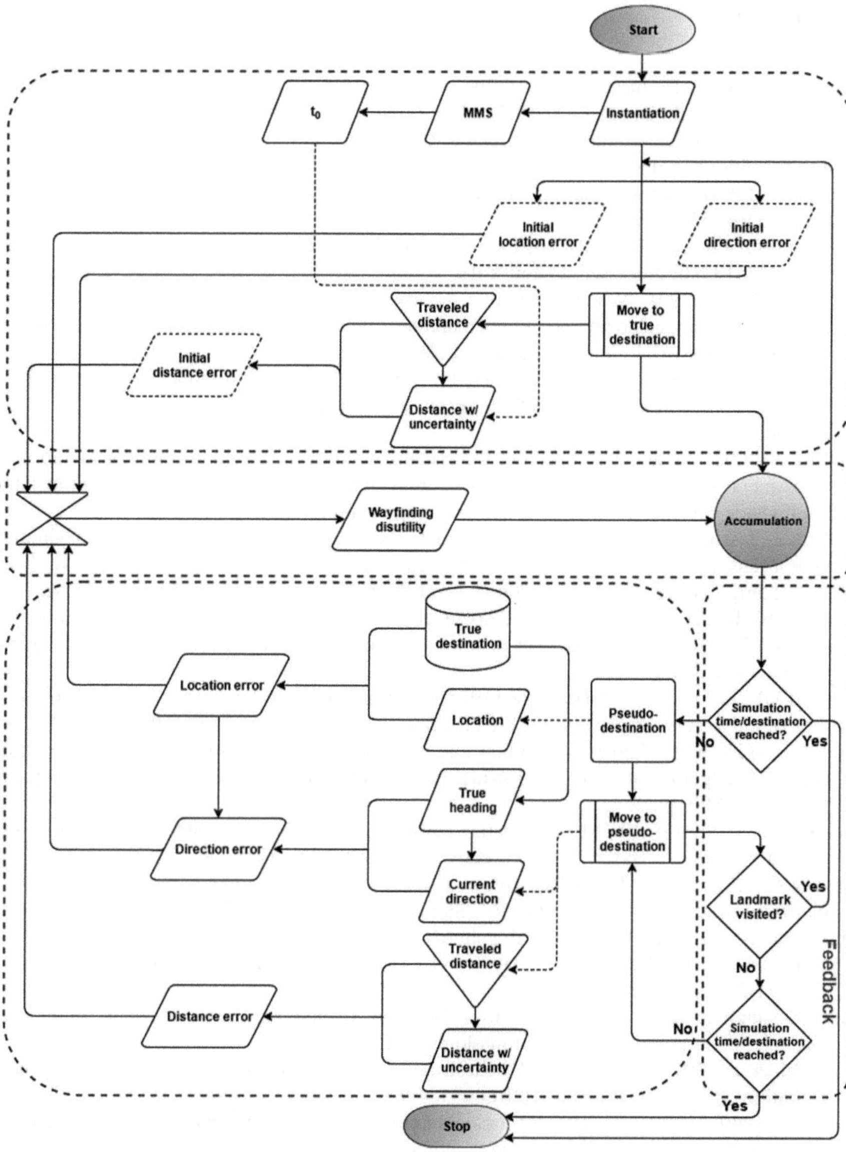

Fig. 3 Agent-based modeling flowchart

References

Algase DL (2003) Biomechanical activity devices to index wandering behaviour in dementia. Am J Alzheimer's Dis Dement 18(2):85–92

Algase DL, Moore D, Vandeweerd C, Gavin-Dreschnack D (2007) Mapping the maze of terms and definitions in dementia-related wandering. Aging Mental Health 11(6):686–698

Allen GL (1999) Cognitive abilities in the service of wayfinding: a functional approach. Prof Geogr 51(4):555–561

Association A (2013) Alzheimer's disease facts and figures. https://www.alz.org/downloads/facts_figures_2013.pdf. Accessed 24 Jun 2016

Amorim M-A (1999) A neurocognitive approach to human navigation. In: Golledge R (ed) Wayfinding behavior: cognitive mapping and other spatial processes. John Hopkings University Press, Baltimore, MD, pp 152–167

Banks CM, Sokolowski JA (2009) Advancing cognitive agent-based modeling: personifying the agents. Proceedings of the 2009 summer simulation multiconference, 13–16 July 2009. Istanbul, Turkey, pp 54–60

Dijkstra J (2008) An agent architecture for visualizing simulated human behavior to support the assessment of design performance. In: 2008 International conference on computational intelligence for modelling control and automation, pp 808–813

Dijkstra J, Jessurun J, de Vries B, Timmermans HJP (2006) Agent architecture for simulating pedestrians in the built environment. In: Bazzan A, Draa B, Klügl F, Ossowski S (eds.) Proceedings of the nineth international workshop of agents and in traffic and transportation, pp 8–16. http://www.ia.urjc.es/ATT/documents/WS28ATT.pdf. Accessed 24 Jun 2016

Dudchenko P (2010) Why people get lost: the psychology and neuroscience of spatial cognition. Oxford, New York

Folstein MF, Folstein SE, McHugh PR (1975) "Mini-mental state": a practical method for grading the cognitive state of patients for the clinician. J Psychiatr Res 12(3):189–198

Golledge RG (1999) Human wayfinding and cognitive maps. In: Golledge R (ed) Wayfinding behavior: cognitive mapping and other spatial processes. John Hopkings University Press, Baltimore, MD, pp 5–45

Golledge RG, Stimson RJ (1997) Spatial cognition, cognitive mapping, and cognitive maps. Spatial behavior: a geographic perspective. Guilford, New York, pp 224–266

Greiner C et al (2007) Feasibility study of the integrated circuit tag monitoring system for dementia residents in Japan. Am J Alzheimer's Dis Dement 22(2):129–136

Makimoto K et al (2008) Temporal patterns of movements in institutionalized elderly with dementia during 12 consecutive days of observation in Seoul, Korea. Am J Alzheimer's Dis Dement 23(2):200–206

Martino-Saltzman D et al (1991) Travel behavior of nursing home residents perceived as wanderers and nonwanderers. Gerontologist 31(5):666–672

Nasir M et al (2014) Prediction of pedestrians routes within a built environment in normal conditions. Expert Syst Appl 41(10):4975–4988

Rieser JJ (1999) Dynamic spatial orientation and the coupling of the representation and action. In: Golledge R (ed) Wayfinding behavior: cognitive mapping and other spatial processes. John Hopkings University Press, Baltimore, MD, pp 168–190

Scarmeas N et al (2009) Disruptive behavior as a predictor in Alzheimer's disease. NIH Public Access 64(12):1755–1761

Shoval N et al (2008) The use of advanced tracking technologies for the analysis of mobility in Alzheimer's disease and related cognitive diseases. BMC Geriatrics 8(1):7

Sposaro F, Danielson J, Tyson G (2010) iWander: an Android application for dementia patients. In: Proceedings of the IEEE engineering in medicine and biology society (EMBC), 2010 annual international conference of the IEEE, Buenos Aires, Argentina, August 31–September 4, pp 3875–3878

Stokes G (1996) Wandering. Winslow Press, Oxon, England

Torrens P, Li X, Griffin WA (2011) Building agent-based walking models by machine-learning on diverse databases of space-time trajectory samples. Trans GIS 15:67–94
Torrens PM et al (2012) An extensible simulation environment and movement metrics for testing walking behavior in agent-based models. Comput Environ Urban Syst 36(1):1–17
Vuong NK et al (2011) Feasibility study of a real-time wandering detection algorithm for dementia patients. In: MobileHealth'11. ACM, New York, pp 11:1–11:4

Design and Validation of Dynamic Hierarchies and Adaptive Layouts Using Spatial Graph Grammars

Kai Liao, Jun Kong, Kang Zhang and Bauke de Vries

Abstract With the thinking paradigm shifting on the evolution of complex adaptive systems, a pattern-based design approach is reviewed and reinterpreted. Although a variety of long-term and lasting explorations on patterns in geographical analysis, environmental planning, and design exist, in-depth investigations into a formalized framework, the process and mechanics of pattern formation, and pattern-based planning and design methodologies are still absent. To face this challenge, our research focuses on spatial cognition-based pattern design (for an intelligent and adaptive/interactive environment in planning and design), inspired by the information theory of complex systems and formal semantics of spatial information. A computational analysis method and design methodology is presented using the spatial graph grammars (SGG) formalism, for the structural complexity of two-dimensional spatial layouts. The proposed model consists of abstract syntax, together with the consistent rules of spatial-semantic compositionality, within a unified and formalized framework. In our model, pattern formation results from dynamic hierarchies and adaptive layouts (driven by complex dynamics and controlled by relevant spatial-semantic specifications) within multiple cognitive levels. Our work demonstrates the application potential of incorporating a novel

K. Liao (✉) · B. de Vries
Departmant of Built Environment, Eindhoven University
of Technology, Postbus 513, 5600 MB Eindhoven, The Netherlands
e-mail: kliao@tue.nl

B. de Vries
e-mail: b.d.vries@tue.nl

J. Kong
Department of Computer Science, North Dakota State University,
Fargo, ND 58108-6050, USA
e-mail: jun.kong@ndsu.edu

K. Zhang
Department of Computer Science, The University of Texas
at Dallas, Richardson, TX 75080-3021, USA
e-mail: kzhang@utdallas.edu

D.A. Griffith et al. (eds.), *Advances in Geocomputation*,
Advances in Geographic Information Science,
DOI 10.1007/978-3-319-22786-3_38

computational tool (in the field of software engineering, data mining, and information visualization/visual analytics) into environmental planning and design.

Keywords Design patterns • Complex adaptive systems • Design cognition and knowledge representation • Spatial graph grammar (SGG) • Spatial-semantic analytics

1 Introduction

The paradigm of pattern-based design has been widely adopted in the fields of software engineering (Gamma et al. 1994; Zhao et al. 2007), spatial information representation and reasoning (Kong and Zhang 2003), and data mining (Qian and Zhang 2004).

The idea of design patterns is rooted in Chomsky's generative grammar and formal language theory (1956, 1957), Simon's hierarchy and complexity (1962, 1973), and Fu's pattern grammars (1971, 1976), to name a few sources. These original and visionary works shed light on the research of theories and methodologies on patterns and pattern-based design with complexity theories, especially the information theory of complex systems (Lindgren 1988, 2003; Andersson et al. 2002).

However, in environmental planning and design, formal pattern-based design methodologies remain absent, although influential early explorations (Stiny and Gips 1972; Yessios 1972, 1987; Alexander et al. 1968) and the most notable long-term and lasting work about patterns and pattern languages by Alexander and his collaborators (Salingaros 1995, 1998; Alexander et al. 1977; Alexander 2003–2005) exist.

This chapter proposes a formalized framework for the process and mechanics of pattern formation in the context of physical planning and architectural design for two-dimensional (2D) spatial layouts. It discusses parsing and generation of such layouts on the basis of complex systems and the related information theory, design cognition, and knowledge representation (de Vries et al. 2005, 2010; Beetz et al. 2005).

2 Theory and Methodology

First, reinterpreting our understanding of pattern and pattern formation with a shift of the thinking paradigm on the evolution of complex adaptive systems is important. This perspective presents us with a profound notion; that is, pattern is actually a fluctuating phenomenon (being or becoming) driven by system dynamics (Prigogine 1977; Prigogine and Stengers 1984; Liao 1996; Liao and Li 1997).

The best example of this kind of thinking paradigm shift might be Alexander's work. It shows an obvious transformation of the understanding of patterns from atomism/reductionism (Alexander and Poyner 1967), structuralism, and utopianism/determinism (Alexander and Eiseman 1982) into complexity (e.g., morphogenetic) (Batty and Longly 1994; Liao 1996; Liao and Han 2005; Knight and Stiny 2001; Alexander 2003–2005), adaption, and uncertainty (Holland 1992; Kelly 1994). However, the potential of complex systems for design patterns in environmental planning and design is still far from being fully explored, especially in the line of formalized and/or computational approaches.

To address this issue, our research focuses on spatial cognition-based pattern design (for an intelligent and adaptive/interactive environment in planning and design), inspired by the information theory of complex systems and formal semantics of spatial information.

2.1 *Dynamic Hierarchies with Emergence*

Simon considers hierarchy to be primary to understanding the organization of complexity, and points out that "almost all the very large systems will have hierarchic organization" (1973, p. 7); therefore, we would be able to approach complex systems "by way of a theory of hierarchy" (1962, p. 481)—i.e., "Near-decomposability" (1962, p. 482) as "reasonable compass" (1962, p. 482).

Holland presented his theory of emergence in (1998). In contrast to Simon's structuralism perspective, Holland regards the process of emergence as crucial, with hierarchical organization being a consequence of this process. Many higher-level "entities" are patterns of organization rather than stable aggregates of lower-level entities.

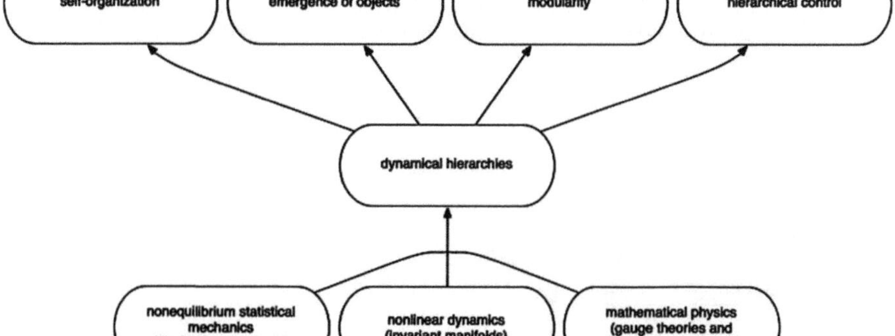

Fig. 1 An overview conceptual framework to model evolving complex adaptive systems (the case of physical-biomolecular systems). [Reproduced from Ericksson (2005, slide 17), courtesy of Martin Nilsson Jacobi]

Referring to Fig. 1 and Holland's theory of "hierarchical organization that emerges in Emergence" with emphasis on bottom-up "building block interactions" and "persistent patterns" with "perpetual novelty" (Holland 1995 pp. 36, 87, (1998), pp. 6, 7, 9, 45; Lane 2006), we consider complex systems as dual organization phenomena determined by both structural and process constraints (Lindgren and Nordahl 1994; Lindgren et al. 1998; Batty 2005; Andersson et al. 2005, 2006; Andersson 2008). Thus, we understand that dynamic hierarchies act as the key to the process and mechanics of pattern formation with emergence.

2.2 Modeling with Multiple Representations

In complex and open systems, the organizational structure and spatial information of patterns would be emergent, along with the phase transitions between equilibrium states and the redistribution of their related structural information.

Pattern formations are critical phenomena during complex systems' evolution with spatial information processing (interaction and adaptation). To model "Evolving Hierarchical Systems" (Salthe 1985), we should not only understand their structures but also model and represent the process (both upward and downward causation) across multilevel hierarchies.

Therefore, our study focuses on the modeling of spatial information processing during a design process across the different design cognition-level hierarchies (at intralevel or/and interlevel; Do and Gross 2001; Haken and Portugali 1995, 2014; Kuipers 2000). It serves the process and mechanics of pattern formation with affluent design cognition and knowledge representations (Beetz and de Vries 2009; Beetz 2014; Niemeijer et al. 2014; Liao et al. 2015) (Fig. 2).

2.3 Adapting to Layout Context with Spatial Semantics

Having spatial information processing with multiple design knowledge representations, the spatial decisions of planning and design correspondingly should knit

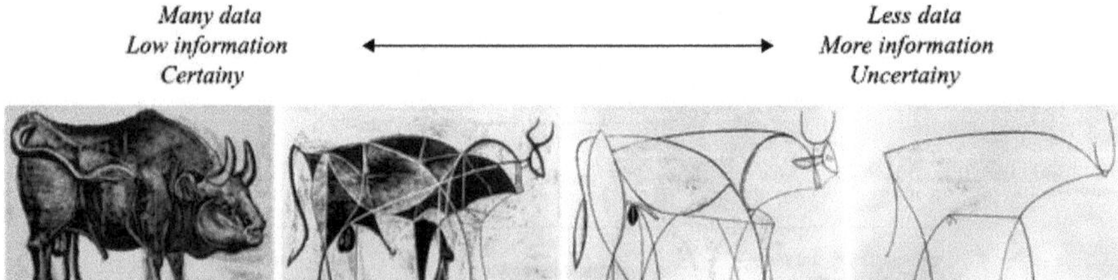

Fig. 2 An illustration of spatial information processing across multiple cognitive levels within multiple design representations. [Redrawn with modifications based on the original figure in Haken and Portugali (2014)]

together across the dynamic hierarchies of scales/levels. This process would form desired patterning and order by interacting with and adapting to a situational and environmental context.

For existing design representation models and computational tools, however, a cognitive gap exists between low-level spatial data/information and high-level design knowledge. The relevant formal studies (especially about architecture-specific, design-oriented formal spatial representation and reasoning, ontologies, and spatial semantics) are limited (Beetz et al. 2005, 2006; Borrmann and Beetz 2010; Egenhofer 2015). This deficiency prevents a direct cross-link between spatial-coordinate information and generic knowledge representation during a design process.

To bridge this gap, our model aims to integrate topological, metric, and semantic features of a 2D space layout design within a unified framework for design-pattern mining, retrieval, (re)configuration, generation, and translation. An approach of spatial-semantic analytics (Zlatev 2010) for dynamic hierarchies and adaptive layout designs is proposed.

3 Implementation

The architectural layouts of the three house projects by Frank Lloyd Wright (Fig. 3) are selected as our case studies. We suggest that a pivotal pavilion (F, the family room area for each) acts as the circulation nexus, social communication hub, and

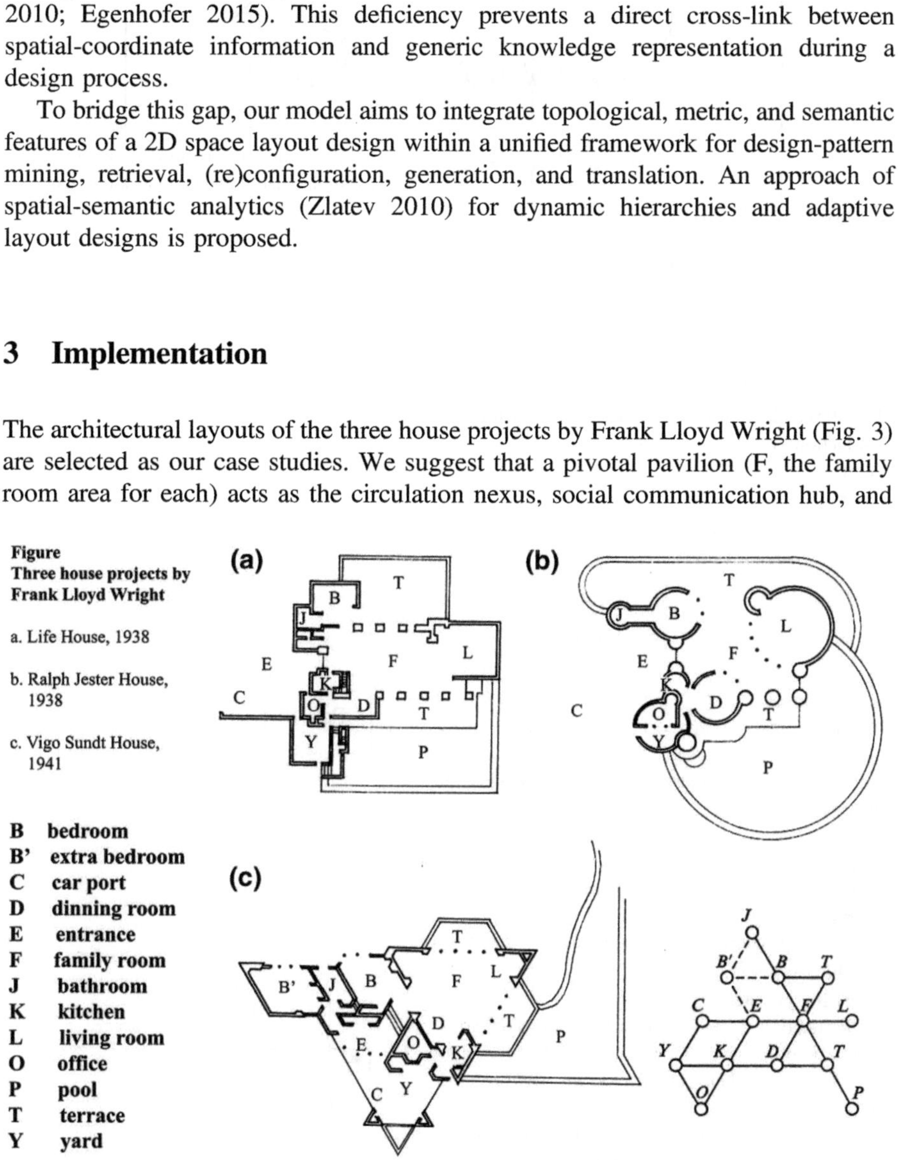

Fig. 3 Three house projects by Frank Lloyd Wright with spatial topological similarity. [Redrawn with modifications based on the original figure in March and Steadman (1974)]

Table 1 An illustration of the graph grammar and spatial-semantic analytic maps for the spatial layouts of three house projects by Frank Lloyd Wright (see Fig. 3)

Event listing of production(s) during the compositional processing	Spatial-semantic analytics, including cardinal direction relation
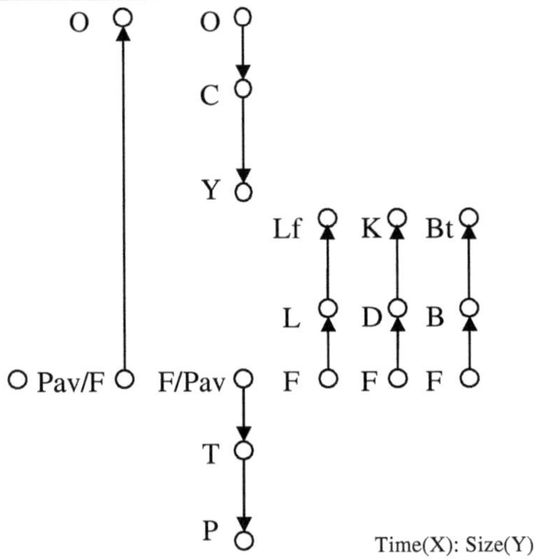 Note: The above diagram demonstrates the following production(s): Pav/F F := F-O F := F-T-P O := O-C-Y F := F-L-Lf F := F-D-K F := F-B-Bt	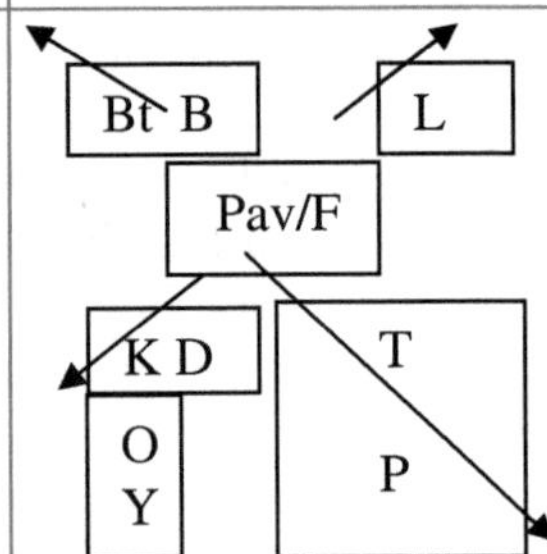 Cardinal direction relations and, for example, size, balance, and alignment
Level I (initial state) (0) The pivotal pavilion = F; (1) F (family room) itself with options (Fig. 3a: rectangle, Fig. 3b: square, then with circles, Fig. 3c: hexagon) Note: Step (0) For parsing, at the top level of the hierarchies, it is the family room itself, with three different geometric/shape options Step (1) The family room acts as the start point of the parsing process	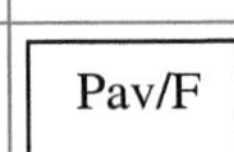
Level II (start state) (2) Production: F := F-O (family room to office) (2.1) The office (Fig. 3a: rectangle with 90° rotation, Fig. 3b: square, then with circles, Fig. 3c: hexagon) Note: Step (2) The O (office) acts as the minor pavilion (which is complemented and is similar to the pivotal pavilion, Pav/F) at the south-western corner of the pivotal pavilion	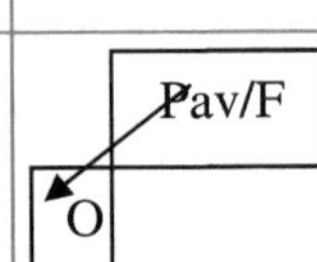

(continued)

Table 1 (continued)

<table>
<tr><td>Level III (continue state)
(3.1) Production(s): F: = (E)-((F)-T)-P
Note:
Step (3.1) shows the large block of F-T-P within "the Grand Courtyard"; i.e., (grand courtyard): = (Entrance - family room - terrace - pool).
Level III (continue state)
(3.2) Production(s): F: = ((F)-O)-Y-C
Note:
Step (3.2) shows the small block of O-Y-C within the tiny courtyard; i.e., (tiny
courtyard): = (office-yard-carport)</td><td>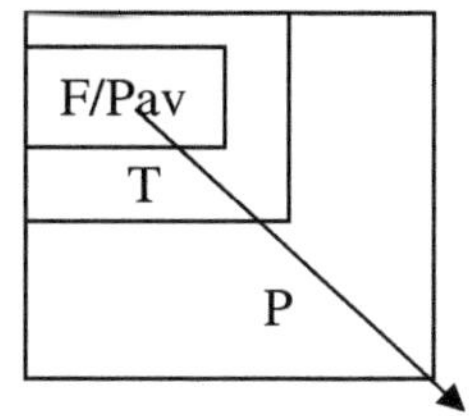

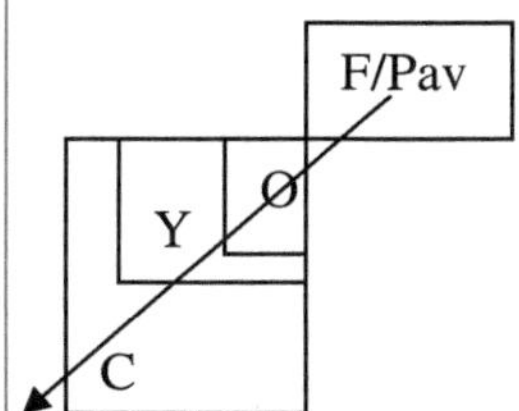
</td></tr>
<tr><td>Level IV (continue state)
(4.1) Production(s): F: = (F)-L-Lf:
Note:
Step (4.1) shows the family room (F), the living room (L), and the fireplace (Lf/fireplace).</td><td>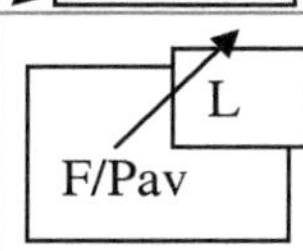
</td></tr>
<tr><td>Level V (end state)
(5.1) Production(s): F := (F)-D-K
Note:
Step (5.1) shows from the family room to the dining area (e.g., D/dining room, K/kitchen, corridor.)
(5.2) Production(s): F := (F)-B-Bt
Note:
Step (5.2) shows from the family room to the bedroom area
(e.g., B/bedroom, Bt/bathroom, corridor)</td><td>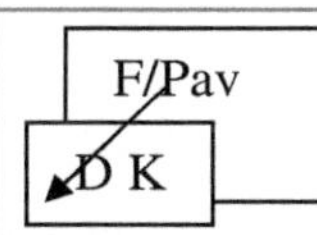

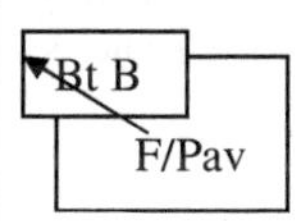
</td></tr>
</table>

visual/compositional crux of the layout design. Respectively, the floor plans in Fig. 3 include (a) rectangles, (b) an underlay square and the above circles, and (c) hexagons.

Setting up the pivotal pavilion (Pav/F) as the origin of a coordinate system and the starting point of a spatial layout, we conduct a parsing operation with the spatial-semantic analytics on the layout design in order to generate the graph grammar and spatial-semantic analytic maps (Table 1).

A spatial graph grammar (SGG) specification, parsing, and induction tool called VEGGIE (Visual Environment of Graph Grammar Induction Engineering), developed by Zhang and (Kong et al. 2006; Zhang 2007; Ates and Zhang 2007; Kong et al. 2008, 2012), is employed for the experimental design. Through parsing with VEGGIE, we are able to represent, retrieve, and regenerate design patterns of 2D spatial layouts automatically (Fig. 4).

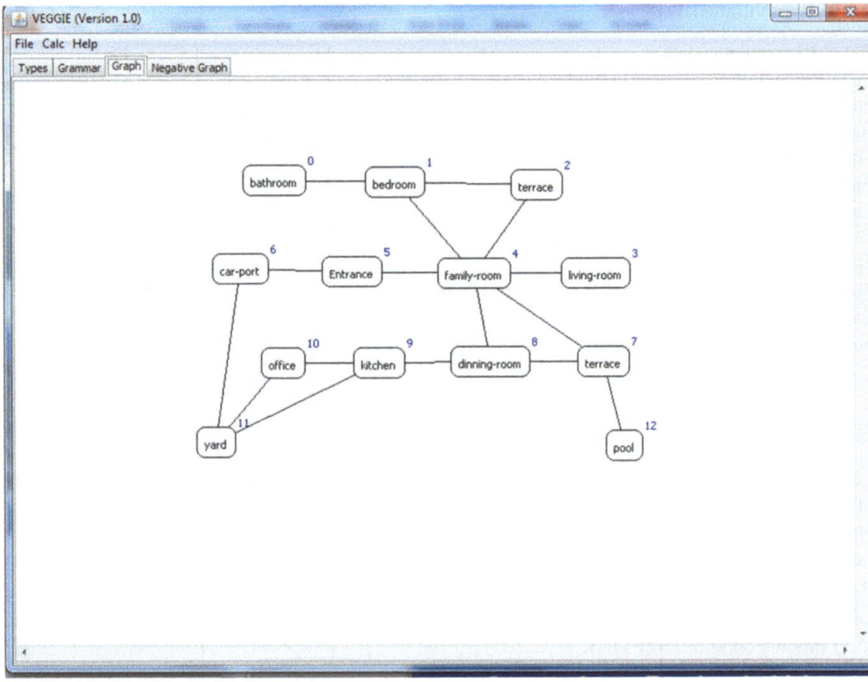

Fig. 4 The screenshot of a VEGGIE-generated, task-specific spatial-semantic analytic map for the spatial layouts of three house projects by Frank Lloyd Wright (see Fig. 3)

4 Conclusion and Discussion

Our work presents a theoretic framework and a conceptual solution for formalized pattern-based design by introducing spatial-semantic analytic intelligence into spatial information processing with dynamic hierarchies and adaptive layouts. The spatial-semantic analytics approach is the crucial contribution of our work, which bridges the gap between low-level data/information of layout configurations, spatial-semantic awareness, and high-level design knowledge, within a unified design cognition-based framework. With our approach, either configurational or compositional information of spatial patterns (out of design cognition within distinct levels/aspects) as a whole in terms of complemented spatial information modeling and design representations would be able to knit to the pattern formation process of spatial layouts (validated by the spatial-semantic compositionality, which is defined by design goals).

Although our work is at a preliminary stage, we demonstrate the application potential of incorporating the novel computational approach of SGG (which is developed for applications in software engineering, data mining, and information visualization/visual analytics) into conventional computation approaches in environmental planning and design. We will continue to develop our computational analysis model and design tool by using SGG.

Acknowledgements Kai Liao wishes to thank his PhD advisor, Prof. Michael Batty; Prof. Paul Longley at CASA, University College London; Prof. Kristian Lindgren, Dr. Claes Andersson at the Complex Systems Group, Chalmers University of Technology, Sweden; Prof. Chia-Yun Han at the University of Cincinnati, USA; and Prof. Deren Li at Wuhan University, China.

References

Alexander C (2003–2005) The nature of order: an essay on the art of building and the nature of the universe. Center for Environmental Structure/Routledge, Berkeley, CA

Alexander C, Eiseman P (1982) Contrasting concepts of harmony in architecture: the 1982 debate between Christopher Alexander and Peter Eisenman—an early discussion of the "New Sciences" of organised complexity in architecture. http://www.katarxis3.com/Alexander_Eisenman_Debate.htm. Accessed 1 Mar 2015

Alexander C, Ishikawa S, Silverstein M (1968) A pattern language which generates multi-service centers. Center for Environmental Structure, Berkeley, CA

Alexander C, Ishikawa S, Silverstein M, Jacobson M, Fiksdahl-King I, Angel S (1977) A pattern language: towns, buildings, construction. Oxford University Press, New York

Alexander C, Poyner B (1967) The atoms of environmental structure. In: Ministry of public building works. Center for Planning and Development Research, Berkeley, CA

Andersson C (2008) Ontogeny and ontology in complex systems modelling. In: Albeverio S, Andrey D, Giordano P, Vancheri A (eds) The dynamics of complex urban systems: an interdisciplinary approach. Springer, Heidelberg, pp 43–58

Andersson C, Frenken K, Hellervik A (2006) A complex network approach to urban growth. Environ Plan A 38(10):1941–1964

Andersson C, Hellervik A, Lindgren K (2005) A spatial network explanation for a hierarchy of urban power laws. Phys A 345(1):227–244

Andersson C, Lindgren K, Rasmussen S, White R (2002) Urban growth simulation from "first principles". Phys Rev E 66(2):026204/1–9

Ates KL, Zhang K (2007) Constructing VEGGIE: machine learning for context-sensitive graph grammars. In: Proceedings of 19th IEEE international conference on tools with artificial intelligence. Los Alamitos, CA, pp 456–463

Batty M (2005) Cities and complexity. The MIT Press, Cambridge

Batty M, Longley P (1994) Fractal cities: a geometry of form and function. Academic Press, London

Beetz J (2014) A scalable network of concept libraries using distributed graph databases. In: Proceedings of the joint ICCCBE and CIB W78 conferences. Orlando, FL, pp 569–576

Beetz J, de Vries B (2009) Building product catalogues on the semantic web. In: Proceedings of managing IT for tomorrow. Taylor & Francis Group, Istanbul, Turkey, pp 221–226

Beetz J, van Leeuwen J, de Vries B (2005) An ontology web language notation of the industry foundation classes. In: Proceedings of the 22nd CIB W78 conference on information technology in construction, Dresden, Germany, pp 670–675

Beetz J, van Leeuwen J, de Vries B (2006) Towards a topological reasoning service for IFC-based building information models in a semantic web context. In: Proceedings of 11th international conference on computing in civil and building engineering, ICCCBE-XI. Montreal, Canada, pp 3426–3435

Borrmann A, Beetz J (2010) Towards spatial reasoning on building information models. In: Proceedings of the 8th European conference on product and process modeling (ECPPM). Taylor & Francis Group, Cork, Ireland, pp 61–67

Chomsky N (1956) Three models for the description of language. IRE Trans Inf Theor 2(3):113–124

Chomsky N (1957) Syntactic structures. Mouton, The Hague

de Vries B, Beetz J, Achten H, Dijkstra J, Jessurun A (2010) Design systems group: knowledge models for design and engineering (2005–2010). In: Achten H, de Vries B, Stappers P (eds) Design research in the Netherlands 2010. Eindhoven University of Technology, Eindhoven, pp 53–63

de Vries B, Jessurun A, Segers N, Achten H (2005) Word graphs in architectural design. Artif Intell Eng Des Anal Manuf 19(4):277–288

Do EYL, Gross MD (2001) Thinking with diagrams in architectural design. Artif Intell Rev 15(1–2):135–149

Egenhofer MJ (2015) Qualitative spatial-relation reasoning for design. In: Gero J (ed) Studying visual and spatial reasoning for design creativity. Springer, Heidelberg, pp 153–177

Ericksson A (2005) Information theory and multi-scale simulations, presentation by the Chalmers Complex Systems Group at the Project EMBIO Kick-off Meeting. Cambridge University, 25–27 July. https://www-embio.ch.cam.ac.uk/meetings/kick-off/Chalmers/AEriksson_Cambridge-2005.ppt. Accessed 27 Jun 2016

Fu KS (1976) Syntactic (linguistic) pattern recognition. In: Fu KS (ed) Digital pattern recognition. Springer, Berlin, pp 95–134

Fu KS, Swain PH (1971) On syntactic pattern recognition. In: Tou JT (ed) Computer and information sciences-1969, Software engineering, vol 2. Academy Press, New York, pp 155–182

Gamma E, Helm R, Johnson R, Vlissides J (1994) Design patterns: elements of reusable object-oriented software. Addison-Wesley, Amsterdam

Haken H, Portugali J (1995) A synergetic approach to the self-organization of cities and settlements. Environ Plan 22(1):35–46

Haken H, Portugali J (2014) Information adaptation: the interplay between Shannon information and semantic information in cognition. Springer series: briefs in complexity, vol XIV. Springer, Berlin

Holland J (1992) Adaptation in natural and artificial systems: an introductory analysis with applications to biology, control, and artificial intelligence. The MIT Press, Cambridge, MA

Holland J (1995) Hidden order: how adaptation builds complexity. Addison-Wesley, Reading, MA

Holland J (1998) Emergence: from chaos to order. Perseus, Reading, MA

Kelly K (1994) Out of control: the new biology of machines, social systems, and the economic world. Addison-Wesley, New York

Knight T, Stiny G (2001) Classical and non-classical computation. Info Tech 5(4):355–372

Kong J, Ates KL, Zhang K, Gu Y (2008) Adaptive mobile interfaces through grammar induction. In: Proceedings of 20th IEEE international conference on tools with artificial intelligence, Herdon, VA, pp 133–140

Kong J, Barkol O, Bergman R, Pnueli A, Schein S, Zhao CY, Zhang K (2012) Web interface interpretation using graph grammars. IEEE Trans SMC—Part C 42(4):590–602

Kong J, Zhang K (2003) Graph-based consistency checking in spatial information systems. In: Proceedings of 2003 IEEE symposium on visual languages and formal methods. IEEE CS Press, Auckland, pp 153–160

Kong J, Zhang K, Zeng X (2006) Spatial graph grammars for graphical user interfaces. ACM Trans Comput-Hum Interact 13(2):268–307

Kuipers B (2000) The spatial semantic hierarchy. Artif Intell 119(1–2):191–233

Lane D (2006) Hierarchy, complexity, society. In: Pumain D (ed) Hierarchy in natural and social sciences. Methodos series, vol 3. Springer, Dordrecht, pp 81–119

Liao K (1996) From Feng-Shui to chaotic/fractal architecture: transformation of urban space design concept of Chinese Shan-Shui (mountain and water, landscape) city. In: Huang G,

Huang T, Huang Y (eds) Proceedings of international symposium on sustainable development of human settlements in mountainous regions. Science Press, Beijing, pp 77–91

Liao K, de Vries B, Kong J, Zhang K (2015) Pattern, cognition and spatial information processing: representations of the spatial layout of architectural design with spatial-semantic analytics. In: Celani G, Sperling D, Franco J (eds) The 16th international conference, the CAAD futures 2015: the next city. Springer, São Paulo, Brazil, pp 547–562

Liao K, Han CY (2005) Collective pavilions: a generative architectural modeling for traditional Chinese pagoda. In: Martens B, Brown A (eds) The CAAD futures 2005: learning from the past. Oesterreichischer Kunst- und Kulturverlag, Vienna, Austria, pp 129–138

Liao K, Li D (1997) An analysis of traditional Chinese architecture and garden design from the viewpoint of chaos theory and fractal geometry. J Wuhan Tech Univ Surv Map (WTUSM) 23-3(Sept):189–203

Lindgren K (1988) Microscopic and macroscopic entropy. Phys Rev A 38(9):4794–4798

Lindgren K (2003) Information theory for complex systems. Lecture notes (Jan 2003, updated in 2014). Department of Physical Resource Theory, Chalmers and Göteborg University

Lindgren K, Nordahl MG (1994) Evolutionary dynamics of spatial games. Physica D 75(1–3):292–309

Lindgren K, Moore C, Nordahl MG (1998) Complexity of two-dimensional patterns. J Stat Phys 91(5–6):909–951

March L, Steadman P (1974) The geometry of environment: an introduction to spatial organization in design. The MIT Press, Cambridge, MA

Niemeijer RA, de Vries B, Beetz J (2014) Freedom through constraints: user-oriented architectural design. Adv Eng Inform 28(1):28–36

Prigogine I (1977) Time, structure, and fluctuations. Nobel lecture 1977. http://www.nobelprize.org/nobel_prizes/chemistry/laureates/1977/prigogine-lecture.pdf. Accessed 1 Mar 2015

Prigogine I, Stengers I (1984) Order out of chaos: man's new dialogue with nature. Bantam Books, New York

Qian Y, Zhang K (2004) Discovering spatial patterns accurately with effective noise removal. In: Proceedings of the 9th ACM SIGMOD workshop on research issues in data mining and knowledge discovery, Paris, France, June 13, pp 43–50

Salingaros NA (1995) The laws of architecture from a physicist's perspective. Phys Essays 8:638–643

Salingaros NA (1998) Theory of the urban web. J Urban Des 3:53–71

Salthe S (1985) Evolving hierarchical systems: their structure and representations. Columbia University Press, New York

Simon H (1962) The architecture of complexity: hierarchic systems. Proc Am Philos Soc 106:467–482

Simon H (1973) Chapter 1: The organization of complex systems. In: Pattee H (ed) Hierarchy theory: the challenge of complex systems. George Braziller, New York, pp 3–27

Stiny G, Gips J (1972) Shape grammars and the generative specification of painting and sculpture. In: Proceedings of IFIP Congress 1971. North Holland Publishing Company, Amsterdam, pp 125–135

Yessios C (1987) A fractal studio. In: Proceedings of the annual conference of the association for computer aided design in architecture (ACADIA). University of North Carolina, pp 169–181

Yessios C (1972) FOSPLAN: a formal space planning language. In: Mitchell WJ (ed) EDRA 3: proceedings of the edra3/ar8 conference, UCLA, vol 23, no 9, pp 1–10

Zhang K (2007) Visual languages and applications. Springer-Verlag, Secaucus, NJ

Zhao C, Kong J, Dong J, Zhang K (2007) Pattern-based design evolution using graph transformation. J Vis Lang Comput 18(4):378–398

Zlatev J (2010) Spatial semantics. In: Cuyckens H, Geeraerts D (eds) The Oxford handbook of cognitive linguistics. Oxford University Press, Oxford, pp 318–350

MIX
Papier aus verantwortungsvollen Quellen
Paper from responsible sources
FSC® C105338

If you have any concerns about our products,
you can contact us on
ProductSafety@springernature.com

In case Publisher is established outside the EU,
the EU authorized representative is:
Springer Nature Customer Service Center GmbH
Europaplatz 3, 69115 Heidelberg, Germany

Printed by Libri Plureos GmbH
in Hamburg, Germany